THE SEA URCHIN: FROM BASIC BIOLOGY TO AQUACULTURE

SEA URCHIN AQUACULTURE, ADVANCED WORKSHOP
20 NOVEMBER - 1 DECEMBER 2000, TORREGRANDE, SARDINIA, ITALY

The Sea Urchin:
From Basic Biology to Aquaculture

Edited by

Yukio Yokota
Department of Applied Information Technology, Aichi Prefectural University

Valeria Matranga
Istituto di BioMedicina e Immunologia Molecolare "Alberto Monroy", Consiglio Nazionale delle Ricerche

Zuzana Smolenicka
International Marine Centre

CRC Press
Taylor & Francis Group
Boca Raton London New York

CRC Press is an imprint of the
Taylor & Francis Group, an **informa** business

CRC Press
Taylor & Francis Group
6000 Broken Sound Parkway NW, Suite 300
Boca Raton, FL 33487-2742

ISBN-13: 978-9-058-09379-0 (hbk)

Publisher's Note

The publisher has gone to great lengths to ensure the quality of this reprint but points out that some imperfections in the original copies may be apparent.

Visit the Taylor & Francis Web site at
http://www.taylorandfrancis.com

and the CRC Press Web site at
http://www.crcpress.com

The Sea Urchin: From Basic Biology to Aquaculture, Yokota, Matranga & Smolenicka (eds)
© Swets & Zeitlinger, Lisse, ISBN 90 5809 379 4

Table of Contents

The Sea Urchin: From Basic Biology to Aquaculture, Yokota, Matranga & Smolenicka (eds)
© Swets & Zeitlinger, Lisse, ISBN 90 5809 379 4

Preface

In 1883 at the International Fisheries Exhibition, London, T. H. Huxley had declared that probably all the great sea fisheries were inexhaustible; consequently any attempt to regulate these fisheries seemed to be useless. Today, nobody agrees with his statement and most people are aware that natural resources are finite and that many species are endangered.

The technique of aquaculture has been established to preserve natural fishery resources while maintaining stable supply of sea products. In fact, we can find various marine fishes and invertebrates from aquaculture in fish markets. In order to develop and improve the technique of aquaculture, to lower the running costs, minimize the impact on the environment, increase the productivity, and cover new species, numerous researches have been carried out. In recent years, the seafood consumption has increased dramatically because of the improvement of transportation and high demands of gourmet food which used to be consumed mainly by local people. Natural sea urchin stocks are globally declining due to over-catching and environmental changes. Aquaculture and systematic natural stock management of sea urchins are recognized as substantial and effective methods to maintain the sustainable sea urchin stocks. The establishment of sea urchin aquaculture requires two types of efforts at least; studies on the practical aspects of aquaculture for these species and researches on the basic sea urchin biology such as development, ecology, physiology and genetics.

This book contains lectures presented at the Advanced Workshop in Honour of Eizo Nakano "Sea Urchin Aquaculture", held at the International Marine Centre, Oristano, Italy, from 20 November to 1 December, 2000, ranging from sea urchin basic biology to social aspects of aquaculture. Prof. Eizo Nakano was one of the founders of the International Marine Centre, a pioneer in the biochemistry of fertilization, a member of Consiglio Scientifico of the Stazione Zoologica Anton Dohrn Napoli, and carried out studies on protein synthesis activation immediately after fertilization in sea urchin eggs by using the technique of radioactive amino acid insertion in eggs. He enormously contributed to the scientific collaboration and friendship between Italy and Japan, and convinced that scientific interaction among young people is indispensable for the development of science.

It should be remembered that a course on sea urchin aquaculture was greatly expected by Eizo Nakano, who started to work at it before his departure. The workshop held in November 2000 materialized his will. Eizo Nakano always showed his high interest in marine aquaculture and his great belief on the importance of the conservation of one of the traditional and classical embryological model systems, such as the sea urchin embryo, with efforts in the maintenance, culture and breeding of live animals. He predicted the importance to develop the sea urchin aquaculture in Europe already in 1988 in view of supporting marine stations, expand and improve international collaborations and establish conditions for aquaculture. In his conclusion in the "International Workshop on Study, Protection and Development of Mediterranean Living Marine Resources" (Proceedings edited by R. Covacci, M.B.A. Djangoz and S. Vallerga, 1988), he put the accent on the importance of collaboration between marine biological laboratories and aquaculture centers. This collaboration, in his mind, would constitute a great advantage both in scientific research as well as in the protection of sea urchin stocks. We think the best way to honor his memory is to work hard and follow him as a model of strength and determination in pursuing new scientific ideas and innovative research fields. Here we can say that his way of working honestly and rigorously should be followed by future generation of scientists. Thanks Eizo for all you gave us.

Needless to say that the workshop would not have been possible without sponsorships. We acknowledge the European Commission, Program Quality of Life, Consiglio Nazionale delle Ricerche, Regione Autonoma della Sardegna, Provincia di Oristano, Commune di Oristano, Italian Embassy in Japan, Carl Zeiss, Zeusaquarium, Alitalia, AVT HORN and International

Marine Centre. Finally, we express special thanks to the colleagues of the Organizing Committee and all the people at the IMC, Dr. Joe Hirano, Dr. Silvana Vallerga, Dr. Renato Covacci, Dr. Maura Baroli, Dr. Rosalba Murgia, Mr. Filippo Angotzi, Dr. Massimiliano Di Bitetto, Ms. Giovanna Sanna, Ms. Donatella Marchi, Mr. Giovanni Piras, Ms. Monica Pinna, Dr. Romana Salaris, Dr. Cristiana Loddo, Dr. Daniele Casu, Dr. Laura Frulio, Dr. Giuseppe Piergallini, Ms. Martina Mussi, Dr. Gianni de Falco, Dr. Didier Gonzales and Dr. Gabriella Tola, whose daily and tireless work was indispensable to this workshop. Here we conclude by repeating our gratitude to the contributors for their co-operation to make the workshop successful and to prepare chapters for this book. The editors are also indebted to Dr. Raffaella Tocco, Dr. Hideto Kasuya and Mr. Takeshi Shinohara for their technical assistance.

Yukio Yokota
Valeria Matranga
Zuzana Smolenicka

The Sea Urchin: From Basic Biology to Aquaculture, Yokota, Matranga & Smolenicka (eds)
© Swets & Zeitlinger, Lisse, ISBN 90 5809 379 4

Introduction to the sea urchin biology

Y. Yokota
Department of Applied Information Technology, Aichi Prefectural University, Nagakute, Japan.

ABSTRACT: Approximately 1,000 species of sea urchins live in different marine habitats, from intertidal to deep-sea environments up to 6,800 m and including the Arctic Sea and the Antarctic Ocean. Sea urchins have been denominated on the basis of their characteristic appearance, which attracted numerous biologists since Aristotle. Many milestones in biological sciences, such as the observation of fertilisation or the isolation of eukaryotic messenger RNA have been carried out using these organisms. Global increase of demand on luxurious and exotic food had a significant impact on sea urchin resources, leading to the evolution of sea urchin aquaculture. In this chapter, biology of the sea urchin is described and basic information is given for those who intend to start studying sea urchins aquaculture.

1. INTRODUCTION

The phylum Echinoderm is divided into 5 classes: Crinoidea, Asteroidea, Holothuroidea, Echinoidea and Ophiuroidea. Although the term 'echinoderm' (meaning 'skin with spines') was derived from the combination of *echinus* (hedgehog, urchin) and *derma* (skin), it was already the original denomination for sea urchins. Indeed, the scientific name of many sea urchin species terminates with – echinus or starts with Echino –. The spines of an echinoderm are essentially different from those of insects, caterpillars, polychaetae and others, from a point of view of their anatomical origin and structure. Being the sea urchin a deuterostome, it is more related to vertebrates and protochordates than to mollusks and arthropods. Freshwater species have not been reported so far.

2. HISTORY OF SEA URCHIN BIOLOGY

In 300s BC, Aristotle (384 - 322 BC) analysed and described the structure of the mouthpart in the sea urchin. According to his description, this organ is called Aristotle's lantern. Although Klein (1734) named the sea urchin as *Echinodermata*, Linné (1758) categorised the sea urchin into Echinus, making part of a larger group named Mollusca. Lamarck (1801, 1816 - 1822) defined a new group named Radiata, and classified the sea urchin into this group together with starfish and sea cucumber, calling them Echinodermata. Anatomical research established the concept of Echinodermata specific body plan. From the end of 19th century to the beginning of the 20th, a large number of investigations on sea urchins were carried out as follows: in Taxonomy, Duncan (1883,1888), Agassiz (1872 - 1874, 1881), Lovén (1874, 1883, 1892), Gregory (1900); in Histology and Morphology, Koehler (1883), Hamann (1887), Prouho (1887); in Physiology, von Uexküll (1896, 1900, 1907) and others. Taxonomic research was further developed by Agassiz & Clark (1907, 1908, 1909), Hawkins (1912, 1920), Clark (1912, 1914, 1917, 1925), Lambert & Theiéry (1909 - 1925), Jackson (1912) and others. Mortensen gathered these results into "A Monograph of the Echinoidea" (1928 - 1951). He described all species, *de facto* as well as extinct ones (in fossil form) based on their taxonomy, morphology, development, ecology etc. A book titled "Treatise on Invertebrate Paleontology (Echinoids)" was published in 1966 by Durham, Fell, Fischer, Kier, Melville, Pawson and Wagner.

The first observation of fertilisation was carried out by Hertwig (1876) in the sea urchin *Paracentrotus lividus*. Descriptive studies on development were performed by McBride (1903, 1914a,b), Gordon (1926), Mortensen (1931, 1937, 1938) and Onoda (1931, 1936, 1938). Shearer (1922), Morgan (1893, 1899, 1909), Hörstadius (1935, 1939) and others contributed to the improvement of experimental embryology of the sea urchin. The idea of Hörstadius and his successors focused not only on the sea urchin development, but also on the understanding of its relevant phenomena, using sea urchin embryos as a model.

We may remember two books, "The Invertebrate: Echinodermata" by Hyman (1955), and "Physiology of Echinodermata" edited by Boolootian (1966).

3. SEA URCHIN MORPHOLOGY

The Echinodermata is an Enterocoelous Coelomate, possessing a pentaradiate body-plan resulting from bilateralism. It does not have a defined head or brain. The calcareous endoskeleton consists of separate plates or pieces, often bearing external spines or protuberances. Its water vascular system of coelomic nature sends numerous small projections (podia) to the exterior. Since Echinoderms, except for Holothuroidea (sea cucumber) do not have neither a head nor a tail, the anterio-posteriority does not appear as clearly defined. Generally, the anus open upwards and the mouth downwards. Therefore, dorso-ventrality cannot be distinguished from the external appearance of a sea urchin. The upper and lower sides are called aboral (or anal) and oral side, respectively.

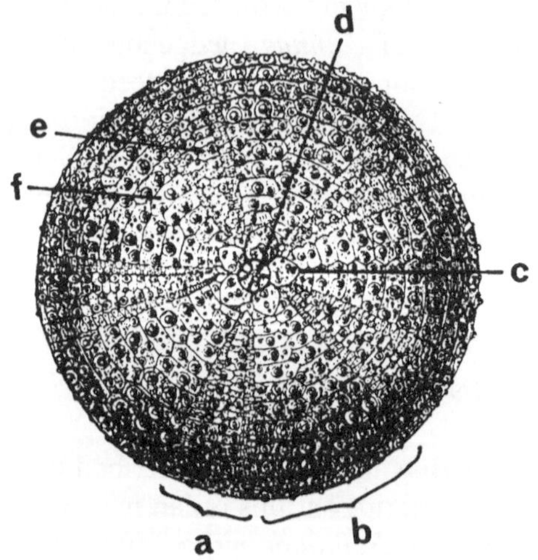

Figure 1. External appearance of regular sea urchins (anal hemisphere)
a: ambulacrum, b: interambulacrum, c: genital pore, d: anus, e: ambulacral plate, f: interambulacral plate.

The Echinoidea (sea urchins) are divided into two structurally different groups: Regularia (regular sea urchin) and Irregularia (irregular sea urchin). Regularia is a group of sea urchins (Fig. 1) having pentagonal structure; Irregularia is composed of sand dollars and heart urchins, which present a symmetrical and antero-posteriority structure. All edible sea urchin species in the world belong to Regularia group. The external morphology of sea urchins belonging to

Regularia is characterised by a test (shell) which is generally hemispheric, with a flat- bottom (oral) half and a curved upper (anal) half (Fig. 2). The body is covered with a huge number of spines and pedicellaria. Sphaeridia are distributed among the spines. These latter are the most conspicuous feature in sea urchins. Spines in the oral and anal areas are shorter than the others. The tube feet (podia), which are flexible and tubular-shaped, are disposed radially, forming 5 double rows of podia. A couple of podia and the region between a pair of rows are called ambulacrum. The area between ambulacra is called interambulacrum, which is usually wider than the ambulacrum. The mouth lies in the centre of the bottom-half surface, where five hard teeth protrude. The mouth is surrounded by a circular peristome, whose texture is that of a soft membrane. The mouth apparatus is called 'Aristotle's lantern'. The anus lies on an eccentric position, almost in the centre of the apex, on the upper half. The anus is surrounded by a membranous structure called periproct, surrounded by a slightly higher area called apical system. Five genital pores are found in this system, corresponding to the interambulacrum. The plate shows a genital pore (genital plate). Spines are a chief characteristic of the sea urchin.

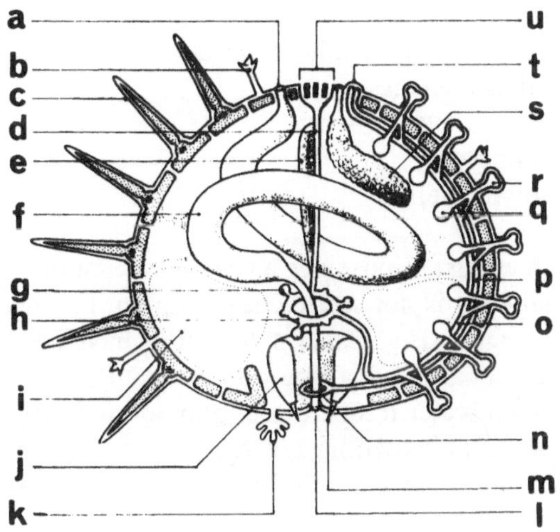

Figure 2 Longitudinal section of the sea urchin.
a: anus, b: pedicellarium, c: spine, d: stone canal, e: axial organ, f: intestine, g: Polian vesicle, h: circular water vessel, i: coelom, j: Aristotle's lantern, k: gill, l: mouth, m: teeth, n: nerve ring, o: radial canal, p: radial nerve, q: ampulla, r: tube foot, s: gonad, t: terminal tentacle, u: madreporite (from Seibutsu Eigo Handbook by Masuda & Noguchi: in Japanese)

According to their size, they are classified into primary, secondary and tertiary spines. Primary (the largest) and secondary ones (smaller) are found in the meridional row, from the peristome to the periproct. Tertiary spines (the smallest) encircle the primary and secondary ones. The difference between primary and secondary spines, though, is not very clear. The sea urchin belonging to genus *Arbacia* has no tertiary spines. There are also spines having different length, such as 30 cm in *Diadema* and 5 mm in *Hemicentrotus*.

Tube foot (podium) is a flexible, tubular structure with a disk on its apex. Podia have five double rows on the ambulacrum, from the peristome to the periproct. Five pairs of buccal podia are generally present in the peristome. The muscle of podia is a smooth muscle. The spicule contained in the connective tissues of podia is one of the classification criteria.
Pedicellaria are minute structures distributed among the spines that can be detected in the peristome. Pedicellaria in *Toxopeneustes* genus are large and white, appearing as a sort of flower. Pedicellaria represent the most remarkable feature on the test surface in *Toxopneustes*.

The body surface is covered with epidermis (except for Cidaroida spines). Variations in the epidermis colour can be observed even in the same species (Example: *Paracentrotus lividus, Echinus esculentus* etc.).

The main part of the test (shell) consists in the coronal system, formed of coronal plates. The periproctal system is encircled by the apical; both are located on top of the test. The peristomial system is situated in the lower part of peristome. Coronal plates are divided into two groups: ambulacral (poriferous, i.e. with ambulacral pores) and interambulacral (non-poriferous). The test consists of 20 rows of plates; five ambulacral areas are formed by two ambulacral plates, with pores for podia passage; five interambulacral areas are two-plated. The interambulacral is usually wider than the ambulacral area.

4. DEVELOPMENT OF THE SEA URCHIN

All species reported so far are marine species and carry out external fertilisation. Since the historical observation by Hertwig (1876), the sea urchin has been employed as a developmental model because of the following reasons:

1) availability of large amounts of eggs and embryos;
2) easy artificial fertilisation
3) possibility of direct observation of phenomena related to fertilisation, thanks to the transparency of eggs in many species;
4) relatively high synchrony of development;
5) easy culturing of early embryos;
6) lack of a hard-egg envelope (shell).

Sea urchins have specific breeding periods once a year. At the moment, the available information on the growth of primordial cells is not sufficient to explain the mechanism of gonad growth. Although investigations on oogenesis and spermatogenesis have been carried out, the molecular mechanism controlling the annual gonadal periodicity is still unclear. What is certain is that gonads maturation depends on water temperature, photoperiodicity and dietary conditions. In addition, gonadal maturation and spawning are reported to be controllable (Bay-Schmith & Pearse 1987, Ito et al. 1989, Sakairi et al. 1989, Masaki & Kawahara 1995, Noguchi et al. 1995, Walder & Lesser 1998).

4.1 *Collection of eggs and spermatozoa*

Natural seawater is suitable for artificial fertilisation and development. As it contains dusts, sands, plankton, microbes and other small particles, though, it is necessary to use filtered seawater at the quality required (Millipore filter (0.45 mm) is the most recommended). Containers for seawater have to be selected carefully. Metal containers should be avoided, as heavy metals could be released into seawater, resulting in the inhibition of fertilisation and cleavage – thus in abnormal development. Plastic containers can promote the formation of exogastrulae, therefore they should be filled with water for a few months prior to use. Glass containers are the most suitable ones.

Concerning the identification of sex and the collection of gametes, we should consider that sea urchins are substantially dioecious, though hermaphrodites can be rarely observed. In this case, a few of five gonads are testes and the others are ovaries. The eggs from a hermaphrodite undergo self-fertilisation, cleavage and development. Sex identification from the outer appearance is impossible in many species. However, identification can be made according to the difference in the shape of genital papilla (Tahara et al. 1958, 1960). Artificial spawning or anatomical observation can help to identify the sex. In order to avoid contamination of spermatozoa in the egg suspension, anatomical instruments (scissors, tweezers etc.) and glassware should be washed with tap water (at least 10 seconds) to avoid spontaneous fertilisation. Spermatozoa are sensitive to low osmotic pressure.

In order to obtain the gametes without any chemicals, a sea urchin has to be opened by inserting a scissor blade into the soft derma around the mouthpart (Aristotle's lantern). To separate the oral hemisphere from the anal one, a transversal cut should be performed. The five gonads are radially attached to the test. A white emulsion (semen) indicates a male, while a red and relatively transparent viscous fluid denotes a female.

Testes are taken out with tweezers or a spoon and put in a small beaker. They can be stored for two days in the refrigerator. Spermatozoa maintain their fertilising capacity in dry sperm. The dry form is also called semen from testes or undiluted sperm.

Ovaries transferred into seawater cause eggs release. Vigorous shaking of ovaries in seawater results in the release of immature eggs with germinal vesicles. Ovaries can be removed by filtration with twice-folded gauze or a plankton-net sheet. Eggs will settle in ten or twenty minutes. The supernatant seawater is discharged by aspiration; then fresh seawater is added to wash the egg suspension. This washing step should be repeated four or five times.

Using the KCl method implies the insertion of a scissor blade into the soft derma around the mouthpart (Aristotle's lantern) in order to remove it together with the coelomic fluid. Pour 0.5 - 0.55 M KCl in to the opened body cavity. In the case of males, KCl must be washed out in seawater and testes are saved. In female animals, the sea urchin is put on a beaker or an Erlenmeyer flask filled with seawater. The eggs spawned appear as disposed in five lines. They are washed four or five times with fresh seawater. Spawning is also induced by the injection of KCl solution through the peristome. In this case, sea urchin can survive after spawning.

Spawning can be also induced by the acetylcholine method. Injection of 1 mM acetylcholine chloride (in sea water or 0.55 M NaCl) into the blastocoel is performed by inserting the injection needle into the peristomal membrane. The advantage of this method is that the sea urchin can survive after spawning, as acetylcholine is hydrolyzed by acetylcholine esterase in the gonads. Acetylcholine solution can be stored at room temperature for approximately 6 months. The injection of a small amount of acetylcholine solution can be useful in distinguishing females from males.

4.2 *Fertilisation and Development*

Eggs must be washed 5 or 6 times before and after fertilisation with Millipore filtered seawater. In this way, microbes (ciliates in coelom), excess of spermatozoa and other polluting agents can be removed. Protozoa and microbes contaminating sea urchin embryo cultures are harmful for embryos and larvae's growth. Embryos are cultured with continuous rotation of seawater at 30 rpm, using a puddle driven by a synchronous motor. The concentration of embryos at the 4-arm pluteus stage should be calibrated as 1 per 5 - 10 ml of seawater. A high concentration of embryos results in delayed development and growth. The aeration of seawater tends to damage the larvae. Seawater must be changed at the interval of 2 or 3 days (1/4 - 1/3). More frequent changes of seawater are recommended. Feeding should take place every 2 or 3 days. Eight to ten ml of nearly full growth culture medium of *Chaetoceros gracilis* or *Dunaliella sp.* is added to one or two liters of sea urchin larvae suspension. In order to avoid culture hypertrophy, washing of algae by centrifugation should be made. The debris formed in the larvae culture gathers in the centre of the bottom by rotating the water; it can be removed using a sucker, or a pipette.

Those who engage themselves in sea urchin aquaculture should familiarise with sea urchin embryos and larvae before being able to judge their conditions by microscopic observations. The developmental table from the unfertilised eggs to the juvenile sea urchin stage after metamorphosis is essential in sea urchin aquaculture. The development of several Japanese species, *Hemicentrotus pulcherrimus*, *Pseudocentrotus depressus*, *Strongylocentrotus intermedius*, *Echinostrephus aciculatus* and *Clypeaster japonicus*, including the structure of skeletons, was illustrated up to the stage of juvenile sea urchin by Ishihara and Noguchi (1996).

Rearing sea urchins from eggs to juveniles requires a specific equipment, such as those listed below:

for the maintenance of adult sea urchins;
- Aquarium with thermo-regulating apparatus

for the culture of algae;
- Autoclave, Sterilising oven, Pipette washer, Illuminated shelf, Air pump with filter, Microscope, Hemocytometer, Cabinet for microbiology

for the culture of sea urchin larvae;
- Filtering system (1 mm or less), Thermo-regulating system for seawater, Microscope (an inverted microscope is better)

for the general objects;
- Drying shelf for glassware, Air conditioner, Distiller, Balance Equipment for glass working.

5. MAINTENANCE AND TRANSPORT OF SEA URCHINS

Materials for aquaria should be selected with care, depending on the circulation system. In a place by the sea, water can be easily supplied by pumping it to the aquarium; the exceeding seawater can be discharged to the sea. The advantage of this open-water circulation system is that fresh seawater is constantly provided. However, an inlet may sometimes be covered with dusts and rubbish (wood fragments, seaweed and others). If water supply stops, there is a risk that all animals in aquarium die. In order to avoid such events, a well can be sunk in the seashore to draw seawater. This could be filtered through the sand. The location to sink a well should also be selected carefully, since underground water can dilute the pumped seawater.

The closed-water circulation system is convenient because marine animals can be kept in a place distant from the sea. The overflowing seawater is not discharged into the sea but recycled by bio-filtration. A pump with a flow rate of 25 - 80 l/min and a lifting capacity of 6 - 9 m can support a container up to 1 t. Seawater management is required to maintain the water

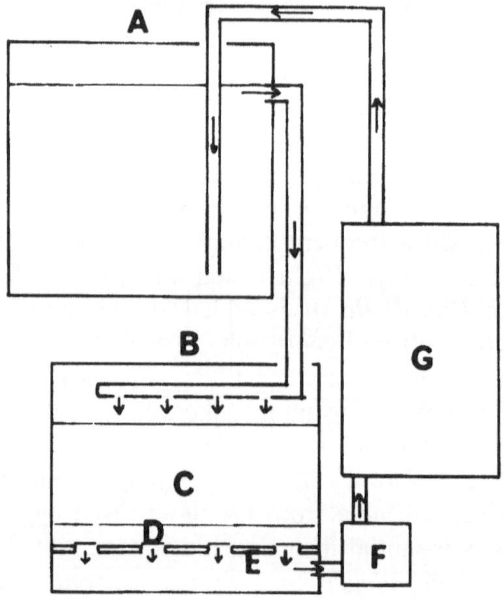

Figure 3. Schematic diagram of closed aquarium system
A: aquarium for sea urchins, B: Filtering system, C: Sand, D: Synthetic or glass fibres, E: Strainer, F: Pump, G: Cooling unit equipped with thermo-regulator. Arrows indicate the flow direction.

quality. The important parameters to consider are pH, salinity, temperature etc. Residual baits and organisms must be removed. A thermo-regulating system and an aeration pump must be provided. The water-cooling system should have a thermostat with a maximum deviation of 1°C. A cooling unit with 1,000 cal/hr is able to regulate the seawater from 500 - 1,000 l. A periodical change of seawater is recommended. De-nitrification by bacteria is essential for the maintenance of water quality. The bacterium is commercially available. The aquarium and container must be covered in order to avoid the rise of salinity due to evaporation. A characteristic system is shown in figure 3.

When sea urchins with mature gametes are kept in aquarium, it must be avoided that all of them are reared in one tank. Spontaneous spawning of one organism stimulates spawning of other individuals in the aquarium. Therefore, mature sea urchins should be separated into groups maintained in an independent vessel.

Transportation of living sea urchins is possible for approximately 24 hours, making sure that they are properly packed in a thermo-insulated box with coolants. Transportation of sea urchins with ripe gonads in seawater is to be avoided, since turbulence promotes spawning. The approximate oxygen air pressure, 159 mmHg, is much higher than that of seawater. Epidermis plays a major role in sea urchin respiration. The animal is able to obtain a sufficient amount of oxygen to survive out of water, when dehydration is avoided. Sea urchins must be kept in a container, taking care that peristomal regions of individuals are not damaged by spines. Figure 4 shows a schematic diagram showing how sea urchins should be packed.

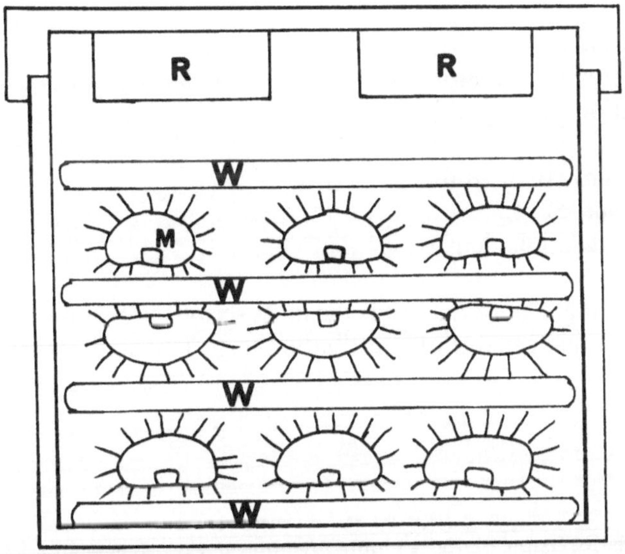

Figure 4. Packaging of sea urchins
R: refrigerant, W: wet materials such as towels or sheets of urethane foam, M: mouth part of sea urchin.

6. FACTORS AFFECTING SEA URCHIN PROLIFERATION AND GROWTH

Although one female sardine spawns millions of eggs, only one or two of those eggs will survive, as it is known. The same could be applied to sea urchins. Many difficulties are found in their development, from the eggs to the mature adults stage. Artificial fertilisation followed by rearing is expected to be an effective aid in recovering sea urchin resources. Artificial propagation of salmon has given successful results. Unfavourable factors for the growth and proliferation of sea urchins are listed below.

Factors decreasing the number of planktonic stage larvae:
Physical: water temperature, current (tide), turbulence of seawater, salinity.
Chemical: pH, heavy metals, hydrocarbons.
Biological: gonads conditions, density, feed, predation, biological properties of settling spots.

Factors affecting the growth of planktonic stage larvae:
Physical: temperature, circulation
Biological: parental gonads conditions, density, feed

Factors decreasing the number of benthic larvae
Physical: characteristics of bottom, temperature, circulation, salinity, waves
Biological: predation

Factors affecting the growth of benthic larvae
Biological: feed

Factors decreasing the juvenile stage larvae:
Physical: characteristics of bottom, temperature, circulation, salinity, wave
Biological: predation

Factors affecting the growth of juvenile stage larvae:
Biological: feed, density, competitive animals

Factors affecting the number of adults
Physical: characteristics of bottom, temperature, circulation, salinity, wave
Biological: predation, fishing, life

Factors affecting the growth of adults
Physical: waves (surge)
Biological: feed, predation, density.

In conclusion, the fact that *Homo sapiens* is the most serious predator of the sea urchin indicates that we are the only animal capable of contributing to sustain and recover the global sea urchin resources.

REFERENCES

Agassiz A. 1872-1874. Revision of the Echini. *Memoir of the Museum of Comparative Zoology, at Harvard University* 3: 1-762.

Agassiz A. 1881. Report on the Echinoidea. Report of Challenger-Exped., 3. pt. 9: 1-321.

Agassiz A. & H. L. Clark 1907, 1908, 1909. Hawaiian and other Pacific Echini. *Memoirs of the Museum of Comparative Zoology, at Harvard University*. 34: 1-204.

Bay-Schmith E. & Pearse J.S. 1987. Effects of fixed daylengths on the photoperiodic regulation of gametogenesis in the sea urchin *Strongylocentrotus purpuratus*. *International Journal of Invertebrate Reproduction and Development* 11: 827-294.

Boolootian R. A. (ed.) 1966. *Physiology of Echinodermata*. 1-822. New York: John Wiley & Sons.

Clark, A. H. 1922. Echinoderms as aberrant arthropods. *Smithsonian Miscellaneous.Collections* 72, 11, 1-20. see also review by Bather F. A. 1922, *Nature* 109: 640-641.

Clark H. L. 1912, 1914, 1917. Hawaiian and other Pacific Echini. *Memoirs of the Museum of Comparative Zoology, at Harvard University* 34: 205-383; 46: 1-284.

Clark. H. L. 1925. *A catalogue of the recent sea urchins in the collection of the British Museum*. 1-250.

Cuvier G. 1817. *Le règne animal*, vol. IV.

Duncan P. 1883. Morphology of the test of the Temnopleuridae. *Journal of the Linnaean Society of London, Zoology* 16.
Duncan P. M. 1888. A revision of the genera and great groups of the Echinoidea. *Journal of the Linnaean Society of London, Zoology* 23: 1-311.
Durham J. W. 1955. Classification of Clypeasteroid Echinoids. *Publications of the University of California at Los Angeles in Biological Sciences* 31: 73-198.
Durham J. W., H. B. Fell, A. G. Fischer, P. M. Kier, R. V. Melville, D. L. Pawson & C, D. Wagner 1966. *Treatise on Invertebrate Paleontology. Echinodermata* 3 (1, 2) (Echinoidea), U211-U633, University of Kansas Press:
Gordon I. 1926. VII. The development of the calcarous test of *Echinus miliaris*. *Philosophical Transaction of the Royal Society of London, Series B, Biological Sciences* 214: 259-310
Gregory J. W. 1900. Echinoidea; in E. R. Lankester (ed.), *A Treatise on Zoology, Part III, the Echinoderma,* 282-332, London: A. & C. Black.
Hamann O. 1887. Beitrag zur Histologie der Echinodermen: Heft 3, Anatomie und Histologie der Echiniden und Spatangiden. *Jenaische Zeitschrift für Naturwissenschaft* 21: 87-266.
Hawkins H. 1912. The classification, morphology, and evolution of the Echinoidea Holectypoida. *Proceedings of the Zoological Society of London* pt. 1: 440-498.
Hawkins H. 1920. The morphology and evolution of the ambulacrum in the Echinoidea Holectypoida. *Philosophical Transactions of the Royal Society of London, Series B, Biological Sciences* 209: 477-480.
Hertwig O. 1876. Beiträg zur Kenntnis der Bildung, Befruchtung und Teilung des tierischen Eies. *Gegenbaurs Morphologisches Jahrbuch* 1: 374-434.
Höerstadius S. 1935. Ueber die Determination im Verlaufe der Eiachse bei Seeigeln. *Publicazioni di Stazione Zoologica di Napoli* 14: 251.
Hörstadius S. 1939. The mechanics of sea urchin development, studied by operative methods. *Biological Reviews* 14: 132.
Hyman L. H. 1955. Echinodermata. *The invertebrates. vol. IV*: 413-589. New York: McGraw-Hill.
Ishihara, K. & Noguchi, M. 1996. Sea urchin. In K. Ishihara (ed.), *Developmental atlas of animals* (In Japanese), 191-220. Tokyo: Kyoritsu-Shuppan.
Ito S., Shibayama M., Kobayakawa A. & Tani Y. 1989. Promotion of Maturation and spawning of sea urchin *Hemicentrotus pulcherrimus* by regulating water temperature. *Nippon Suisan Gakkaishi* 55: 757-763. (in Japanese)
Jackson R. T. 1912. Phylogeny of the Echini, with a revision of palaeozoic species. *Memoirs of the Boston Society of Natural History* 7: 1-491.
Klein J. T. 1734. Naturalis dispositio echinodermatium.
Koehler R. 1883. Recherches sur les echinides des cotes de Provence. *Annales du Musée d'Histoire Naturelle de Marseilles, Zool.* 1: mem. 3.
Lambert, J. & P. Theiéry 1909-1925. *Essai de Nomenclature Raisonnée des Echinides.* 1-607, Chaumont: Libraries Septum Ferrari.
Lamarck J. B. 1801. *System des Animaux sans Vertebres.*
Lamarck J. B. 1816-1822. *Histoire Naturelle des Animaux sans Vertebres.*
Linné C. 1758. *System Naturae*
Lovén S. 1874. Etudes sur les echinoidees. *Konglia Svenska Vetenskaps Akademiens Handlligar. Ny foljd.* 11: no. 7.
Lovén S. 1883. On Pourtalesia. *Konglia Svenska Vetenskaps Akademiens Handlligar. Ny foljd.* 19: no. 7
Lovén S. 1892. Echinologica. *Konglia Svenska Vetenskaps Akademiens Handlligar. Ny foljd.* 18, Afd. 4, no.1.
MacBride E. W. 1903. VI. The development of *Echinus esculentus* together with some points in the development of *E. miliaris* and *E. acutus*. *Philosophical Transactions of the Royal Society of London, Series B, Biological Sciences* 195: 285-327.
MacBride E. W. 1914a. The development of *Echinocardium cordatum*. Part 1. The external feature of the development. *Quarterly Journal of Microscopical Science* 59.
MacBride E. W. 1914b Text-book of Embryology. Invertebrata Vol. I., pp. 504-529. London: Macmillan.
Masaki K. & Kawahara I. 1995. Promotion of gonadal maturation by regulating water temperature in the sea urchin *Pseudocentrotus depressus* –I. *Bulletin of Saga Prefectural Sea Farming Center* 4: 93-100.

Morgan T. H. 1893. Experimental studies on echinoderm eggs. *Anatomischer Anzeiger* 9: 141.
Morgan T. H. 1899. The action of salt solution on the unfertilized and fertilized eggs of *Arbacia*, and of other animals. *Wilhelm Roux' Archiv für Entwicklungsmechanik der Organismen* 8: 448.
Morgan T. H. 1909. The effects produced by centrifuging eggs before and during development. *Anatomical Record* 3: 155.
Mortensen T. 1927, *Handbook of the echinoderms of the British Isles. Humphrey Milford*, London: Oxford University Press.
Mortensen T. 1928-1951. *Monograph of the Echinoidea*. Copenhagen: C. A. Reitzel.
Mortensen T. 1931. Contributions to the study of the development and larval forms of echinoderms. I. The development and larval forms of some tropical echinoderms. II. Observations on some Scandinavian echinoderm larvae. *Memoir of the Royal Academy of Science Letter Denmark*, 4: 1-39.
Mortensen T. 1937. Contributions to the study of the development and larval forms of echinoderms. III. *Memoir of the Royal Academy of Science Letter Denmark*, 7: 1-65.
Mortensen T. 1938. Contributions to the study of the development and larval forms of echinoderms. IV. *Memoir of the Royal Academy of Science Letter Denmark*, 7: 1-59.
Noguchi H., Kawahara I., Goto M. & Masaki K. 1995. Promotion of gonadal maturation by regulating water temperature in the sea urchin *Pseudocentrotus depressus* – II. *Bulletin of Saga Prefectural Sea Farming Center* 4: 101-110.
Onoda K. 1931. Notes on the development of *Heliocidaris crassispina* with special reference to the structure of the larval body. *Memoirs of the College of Science, Kyoto Imperial University Ser. B*, 7: 103-134.
Onoda K. 1936. Notes on the development of some Japanese Echinoids with special reference to the structure of the larval body. *Japanese Journal of Zoology* 6: 637-654.
Onoda K. 1938. Notes on the development of some Japanese Echinoids with special reference to the structure of the larval body Report III. *Japanese Journal of Zoology* 8: 1-13.
Prouho H. 1887. Recherches sur le *Drococidaris papillata* et quelques autres Echinides de la Méditerranée. *Archives de Zoologie Experimentale et Générale. ser. 2*, 5: 289-380.
Sakairi K., Yamamoto M., Ohtsu K. & Yoshida M. 1989. Environmental control of gonadal maturation in laboratory-reared sea urchins, *Anthocidaris crassispina* and *Hemicentrotus pulcherrimus*. Zoological Science 6: 721-730.
Shearer C. 1922a. On the oxidation processes of the echinoderm egg during fertilisation. *Proceedings of the Royal Society of London, Series B. Biological Sciences* 93: 213.
Shearer C. 1922b. On the heat production and oxidation processes of the echinoderm egg during fertilisation and early development. *Proceedings of the Royal Society of London, Series B. Biological Sciences* 93: 410.
Tahara Y., Okada M. & Kobayashi N. 1958. Secondary sexual characters in Japanese sea urchins.*Publications of Seto Marine Biological Laboratory* 7: 165-172.
Tahara Y., Okada M. & Kobayashi N. 1960. Further notes on the sexual dimorphism in Japanese sea urchins. *Publications of Seto Marine Biological Laboratory* 8: 183-189.
Uexkull J. von 1896. Ueber Reflexe bei den Seeigeln. *Zeitschrift für Biologie* 34: 298-318.
Uexkull J. von 1900. Die Physiologie des Seeigelstachels. *Zeitschrift für Biologie* 39, 73-112.
Uexkull J. von 1907. Studien uber den Tonus. IV. Die Herzigel. *Zeitschrift für Biologie* 49: 307-332.
Walder C.W. & Lesser M.P. 1998. Manipulation of food and photoperiod promotes out-of-season gametogenesis in the green sea urchin, *Strongylocentrotus droebachiensis*: Implication for aquaculture. Marine Biology 132: 663-676.

The Sea Urchin: From Basic Biology to Aquaculture, Yokota, Matranga & Smolenicka (eds)
© Swets & Zeitlinger, Lisse, ISBN 90 5809 379 4

Imaging in the microscope, phase contrast and DIC-instrumentation and applications in cell, developmental and marine biology

H. Gundlach
Carl Zeiss, Division of Research and Technology, Oberkochen, Germany

ABSTRACT: The exact design of the microscope was established from a theory for image formation developed by Ernst Abbe in 1873. With the introduction of apochromate lenses and oil immersion, the theoretical resolution limit of 200 nm became available at this early stage. Light microscopy advanced further by new lens design, optical contrast enhancement methods and application of fluorescent and immuno-fluorescent dyes, also in combination with inverted microscopes. Illumination and contrast are the central points in the discussion of imaging performance. Without sufficient contrast, neither the required magnification nor the maximal resolution can be obtained. Köhler (1893) developed optical illumination for microscopy. The main advantage of Köhler illumination is that in spite of the small filament of the light bulb, the entire area of the illuminated field diaphragm, as well as the illuminated part of the specimen has a uniform luminance. Nomarski's differential interference contrast (DIC) microscopy is discussed in comparison to Zernike's phase contrast (PhC) microscopy. The possibilities and limits of both are demonstrated by various applications. The high contrast and the use of the full numerical aperture of the DIC optics makes it possible to obtain a series of "optical sections" through rather thick living specimens (e.g. head of water flea, salivary gland of *Drosophila*, *Xenopus* nucleolus, sea urchin egg, mouse embryo etc). PhC and DIC optics are today available for high resolution light microscopy until N.A. 1.4 Oil as well as for long working distance (LWD) optics, mainly combined with inverted biological microscopes.

1. INTRODUCTION

The last 15 years was a renaissance of light microscopy which was not expected in such a dimension. The progress in modern light microscopy is produced by means of new design in optics and instrumentation, improvement of optical contrast enhancement techniques, e.g. phase- and DIC contrast, polarisation and reflection contrast also in combination with epi-fluorescence and in combination with inverted microscopes.

Multi-colour fluorescence microscopy is of great importance by nonisotopic *in situ* hybridisation for the simultaneous detection of multiple nucleic acid sequences for the analysis of cell components, genes and chromosomes, also in combination with transmitted light contrast enhancement techniques.

A new approach in light microscopy is given by means of opto-electronic devices and instrumentation for digital imaging. These technologies open a new dimension in 2 D and 3 D imaging and analysis and create the bridge between light microscopy and the low magnification range of electron microscopy but with the importance of using living specimens.

In this paper we will focus mainly on the differential interference contrast microscopy according to Professor G. Nomarski in comparison to phase contrast microscopy by Zernike. The possibilities and limits of both will be demonstrated by various applications.

2. IMPROVEMENT OF IMAGING

For the image formation in the microscope, three main factors are of the utmost importance in obtaining high-quality optics:

- the types of the glass used
- the design of the optical system
- the high precision manufacture of the optical and mechanical system

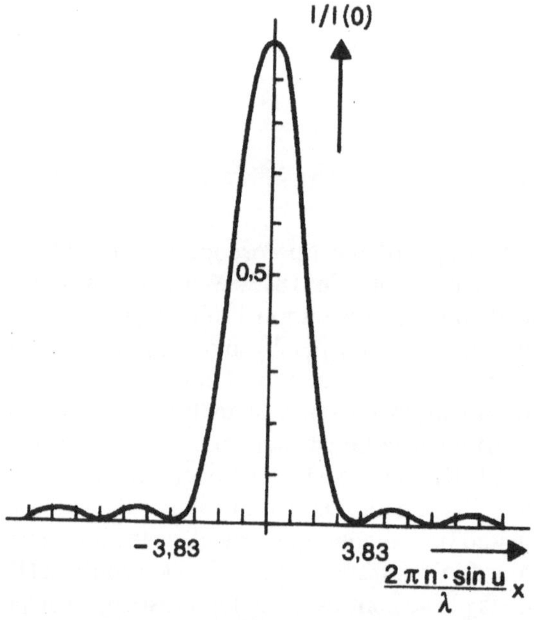

Figure 1. Intensity distribution in the image of a luminous point.

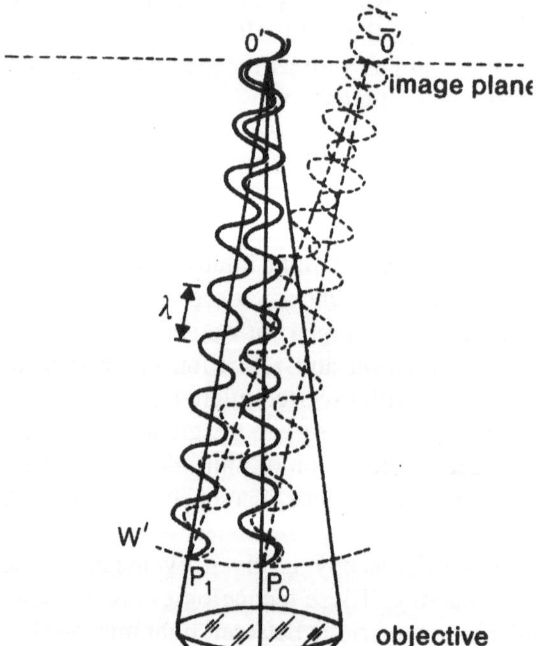

Figure 2. Intensity distribution in the image of a luminous point.

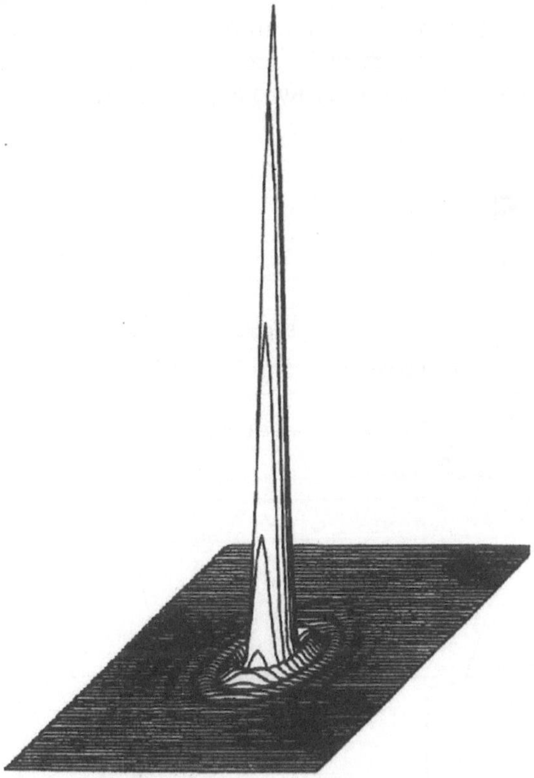

Figure 3. Computed intensity distribution.

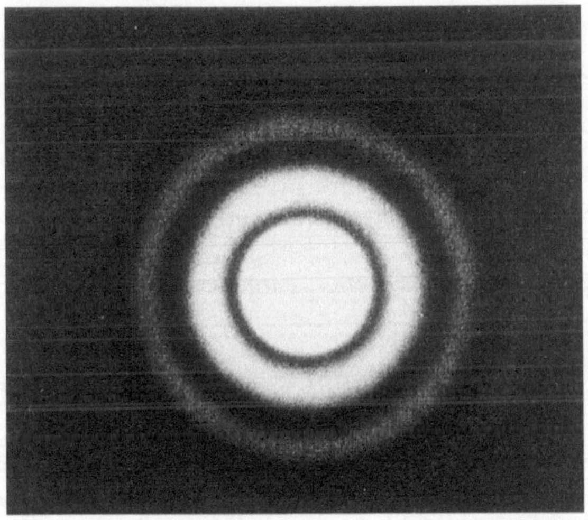

Figure 4. Airy disk.

Large-scale computer systems now permit the use of all theoretically possible mathematical methods in the design of optical system, rapid computation of the point spread function and thus the realistic simulation of the process of image formation in the microscope (Fig. 1-4).

This Infinity Colour Corrected System (ICS, introduced 1986 by Carl Zeiss) objectives which have been optimised using these methods in particular provide maximum image quality. The primary intermediate image has been corrected for monochromatic and chromatic aberrations and "S" indicates that a new correction system is formed by the objective and the tube lens (Fig. 5). One of the most important features of the ICS optics is that the space of the parallel beam path between objective and the tube lens can be optimally utilised to accommodate analyser. DIC prism or reflector slider for incident-light fluorescence works without any need for further optics. The line of Plan-Neofluar objectives are suitable for all illumination- and contrast methods for the large field of view 25 and a chromatic correction which almost matches that of Plan-Apochromat objectives and very well recommended of Nomarski DIC and polarisation because of special Schott FK 51 (= fluoro crown) glass instead of fluorite.

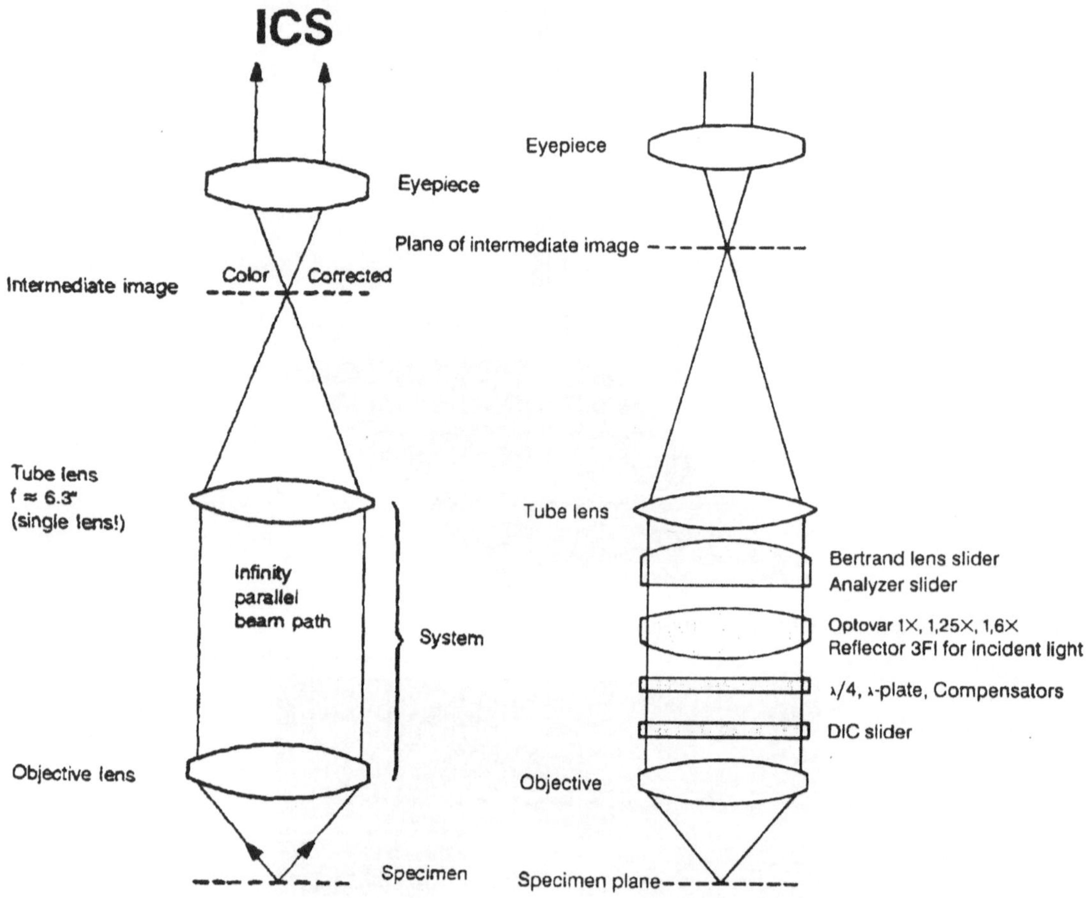

Figure 5. Scheme imaging light path of ICS optics. Introduced 1986 by Carl Zeiss.

3. NOMARSKI DOUBLE-BEAM-INTERFERENCE CONTRAST MICROSCOPE

Figure 6 shows a diagram of a double-beam-interference contrast light beam according to Smith with two Wollaston prisms which are located in the lamp side focal plane of the condenser and the back focal plane of the objective. This system is limited to a small number of objectives and condensers.

A very neat and technically feasible solution to this problem was the introduction of Nomarski's modified Wollaston prism, which in the following will be briefly called „Nomarski prism".(Fig. 7). As a result, the interference plane is outside the compound prism, although the Nomarski prism itself is located at a relatively long distance from the objective. This is of considerable importance for the design of a DIC-microscope. Today, a wide selection of objectives and condensers are available, from long working distance optics (= LD) in combination with inverted microscopes until high magnification/high aperture optics with the maximum of n.A. of 1.4 oil immersion system.

Another advantage is the possibility to compare DIC and qualitative polarisation microscopy without changing objectives. For routine applications phase contrast objectives can also be used for DIC microscopy.

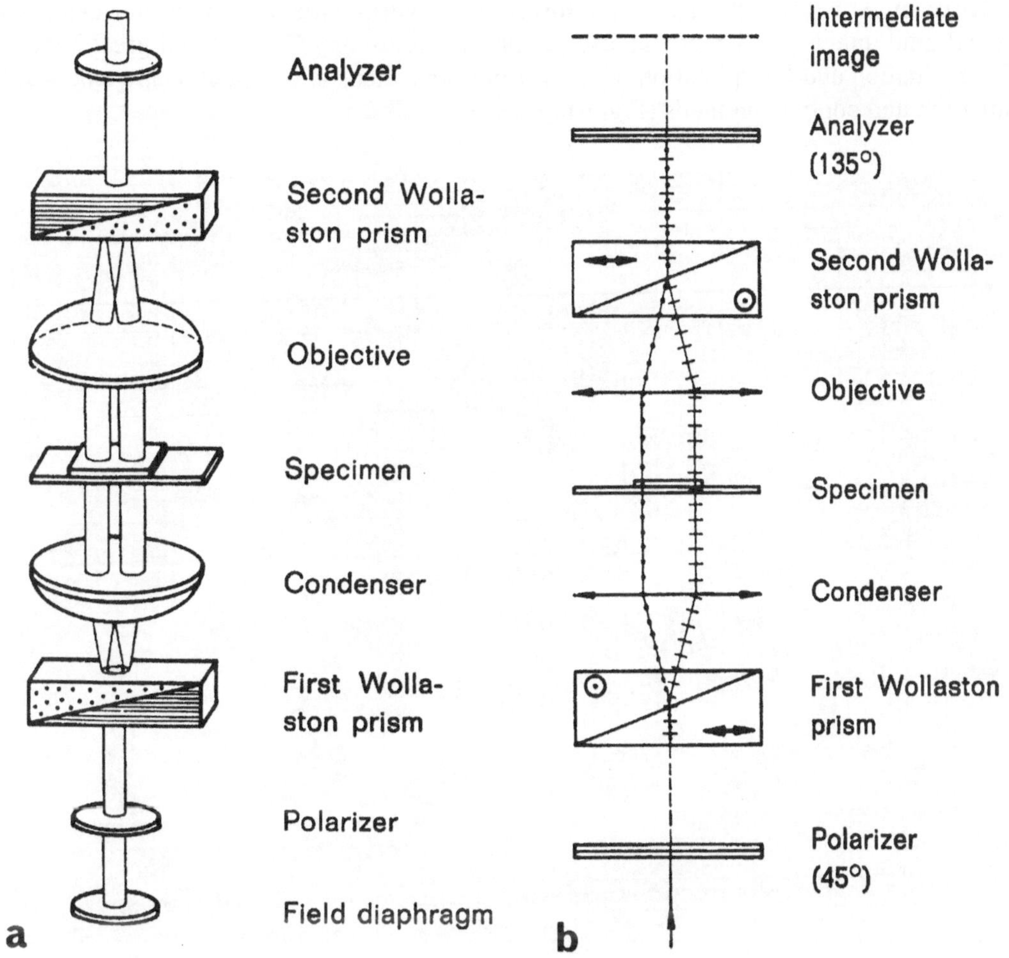

Figure 6. a) Diagram of a double beam interference contrast microscope according to Smith.
b) Principle of a double beam interference contrast microscope with symbols used for the optic axes.

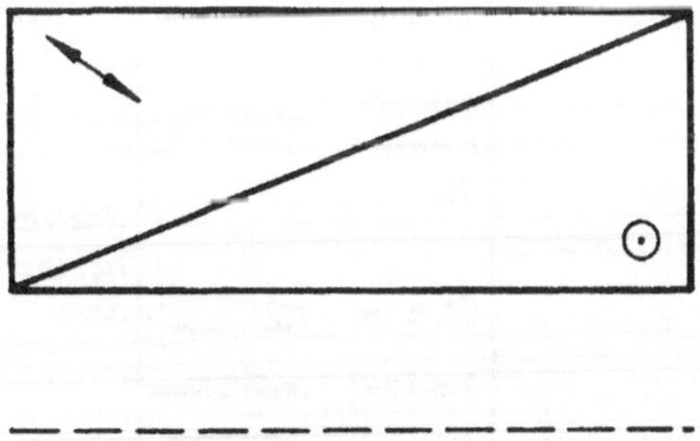

Figure 7. Nomarski prism with direction of optic axes and position of interference plane.

4. APPLICATIONS

Before DIC Nomarski in 1956 was introduced, phase contrast and darkfield are the suitable methods for the evaluation of unstained and living material. Figure 8 is the first photomicrograph of living giant chromosomes in a nucleus of *Chironomus* taken in 1941 by Michel, who has also produced first movie film of the meiosis and mitosis.

This specimen is used in many laboratories today world wide as a model for cell research also for digital imaging and confocal laser scanning microscopy (Fig. 19). Of great importance for the evaluation and interpretation of the various specimens is the quick change between the illumination and contrast methods (Fig. 9).

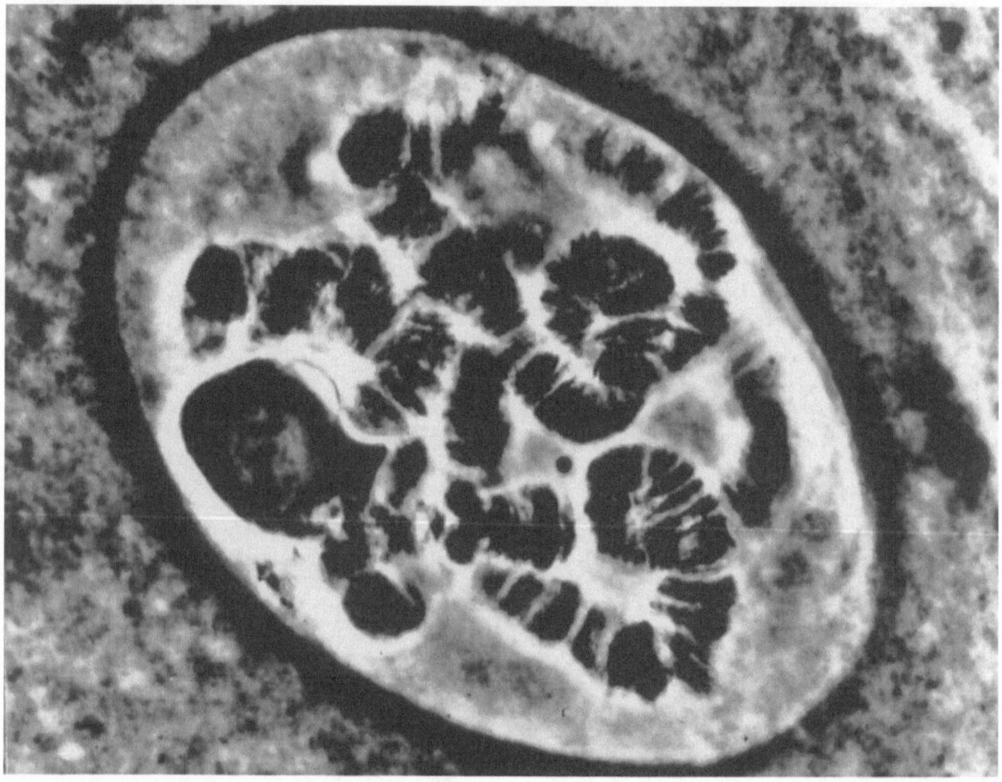

Figure 8. First photomicrograph of living nucleus with giant Chromosomes of *Chironomus*. Dr. Kurt Michel

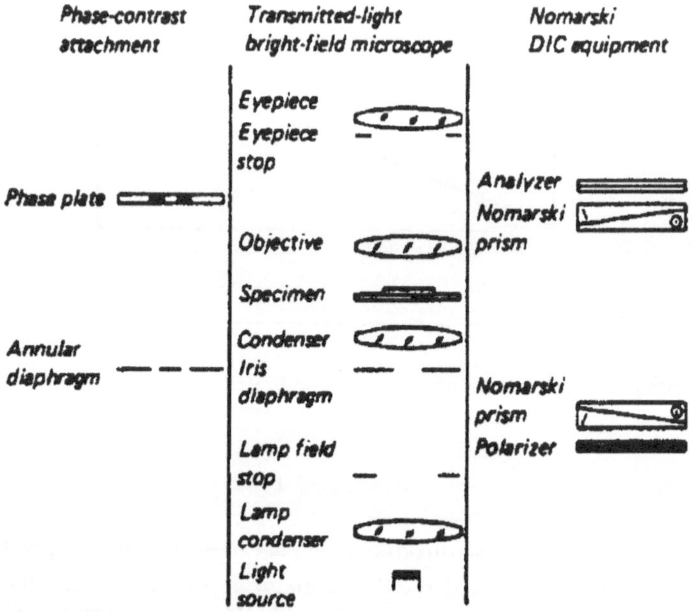

Figure 9. Diagram bright-field, DIC and Phase contrast

Figure 10 shows a comparison between phase and DIC from a head of a water flea and there is no doubt which method is the most suitable one. Mainly if the specimen has a certain thickness, the DIC image is much more sharper and more brilliant because of no "halo" effect and smaller depth of focus. The nerves and the muscles are appearing in a crisp and three dimensional image. Using phase contrast the image becomes indistinct caused by a „halo" and the greater depth of focus.

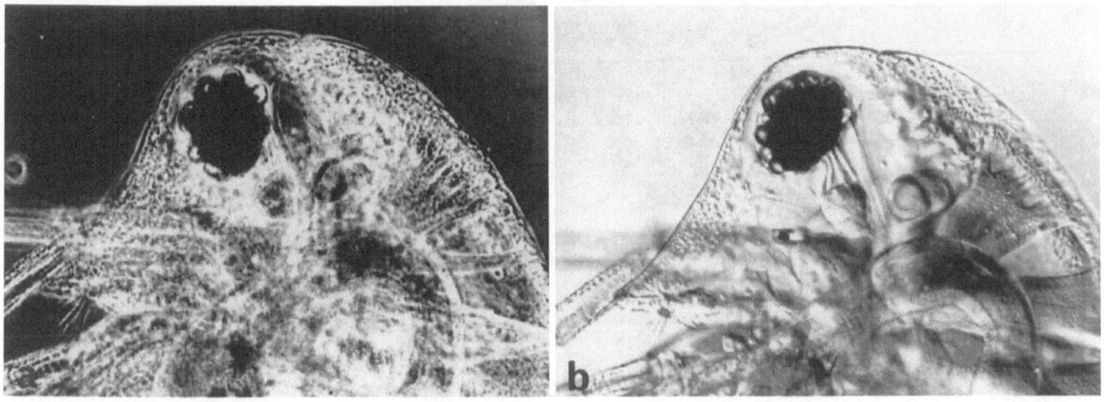

Figure 10. Head water flea, a) phase contrast, b) Head water flea, DIC Nomarski.

In DIC the ratio of focus and resolution is variable. Working with high numerical aperture the depth of focus is small. In this case, optical sectioning is possible (Fig. 11). Six different focus planes are imaged, starting from the top to the bottom of salivary gland, ca. 70 μm thickness. That was the first time that such organs could be evaluated in living state. High resolution DIC was produced in the same specimen with an oil immersion objective 100x for imaging single chromosomes in the interphase nucleus including nucleoli and chromosome banding (Fig. 12).

In 1875 Oskar Hertwig discovered in a sea urchin the fertilisation of an egg, one of the fundamental biological processes. The results in Figure 13 are taken during a workshop in Villefranche-sur-Mer, France in 1976. Only by means of Nomarski optics the chromosomes inside of the cell could be made visible for evaluation and analysis.

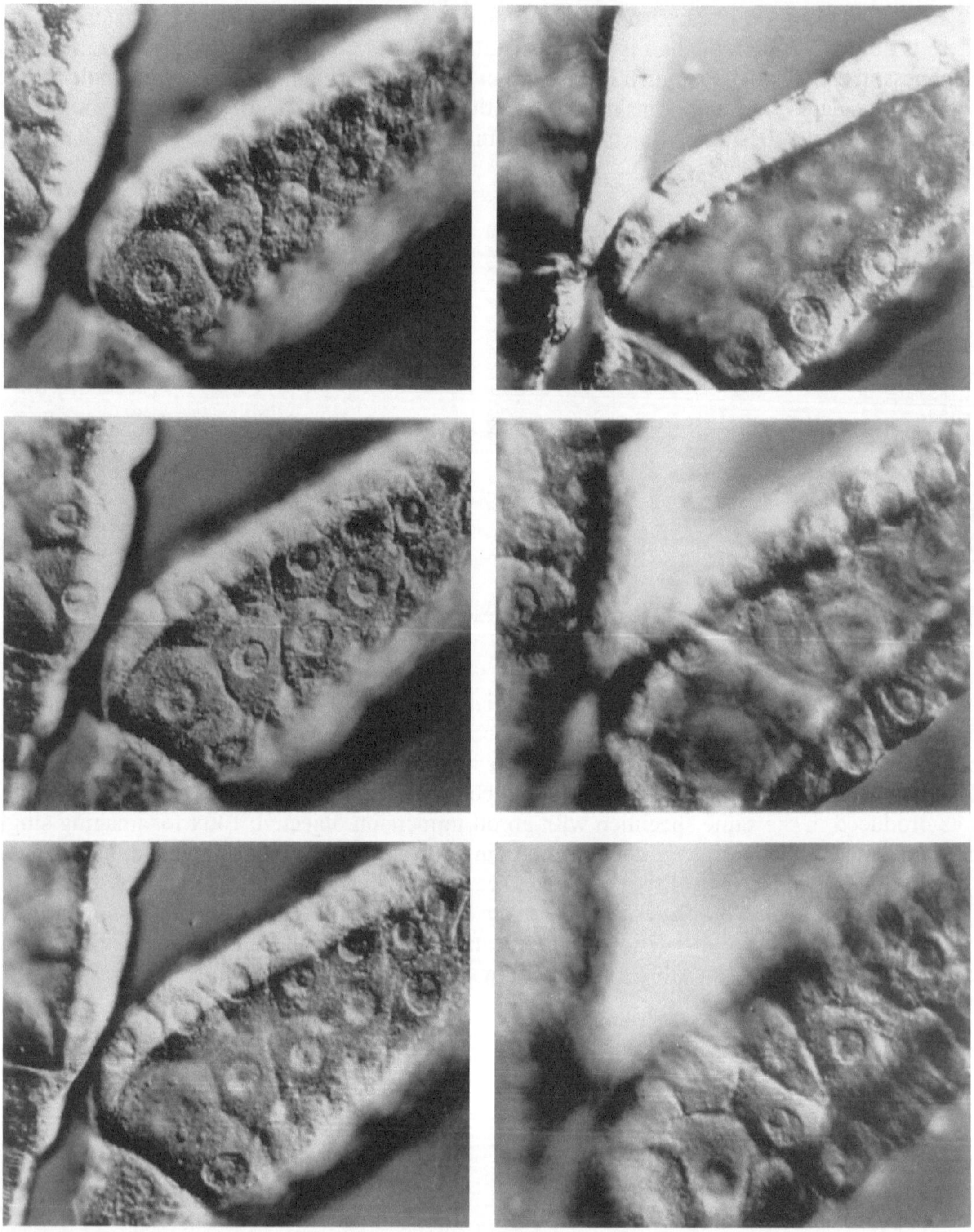

Figure 11. Salivary gland of *Drosophila*. Optical sectioning in DIC. Objective 40x. Specimen Prof. L. Rensing, Zoological Institute, University Bremen.

Figure 12. Nucleus of salivary gland DIC.

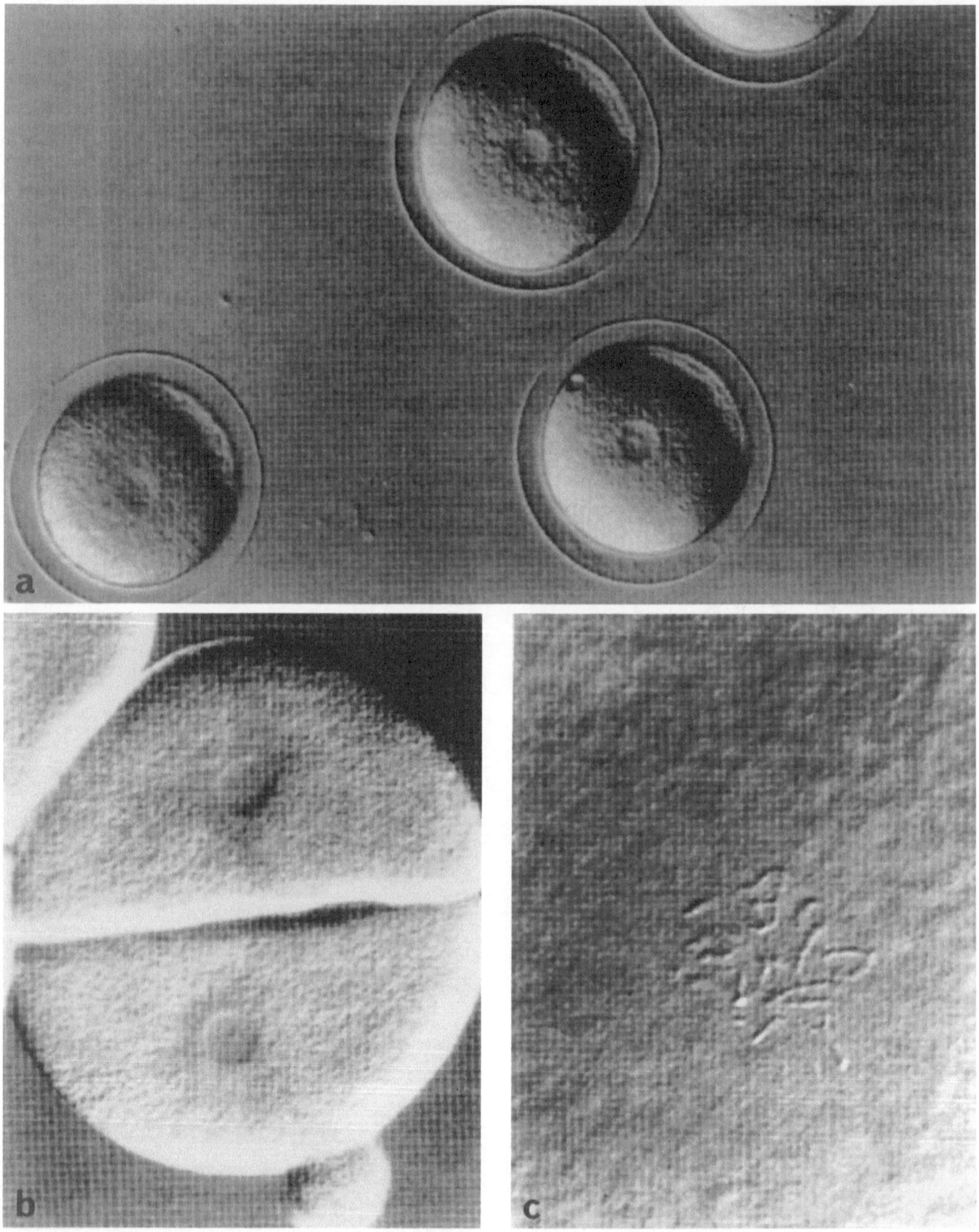

Figure 13. a) Eggs of the sea urchin after insemination. Objective 16x, b) Cell division, Objective 40x, c) Chromosomes, Objective 63/1,4x.

5. INVERTED MICROSCOPY

Inverted microscopes with rigid stages introduced 1976 are great advantage for investigations of cell biology and cell physiology experimenting in living cells with micro pipettes or microelectrodes under an optical observation. Special LD condensers and LD objectives (= long working distance), permit manipulations on living specimens in culture chambers. Figure 14 shows the development of the rabbit embryo in vitro in various magnifications. Because of the thickness of the embryo only DIC can be applied.

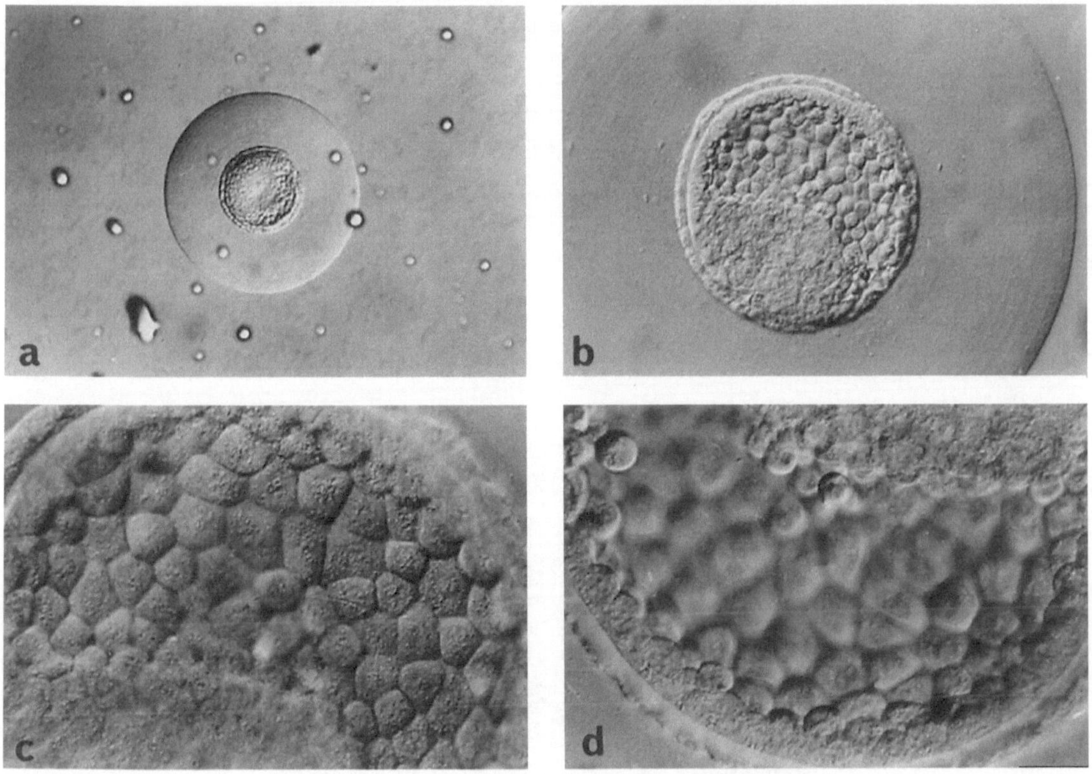

Figure 14. Embryo of the rabbit *in vitro*, ca. 3 days old. a) Objective 6.3x b) Objective 16x c) Objective LD 40x Fig. 14 d) Objective LD 16x Optical sectioning.

Identification and selective micro preparation of live nuclear components in oocytes of *Xenopus laevis* are demonstrated in the following example. Freshly prepared ovaries of the leopard frog contain growing oocytes of different size; mature oocytes can be up to 1.8 mm in diameter. The *in situ*-organisation of an amplified nucleolus of the freshly isolated oocyte nucleus is studied at first in DIC (Fig. 15-17), which permits high resolution focusing in different planes of the specimen in that extraordinary large nuclei (diameter up to 700 μm). The oocyte chromosomes are recognisable in a centre of the sphere (Fig. 16). The DIC method facilitates high resolution optical microscopy of living cell components which, furthermore can be isolated by selective micro preparation for additional analysis in the electron microscope.

Such investigations are becoming increasingly important because they allow a precise description of gene activity of the nuclear DNA contained in the nuclei and an analysis of the interaction between special protein components and gene activity.

The present state of the art in inverted microscopy is the combination of epi-fluorescence, also in multi-parameter mode and DIC or phase contrast simultaneously. By means of this procedure biochemical information can be directly correlated to the morphology and nuclear structures.

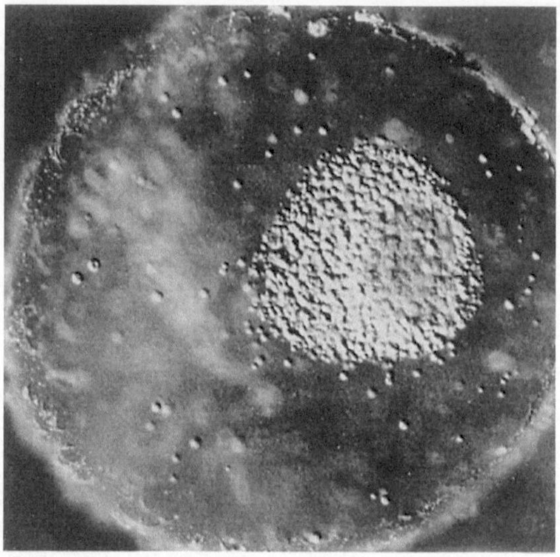

Figure 15. Freshly isolated nucleus.

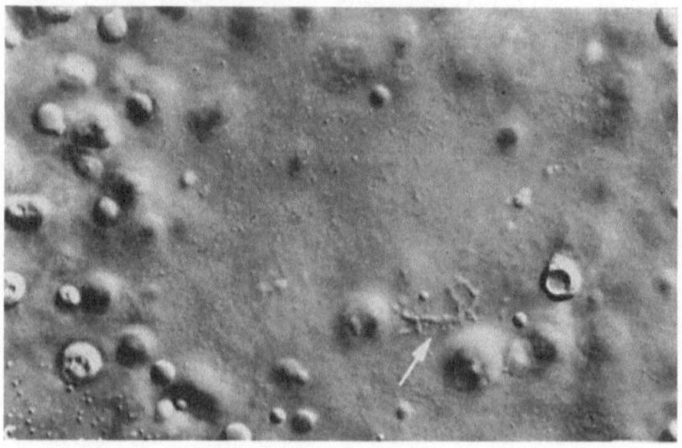

Figure 16. Optical section through the nucleolar sphere with chromosomes (arrow).

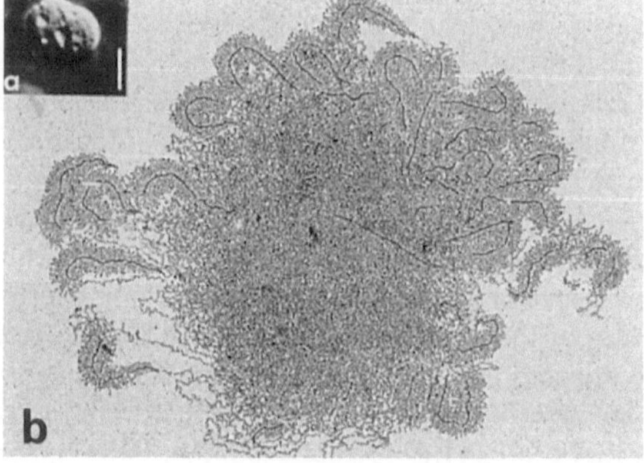

Figure 17. a) Isolated nucleolus in DIC. Scale 10 μm, b) Electron photomicrograph of a spreading preparation of nuclei.

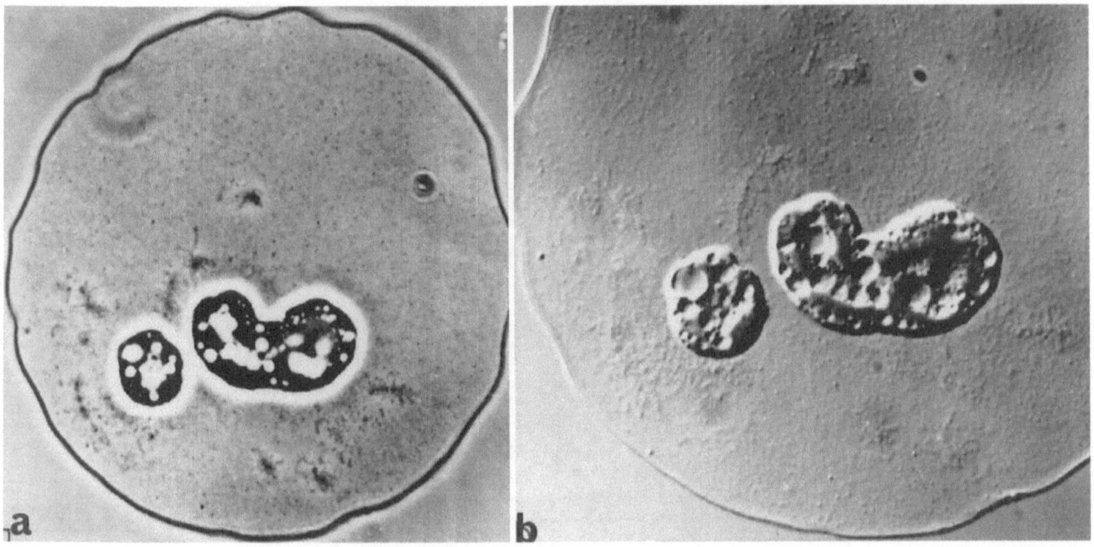

Figure 18. Isolated nucleus, 2 µm thick section. Planapo 63/1,4 Oil, a) phase contrast b) DIC Nomarski.

The life cycle of the green alga *Acetabularia* is characterised by a vegetative phase in which the zygote produces a large and highly differential cell body and a giant ("primary") nucleus with many nucleoli which can be imaged ideal by Nomarski optics. Using oil immersion 100x, a part of the endo plasmatic reticulum could be made visible. Figure 18 shows a 2 µm thick section of an isolated nucleus, a comparison between phase and DIC. Several lampbrush chromosomes are visible; the association of one such chromosome with a nuclear unit is only visible in the DIC image because of the "halo" disturbance in the phase contrast image.

A suitable method for the evaluation of the adhesion mechanism of cultured cells to glass surface during mitosis is the combination of reflection contrast, phase and DIC. The contact areas of the mitotic cell are revealed by reflection contrast while the cell body and a nucleus is imaged by phase or DIC.

6. OPTOELECTRONIC LIGHT MICROSCOPY

Especially new techniques like video- and laser scanning microscopy expanding the field of classical light microscopy into a new dimension, overlapping the low range of electron microscopy but with the importance of using living specimens. The rapid progress in image processing and image analysis has considerable widened the range of light microscopy and permits hitherto in achievable new insights. By means of video enhanced contrast techniques (= VEC), introduced beginning 1980 by Allen & Inoue (1980), structures can be made visible which are ca. 10x smaller than the theoretical limit of resolution in a light microscope (e.g. living single micro-tubules). In a laser scanning microscope also all optical contrast enhancement techniques can be used. High magnifications are possible by means of scan zoom without loss of brightness and contrast (Fig. 19) as well as 3 D imaging and 3 D reconstruction.

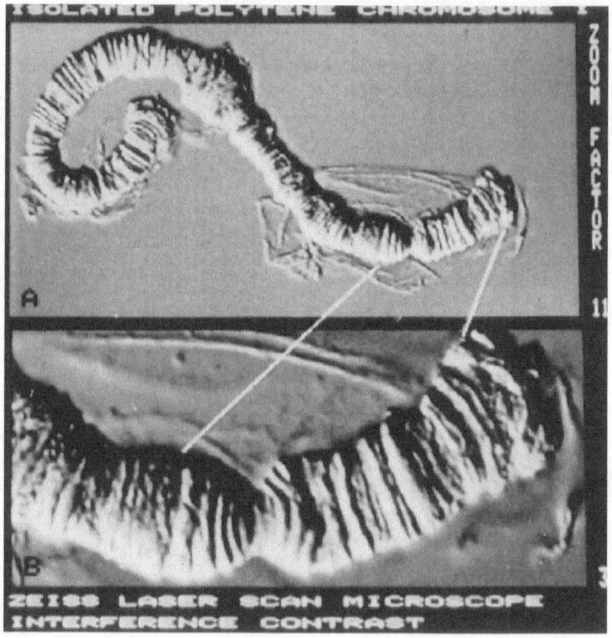

Figure 19. Giant chromosome of *Chironomus* in DIC (A) and LSM zooming (B).

REFERENCES

Abbe, E. 1873. Beiträge zur Theorie des Mikroskops und der mikroskopischen Wahrnehmung. *Archiv für Mikroskopische Anatomie.*9, 413.

Allen, R.D., David, G.D. Nomarski, G. 1969. The Zeiss – Nomarski differential interference equipment for transmitted-light microscopy. *Zeitschrift für wissenschaftliche Mikroskopie und mikroskopische Technik* 69, 4, 193-221.

Allen, R.D., Allen, N.S. & Travis, J.L. 1981. Video-enhanced contrast, differential interference contrast (AVEC-DIC) microscopy. A new method capable of analyzing microtuble-related motility in the reticulopodial net work of *Allogromia laticollaris. Cell Motility* 1, 291-302.

Arndt-Jovin D.J., Robert-Nicoud, M., Kaufman, S.J. & Jovin, T.M. 1985. Fluorescence digital imaging microscopy in cell biology. *Science* 230, 247-256.

Gundlach, H. 1978. The egg of the sea urchin – a typical in cell biology. *Zeiss Information* 87, 63.

Gundlach, H., Trendelenburg M.F. 1980. Identification and selective micro preparation of live nuclear components with the Zeiss IM 35 inverted microsocpe. *Zeiss Information* 91, 36-40.

Köhler, A. 1893. Ein neues Beleuchtungsverfahren für mikrophotographische Zwecke. *Zeitschrift für wissenschaftliche Mikroskopie und mikroskopische Technik* 10, 433 -440.

Lang, W. 1968. Nomarski differentical inferference-contrast microscopy. *Zeiss Information* 70, 114 ff.

Muchel, F. 1988. ICS – a new principle in optics. *Zeiss Information* 30, 20-27.

Spring, H, Grierson, D., Hemleben, V., Stohr, M., Krohne, G., Stadler, J, & Franke, W.W. 1978. DNAcontents and numbers of nucleoli and pre – r RNA-genes in nuclei of gametes and vegetative cells of Acetabularia Mediterranea. *Experimental Cell Research* 114, 203-215.

Zernike, F. 1935. Das Phasenkontrastverfahren bei der mikroskopischen Beobachtung. *Physikalische Zeitschrift* 36, 848.

The Sea Urchin: From Basic Biology to Aquaculture, Yokota, Matranga & Smolenicka (eds)
© Swets & Zeitlinger, Lisse, ISBN 90 5809 379 4

Morphology of *Paracentrotus lividus* embryological development

G. Giudice & I. Albanese
Dipartimento di Biologia Cellulare e dello Sviluppo, Università di Palermo, Palermo, Italia

The morphology of development of the sea urchin *Paracentrotus lividus* from the stage of oocyte till that of pluteus is described with the aid of optical microscopic photographs.

Sea urchin development represent one of the main models used in developmental biology for a series of reasons: First, you can easily fertilize millions and even billions of eggs and observe the various steps of fertilization. They can be also parthenogenetically activated by a variety of means. Second, immediately after fertilization they start dividing at a very fast pace, recalling that of bacteria, in a way that is synchronous enough to permit the biochemical and molecular study of a large number of individuals all at the same developmental stage. Third, they are permeable to precursors of macromolecules, so to enable one to easily study nucleic acid and protein synthesis. Fourth, they can be microinjected and micromanipulated with reasonably little effort, so that one can study the effect of injected DNA or RNA constructs and carry out microtransplantation experiments. Lastly, they can be raised trough metamorphosis in order to obtain an offspring so that genetic studies can be carried out, and also, which is one of the main purposes of this course, commercially valuable adult animals can be cultured. (Giudice 1973, 1986, 1999)

The development of the Mediterranean species *Paracentrotus lividus* will be shortly described, which applies with very little modifications also to many other sea urchin species.

The unfertilized egg (figure 1b) has different properties along its axes: namely, the pole in which you can see the pronucleus is called the animal pole, and the opposite the vegetal pole. The sperm, attracted toward homologous eggs and activated by a species-specific peptide released by the jelly coat surrounding the egg, undergoes the acrosome reaction and exposes on its surface a protein, called bindin, which interacts, again species-specifically, with an egg receptor. This event triggers almost instantaneously several changes within the egg, whose membrane potential decreases transiently, thus providing a fast block to polyspermy. Moreover, a slight increase in the intracellular pH and a rise in free calcium ions occur, the latter induced at least in part by nitric oxide (NO) generated by a NO synthase present in the sperm. The calcium increase activates several metabolic processes, and causes the exocytosis of cortical granules and the consequent elevation of a fertilization envelope, responsible for the permanent block to polyspermy. Within about 45 minutes following fertilization the egg divides along the animal vegetal-axis into two equivalent cells (figure 1c and d). About half an hour later the egg divides again along this axis, but in a plane perpendicular to the first one, giving rise to four again equivalent cells (figure 1e). The third division occurs along an equatorial plane originating eight cells, four belonging to the animal territory and four to the vegetal one (figure 1f). The fourth cell division divides the four animal blastomeres again meridionally, originating eight cells called mesomeres, while the four vegetal cells are unequally divided along a subequatorial plane into four small cells close to the vegetal pole, called micromeres, and four large cells above them, called macromeres. This is the stage of 16 cells (figure 1g). The micromeres are especially important because they will divide again originating four larger micromeres and four smaller ones; the larger ones will divide at least three more times and will give rise to the primary mesenchyme cells, which will originate the larval skeleton; the smaller ones will divide only once, at least for the moment, and will then give rise to the coelomic pouches from which through metamorphosis the adult will be originated. The fifth division produces the stage of 32 cells. The next cell division produces 60 cells because of the small micromeres lag behind with respect to the large ones

With the following divisions a blastula is originated, made of a hollow sphere lined by one layer of epithelial ciliated cells surrounding a cavity, which is called blastocoel (figure 1i-l). Till this stage cell divisions have followed each other at about half an hour intervals. When the blastula has formed cell divisions slow down and a pause is observed, which will last till the beginning of gastrulation. The blastula rotates within the fertilization envelope and then produces a hatching enzyme (figure 1j and k), which dissolves the fertilization envelope allowing the blastula to freely swim in the seawater, stage of swimming blastula (figure1l). At this point some cells of the vegetal pole deriving from the large micromeres start to glide inside the blastocoel to form the primary mesenchyme, stage of mesenchyme blastula (figure 1l and m). Shortly before the primary mesenchyme cell ingression the most vegetal part of the blastocoelic wall becomes thicker and flat, to form the so called vegetal plate This then starts to invaginate to form the first intestine, stage of early gastrula (figure 1m). At this stage the primary mesenchyme cells surround the point of intestine ingression and at one side, the future oral side of the embryo, they form two little columns, one per side inside which the cells join to each other through pseudopodia (figure 1n and o). Within these cells the first rudiment of the embryonic skeleton is deposited in the form of triradiated spiculae. The intestine elongates toward the animal pole, stages of mid gastrula and late gastrula (figure 1n and o) helped also by the action of the secondary mesenchyme cells delaminating from the tip of the archenteron. At this point the intestine bends toward one side, the oral one, where it attaches and opens a mouth, stage of prism (figure 1p). The intestine now starts to show two constrictions dividing it into an esophagus, close to the mouth, a stomach and an intestine. From the part of the intestine forming the mouth two pouches, one per side, originate; from one of these, usually the left one, the adult will be formed through metamorphosis. In the meanwhile the skeleton has elongated to form two anal rods, laying on the plane of the intestine ingression, which represents the anus, two oral rods, above the mouth, two body rods which grow toward each other and join to form the apex of an acute angle at the site opposite to the mouth, and two transverse rods growing toward each other on the anal plane. The anal rods and the oral rods, accompanied and lined by epithelium, form respectively the anal arms and the oral arms of the larva, which at this stage is called pluteus (figure 1q, r and s).

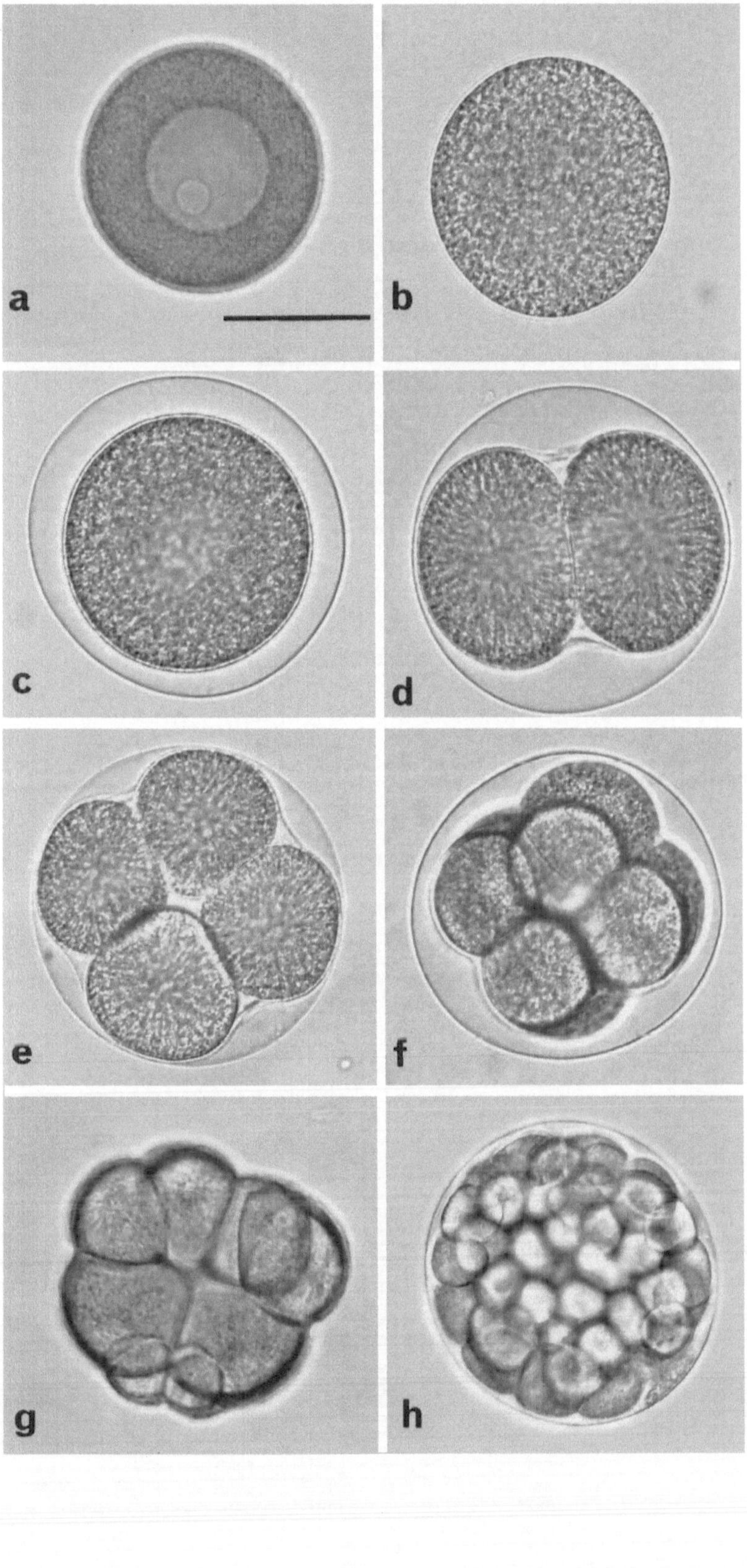
a
b
c
d
e
f
g
h

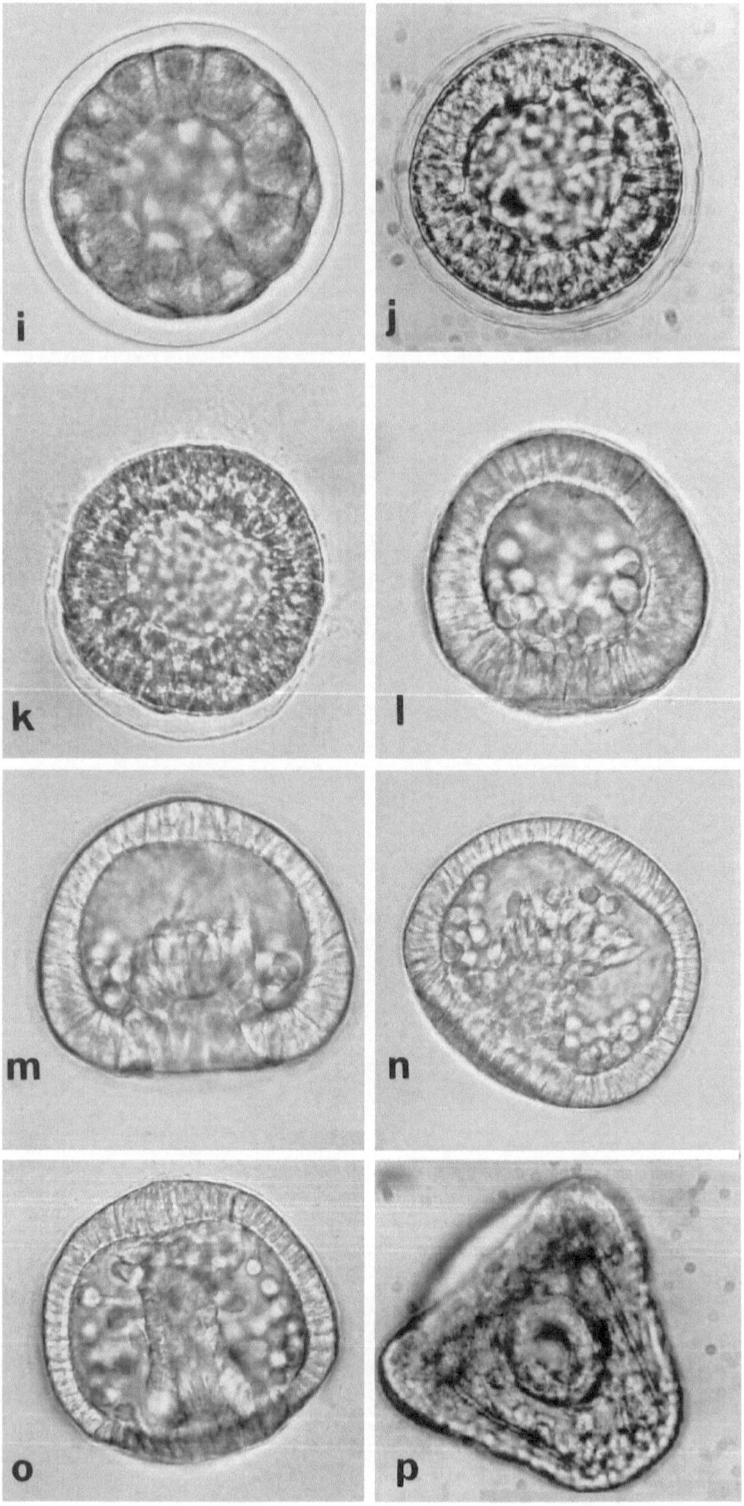
i
j
k
l
m
n
o
p

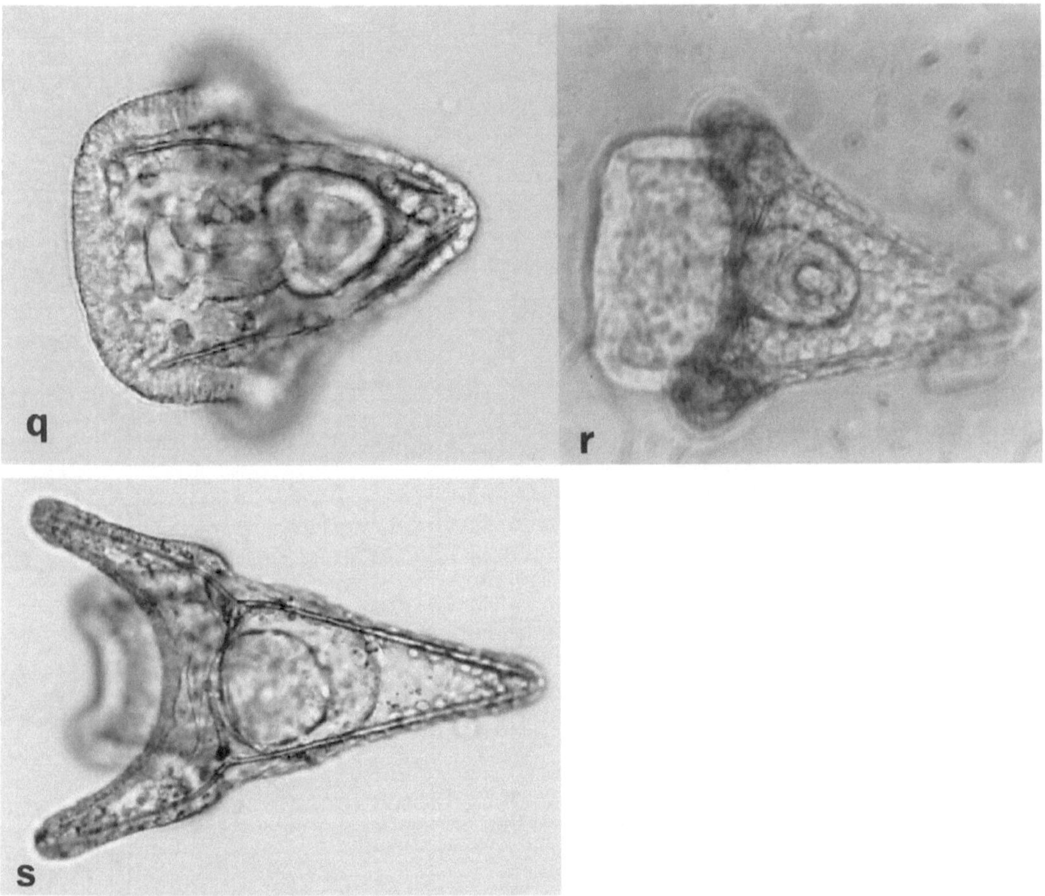

Figure 1. Development of *Paracentrotus lividus:* a oocyte; b unfertilized egg; c fertilized egg, d the 2-cell stage; e the 4-cell stage; f the 8-cell stage; g the 16-cell stage; h the morula stage; i the early blastula stage; j the blastula membrane starts to be digested; k the hatching blastula stage; l the mesenchyme blastula stage; m the onset of gastrulation; n the mid-gastrula stage; o the late-gastrula stage; p a prism viewed from the vegetal face; q a more advanced prism viewed from the same face but slightly rotated toward the oral face; r early pluteus viewed from the same side as in *q*; s late pluteus, viewed from the same face. j and p x 550; r x 360 (Giudice, 1973), a-i, k-o, q and s (bar = 50 mm), by courtesy of Dr. V. Matranga.

REFERENCES:

Giudice, G. 1973. *Developmental Biology of the Sea Urchin Embryo.* New York and London: Academic Press.

Giudice, G. 1986. *The Sea Urchin Embryo. A Developmental Biological System.* Berlin, Heidelberg, New York, Tokyo. Springer-Verlag.

Giudice, G. 1999. Genes and Their Products in Sea Urchin Development in R.A.Pedersen & G.P. Schatten (Eds), *Current Topics in Developmental Biology* 45: 41-106. S. Diego, London, Boston, New York, Sidney, Tokyo, Toronto: Academic Press.

The Sea Urchin: From Basic Biology to Aquaculture, Yokota, Matranga & Smolenicka (eds)
© Swets & Zeitlinger, Lisse, ISBN 90 5809 379 4

Sea urchin developmental genetics and culture

R.A. Cameron
Division of Biology 156-29, California Institute of Technology, Pasadena, California, USA

1. INTRODUCTION

The purpose of this essay is to survey the genetic basis for elements of the biology of sea urchins that are germane to the subject of culture. Due in no small part to their easy availability and clear transparent eggs and embryos, sea urchins have been a favorite research model for cell and developmental biology for the last 150 years (Giudice, 1973, 1986; Czihak, 1975; Davidson, 1976, 2001). Much is known about the early developmental phases including cell lineage (Cameron & Davidson, 1991) and gene expression (reviewed in Davidson et al, 1998). Elements of reproduction including environmental effects on gametogenesis are well studied (Pearse & Cameron, 1991). This extensive collection of information on development through the life cycle and on reproduction can be used to inform us on the subject of husbandry and culture.

In modern terms, development of sea urchins as well as other animals can be viewed as the outcome of a complicated spatial and temporal series of differential gene expressions that occurs in a group of rapidly dividing cells. The control of these expressions must depend on a complex hardwired regulatory program built into the genome. The automated methodologies of molecular biology are just now rendering these regulatory programs accessible in sea urchin embryos, an ideal research model for the analysis of gene regulatory networks. Furthermore, sea urchin eggs lend themselves easily to gene transfer methods, a key feature necessary for the study of the regulation of transcription and the function of signaling systems in development. While these areas of research don't pertain directly to culture, the information gained here is being extrapolated to explain basic aspects of sea urchin reproduction.

I will first consider some data on inbreeding and the problem of recessive lethals; then I will describe some of our results on twinning and sex determination. The remarkable advances made in the understanding of the molecular basis of sea urchin development will fill the remaining space where we will examine the transcriptional control of developmentally important genes and the results of the sea urchin genome project.

2. INBREEDING EFFECTS

Broadcast spawning animals like sea urchins afford an opportunity for maximal gamete mixing. Thus the chance for deleterious effects through inbreeding is avoided. Furthermore, the high DNA polymorphism and very large population size of the purple sea urchin, *Strongylocentrotus purpuratus* implies a large and evolutionarily old gene pool. However, as a limited broodstock is chosen for culture the extent of normal heterozygosity in sea urchin populations becomes important. An inbred line of the purple sea urchin that we have been maintained for more than 7 generations offers an opportunity to examine this genetic feature of sea urchin populations (Leahy et al, 1994). Our inbred line is descended from a single pair of wild animals and then continued through sibling matings (Figure 1). In our laboratory cultures, metamorphosis occurs after about five weeks post-fertilization and juveniles are capable of shedding gametes at 11-12 months of age. We conducted a series of simulations to estimate the degree of lethality expected in our crosses at different levels of recessive lethals per genome (Leahy et al, 1994). As expected the fraction of crosses that produce lethality is highest early in the inbreeding process, since the lethal genes are gradually lost as inbreeding proceeds. This phenomenon was taken into account when we chose animals from the F2 and F3 generations rather than older ones on which to perform a series of crosses. Systematic observations were also made on self-fertilized cultures derived from several naturally occurring hermaphrodites. Morphological development, survival efficiency, and expression of three territorial embryonic markers were assayed in the embryos developing from these crosses. Unexpectedly, out of >90 controlled crosses, we observed no developmental failures whatsoever up to the end of embryogenesis (i.e., onset of feeding) that could be attributed to homozygous, zygotically acting recessive genes. However, during postembryonic larval development, lethality, morphological malformation, and arrest are observed in inbred cultures at a high frequency. In an unsorted sample of larvae we saw individuals at many different stages and degrees of abnormal morphology ranging from 4 arm

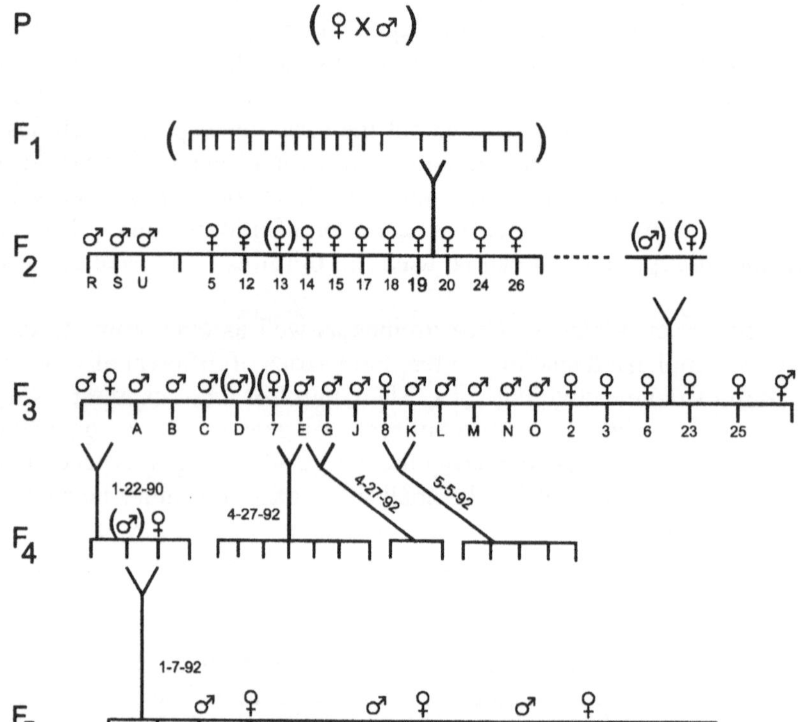

Figure 1. The pedigree of the inbred line of the purple sea urchin from inception to 7th generation. Each subsequent cross is indicated by joined vertical lines and the progeny as a horizontal line. Dead animals are indicated in parentheses. The F7 generation has not been sexed yet. so individuals are shown only as vertical lines.

larvae like the morphology seen at the onset of feeding up to individuals with 8 arms and epaulets, a later appearing locomotory organ. The incidence of these zygotic developmental failures is such that it appears that there is at least one recessive genetic defect affecting larval development per haploid parental genome. The relative imperviousness of the basic embryonic process to defects arising from homozygosity is consistent with other evidence implying that territorial specification in sea urchin embryogenesis is controlled by maternally rather than zygotically expressed gene products.

The degree of heterozygosity in sea urchins generally and in the inbred line is of interest now that the puzzling lack of apparent embryonic lethals has come to light. How much of the average genomic variation revealed by hybridization kinetic studies is represented in differences between alleles of individual genes? Genomic variation has been estimated from restriction fragment length polymorphisms using single open reading frame probes and individual genomic DNA. Some 24 restriction enzyme-probe combinations that are useful for this estimate had been reported in the literature for *S. purpuratus* up to 1996. These include genes coding for the five cytoskeletal actins; muscle actin; a major spicule matrix protein; bindin; a single copy region of a maternal transcript with interspersed repeats; a calcium binding protein; a beta-tubulin; and a single copy histone gene. The heterozygosity for these genes as calculated from the standard formula: $H=1-\Sigma(c_i)^2$, varies between 0.15 and 0.83. The mean value for heterozygosity, 0.55, agrees closely with that calculated from the polymorphism as measured from single copy melting curves (Britten et al, 1978). Another way to measure genomic variation that is independent of genes under selection is simple sequence repeats or micro-satellites. These are repeat units of 2,3,4 or more nucleotides that are repeated many times. Micro-satellite sequences can be treated as loci with polymorphic alleles that obey the rules of Mendelian genetics but since they may not be included in coding regions they are probably not under selection.

We characterized 8 microsatellite loci, 6 screened from a genomic DNA library and 2 that had previously been found in unrelated sequencing projects. Of these 8, five were simple GA repeats of 10-14 iterations and the others had at least 10 repeat stretches of 2 or 4 nucleotide repeats with intervening stretches of up to 20 non-repeating nucleotides. Taken together the primers chosen for these loci produce fragments that varied in length from 90-108 base pairs for the shortest locus to a range of 354-415 base pairs for the longest. Since this work was done, a BAc-end sequencing project conducted as part of the Sea Urchin Genome project has produced more that 4000 simple sequence repeats (see below).

We were able to successfully carry out micro-satellite genotyping using our 8 primer sets with twenty-seven sea urchins taken from field collections or from first generation crosses of animals maintained in our culture system apart from the inbred line. There were at least five of the primer sets that worked for any one individual and at several loci almost all of the individuals yielded results. The mean homozygosity over the entire data set was 71%, somewhat higher that estimates from known genes or the melting curve data (see above). The one locus for which only homozygotes could be found still presented 6 different alleles in this randomly selected population. The significance of this peculiar distribution of homozygotes for different alleles is yet to be determined. We assume that a single band reflects a diploid homozygote.

However, it is likely that sea urchins have a chromosomally based sex determination mechanism (Ransick et al, 1996, see below). Thus, this microsatellite locus could reside on the male chromosome and only yield a single product rather than two of the same size. Only one female of the 8 analyzed yielded a PCR product at this locus. For the inbred line, no more than 4 alleles at any locus are expected started from a sibling mating. We genotyped individuals from the F2 and F3 generation which overlapped with the recessive lethal study above. Here again at least five loci amplified for each individual. Among the individuals measured, only one of the 8 loci retained all four alleles. Between these two generations, an allele was lost from the measured individuals at 4 of the 8 loci analyzed. At the 14_6 locus where all of the F3 individuals yielded a PCR product, the loss of alleles is irrefutable. In the other cases enough genotypes are missing to render the loss of an allele ambiguous. The average homozygosity measured from these allele frequencies increased from 63% at the F2 to 78% at the F3. One locus shows no change in homozygosity while the others exhibit significant increases. Locus 16-11 seems fixed in the F3 generation and three other loci are completely homozygous but for different alleles. Only locus S4 shows no change in the degree of homozygosity between the two generations. This sequence is the only one known to be expressed.

3. TWINNING, SEX DETERMINATION AND SMALL MICROMERES

A few reports scattered through the literature on sea urchin reproduction (reviewed in Pearse and Cameron, 1991) raise the possibility that sex determination has an environmental component. Consequently, it is useful to know if sea urchins have sex chromosomes since this implies that sex determination is independent of environment, per se. The only karyograms done on the purple sea urchin show sets of 42 small telocentric chromosomes between which there is no recognizable dimorphism (Gerhart, 1983). We applied the techniques of experimental biology to address this question. If sex determination is autosomal, then twins raised in pairs might occasionally have different sexes. We separated the first two blastomeres of the purple sea urchin and the reared the resulting twins in pairs through larval life and then to sexual maturity (Ransick et al, 1996). Out of 8 pairs in this study and one from the literature (Marcus, 1979), each set of twins was always the same sex. Four pairs were female and 5 were male. We derive two conclusions for this study: changes in the pattern of specification of embryonic cell fates following blastomere separation result in the creation of two sets of fully functional coelomic pouches and adult rudiments (the source of the gonads, see below), and sex determination in sea urchins must be chromosomal, since the pairs of twins were always of the same sex.

The four small micromeres result from the unequal division of the micromeres which themselves arise from an unequal division at the 4th cleavage of the sea urchin embryo. The small micromeres only contribute to the coelomic pouches, which produce major components of the adult body plan during post-embryonic development (Cameron et al, 1991; reviewed in Cameron & Davidson, 1991). The slowed pace of division in the small micromeres and their apparent quiescence during late embryonic development suggested to some that these cells might be the precursors of primordial germ cells in the sea urchin. To test the proposition that the small micromeres are the definitive primordial germ cell lineage of the sea urchin, we deleted the 4th cleavage parents, and raised the deleted embryos through larval life and metamorphosis to sexual maturity (Ransick et al, 1996). Almost all of the experimental animals produced functional gametes excluding the possibility that the germ line arises exclusively and obligatorily from descendants of the small micromeres; but rather the germ cell lineage becomes specified during the post-embryonic development of the adult body. A survey of the literature indicates that there is no known case of an embryonic primordial cell lineage in a bilaterian species that displays maximal indirect development.

4. TRANSCRIPTIONAL CONTROL AND THE GENOME PROJECT

The sea urchin embryo has been an especially useful model for the analysis of the *cis*-regulatory elements and the gene regulatory networks. *cis*-Regulatory modules are the hardwired, primary building blocks of the gene regulatory networks that control the developmental process. Operationally, a *cis*-regulatory module is a fragment of genomic DNA that when fused to a reporter gene and introduced into a zygote will recapitulate some aspect of the expression of the gene from which the fragment was isolated. The array of sites in the DNA fragment to which transcription factors bind contain the heritable genomic sequence code necessary and sufficient for the function of that module (reviewed by Arnone & Davidson, 1997). *cis*-Regulatory modules have been likened to computational machines that integrate at the gene a variety of input information, presented in the form of activated transcription factors (see Davidson, 2001 for review). The information carried to the module includes information regarding signaling activities from neighboring cells and intracellular information like position in the cell cycle or its lineage origin. The most thoroughly studied sea urchin developmental regulatory gene, *endo16*, is an endoderm differentiation gene (Yuh *et al.*, 1998, 2001). The *endo16* gene serves as a paradigm illustrating the anatomy of *cis*-regulatory elements and their functional interactions (Yuh *et al.*, 1998; 2001). A model of the *cis*-regulatory structure of *endo16* represents the internal regulatory architecture that enables this gene to function appropriately in all cells of the organism throughout its life cycle.

There is a fair amount of genomic information available for the purple sea urchin. Its collection was motivated by the need to identify *cis*-regulatory regions, but it is also useful for other approaches such as the microsatellite genotyping mentioned above. The mass of the diploid genome is 1.8 pg (Hinegardner, 1974) contained in 42 chromosomes (Gerhart, 1983). This corresponds to about 800 megabases of sequence. The sea urchin displays a pattern of short period interspersion of single copy sequences such that about 25% of the genome is represented in many thousands of repetitive element families (Davidson, 1976) and simple sequence repeats (Cameron *et al.*, 1999). At a computational criterion of sequence similarity that is equivalent to the standard experimental hybridization conditions performed earlier there are in this genome about 2,700 families of repeats of ≥ 70 copies per genome (Cameron *et al.*, 2000).

An advantage of the sea urchin genome for *cis*-regulatory analysis is its relative simplicity. Major blocks of genes have been duplicated during vertebrate evolution. For example, sea urchins and the cephalochordate, amphioxus, have one cluster of *hox* genes, whereas mammals have four (Do & Lonai, 1988; Scott, 1992; Bailey *et al.*, 1997). The absence of duplicated genomes and genome regions that may make the selection of orthologous genes problematical in vertebrates is an advantage.

Another metric of genomic simplicity is gene number. Estimates of gene number in the genome of *S. purpuratus* support this conclusion. An estimate from a BAC-end sequence search against the Genpept database (Cameron *et al.*, 2000) yields an estimate of about 25,000 genes. However given the assumptions in the calculation this is likely a rough estimate. In a small sample of the genome totaling a little over 3.3 mb (0.41% of the genome) for which complete annotated genome sequence is available, the apparent gene density extrapolates to 20,200 genes (Cameron *et al.*, 2002). So, provisionally, the best guess at present is about $22 \pm 5 \times 10^3$ genes.

The Sea Urchin Genome Project utilizing *S. purpuratus*, the "California purple sea urchin" is the major repository for the genomic information mentioned above. A description of the Project resources can be found on the Sea Urchin Genome Project Web site at: http://sea-urchin.caltech.edu/genome. The Project has produced a virtual BAC-end sequence tag (STC) map of the genome, resulted in a data set consisting of about 76,020 BAC-end sequences or about 5% of the genomic sequence length. Another completed goal of the Sea Urchin Genome Project is the generation and printing of large arrayed cDNA libraries representing the stages of development and many adult tissues. For an organism such as the purple sea urchin where there is as yet a relatively small EST data base the main use of arrayed libraries is gene discovery. From these libraries an expandable EST data base from cDNA sequencing projects is being compiled. At present there are about 12,000 ESTs in the data base. The information amassed on the sea urchin genome could come into play as the approaches and strategies of sea urchin culture become more technologically advanced. It is possible to produce with the information available today a carefully bred cultured line of animals to the specifications desired.

5. CONCLUSIONS

It has been said that today is the dawn of the Biological Age. The advances in our understanding of how animals work since the advent of molecular approaches have been rapid and exciting. The developmental genetics that are detailed in this essay demonstrate that sea urchins are a useful model for this advancement and that the information can be brought to bear on the questions of husbandry and culture as well.

REFERENCES

Arnone, M. & Davidson, E.H. 1997. The hardwiring of development: Organization and function of genomic regulatory systems. *Development* 124: 1851-1864.

Bailey, W.J., Kim, J., Wagner, G.P. & Ruddle, F.H. 1997. Phylogenetic reconstruction of certebrate Hox cluster duplications. *Molecular Biology of Evolution* 14: 843-853.

Britten, R.J., Cetta, A., & Davidson, E.H. 1978. The single-copy DNA sequence polymorphism of the sea urchin *Strongylocentrotus purpuratus. Cell* 15: 1175-1186.

Cameron, R.A. & Davidson, E. H. 1991. Cell type specification during sea urchin development. *Trends in Genetics* 7: 212-218.

Cameron, R.A., Fraser, S.E., Britten, R.J., & Davidson, E.H. 1991. Macromere cell fates during sea urchin development. *Development* 113: 1085-1091.

Cameron, R.A., Leahy, P.S., Britten, R.J. & Davidson, E.H. 1999. Microsatellite loci in wild-type and inbred *Strongylocentrotus purpuratus. Developmental Biology* 208: 255-264.

Cameron, R.A., Mahairas, G., Rast, J.P., Martinez, P., Biondi, T.R., Swartzell, S., Wallace, J.C., Poustka, A.J., Livingston, B.T., Wray, G.A., Ettensohn, C.A., Lehrach, H., Britten, R.J., Davidson, E.H. & Hood, L. 2000. A sea urchin genome project: Sequence scan, virtual map, and additional resources. *Proceedings of National Academy of Sciences USA* 97: 9514-9518.

Cameron, R.A., Oliveri, P., Wyllie, J. & Davidson, E.H. 2002. *cis*-Regulatory Activity of Randomly Chosen Genomic Fragments from the Sea Urchin. submitted to *Proceedings of National Academy of Sciences USA*.

Czihak, G., ed. 1975. *The sea urchin embryo--biochemistry and morphogenesis*. Berlin, New York:Springer-Verlag.

Davidson, E. H. 1976. *Gene Activity in Early Development*, Second Edition. New York: Academic Press.

Davidson, E.H., Cameron, R.A. & Ransick, A. 1998. Specification of cell fate in the sea urchin embyos:summary and some proposed mechanisms. *Development* 125: 3269-3290.

Davidson, E.H. 2001. *Genomic Regulatory Systems. Development and Evolution.* San Diego: Academic Press.

Do, M.S. & Lonai, P. 1988. Gene organization of murine homeobox-containing gene clusters. *Genomics*. 3: 195-200.

Gerhart, S.G. 1983. Sea urchin cytogenetics. Thesis, Dept. Medical Sciences, University of Calgary, Canada.

Giudice, G. 1973. *The Developmental Biology of the Sea Urchin Embryo*. New York: Academic Press.

Giudice, G. 1986. *The Sea Urchin Embryo. A Developmental Biological System.* Berlin: Springer-Verlag.

Hinegardner, R. 1974. Cellular DNA content of the echinodermata. *Comparative Biochemistry and Physiology*. 49B: 219-226.

Leahy, P.S., Cameron, R.A., Knox, M.A., Britten, R.J. & Davidson, E.H. 1994. Development of sibling inbred sea urchins: Normal embryogenesis, but frequent postembryonic malformation, arrest and lethality. *Mechanisms of Development* 45: 255-268.

Marcus, N.H. 1979. Developmental aberrations associated with twinning in laboratory-reared sea urchins. *Developmental Biology* 70: 274-277.

Pearse, J.S. & Cameron, R A. 1991. Echinodermata:Echinoidea. In A.C. Giese, J.S. Pearse & V.B.

Pearse (eds.), *Reproduction of Marine Invertebrates, Vol. 6*: 514-662. Pacific Grove: Boxwood Press.

Ransick, A., Cameron, R.A. & Davidson, E.H. 1996. Post-embryonic segregation of the germ line in sea urchins, in relation to indirect development. *Proceedings of National Academy of Sciences USA* 93: 6759-6763.

Scott, M.P. 1992. Vertebrate homeobox gene nomenclature *Cell* 71: 551-553.

Yuh, C.-H., Bolouri, H. & Davidson, E.H. 1998. Genomic *cis*-regulatory logic: Functional analysis and computational model of a sea urchin gene control system. *Science* 279: 1896-1902.

Yuh, C.-H., Bolouri, H. & Davidson, E.H. 2001. *cis*-Regulatory logic in the *endo16* gene: Switching from a specification to a differentiation mode of control. *Development* 128: 617-628.

The Sea Urchin: From Basic Biology to Aquaculture, Yokota, Matranga & Smolenicka (eds)
© Swets & Zeitlinger, Lisse, ISBN 90 5809 379 4

Developmental modes and rudiment formation in sea urchins

S. Amemiya
Department of Integrated Biosciences, Graduate School of Frontier Sciences, University of Tokyo, Japan

ABSTRACT: After completion of gastrulation, echinoid embryos develop into 4-armed pluteus larvae. The larvae can be raised to this stage without feeding, but they require food to develop further. When 4-armed plutei feed on diatom, they develop through a 6-armed stage into 8-armed larvae. Morphologically, plutei are bilaterally symmetrical, whereas adult sea urchins possess pentaradiate symmetry. Essential parts of the adult structure with five-fold radial symmetry are derived from the left coelomic pouch initially formed on the archenteron tip at the late gastrula stage. The developmental mode in which the adult structure is established through feeding pluteus is referred to as the "indirect development". Direct developing species, which skip some of the steps in the feeding larval stages, have evolved in several phyletic lineages of echinoids. Various degree of elimination of larval traits and of changes in relative timing of expression of larval traits are found in these species, although the larvae are commonly devoid of stomodeum and functional gut.

1. INTRODUCTION

The sea urchin (echinoid) is a good developmental model, because they are easily obtained, and because the gametes can be obtained throughout the year by choosing appropriate species. Various species of sea urchins are found near the seashore along the Japanese archipelago. These are *Anthocidaris crassispina*, whose Japanese name is purple urchin, *Hemicentrotus pulcherrimus* or horse dung urchin, *Strongylocentrotus intermedius*, *Strongylocentrotus nudus*, *Pseudocentrotus depressus* or red urchin, *Toxopneustes pileolus* or trumpet urchin, *Mespilia glubulus*, *Diadema setosum*, *Heterocentrotus mammillatus* or pipe urchin, *Asthenosoma ijimai* and *Phyllacanthus dubius* or the Japanese bomb urchin. Some species of irregular urchins, or sand dollars, are also found on coast regions of the Japanese mainland, such as *Clypeaster japonicus* or the Japanese octopus pillow, *Scaphechinus mirabilis*, *Astryclypeus manni* and *Peronella japonica*. Among them, *Strongylocentrotus intermedius* and *Pseudocentrotus depressus* are commercially important for food. *Hemicentrotus pulcherrimus*, *Anthocidaris crassispina*, *Clypeaster japonicus* and *Scaphechinus mirabilis* are most commonly used as the experimental materials for the developmental biology.

The sea urchins, along with starfish, belong to the Phylum echinodermata, which is characterized by the establishment of pentameral symmetry in adults. This pentamerous symmetry is not readily apparent in the sea urchin (Fig. 1a), but is clearly visible in starfish. However, the penta radiate symmetry in sea urchins can be more clearly shown if the spines and other surface structures are removed (Fig. 1b). The larvae of typical sea urchins are able to develop into 4-armed pluteus without taking food. The larvae have left-right (LR) symmetry, a completely different morphology from that of adults. How is the pentamerous symmetry in the adult established from larvae with LR symmetry? This paper deals with the process for establishing pentaradiate symmetrical morphology from larvae with LR symmetry through metamorphosis.

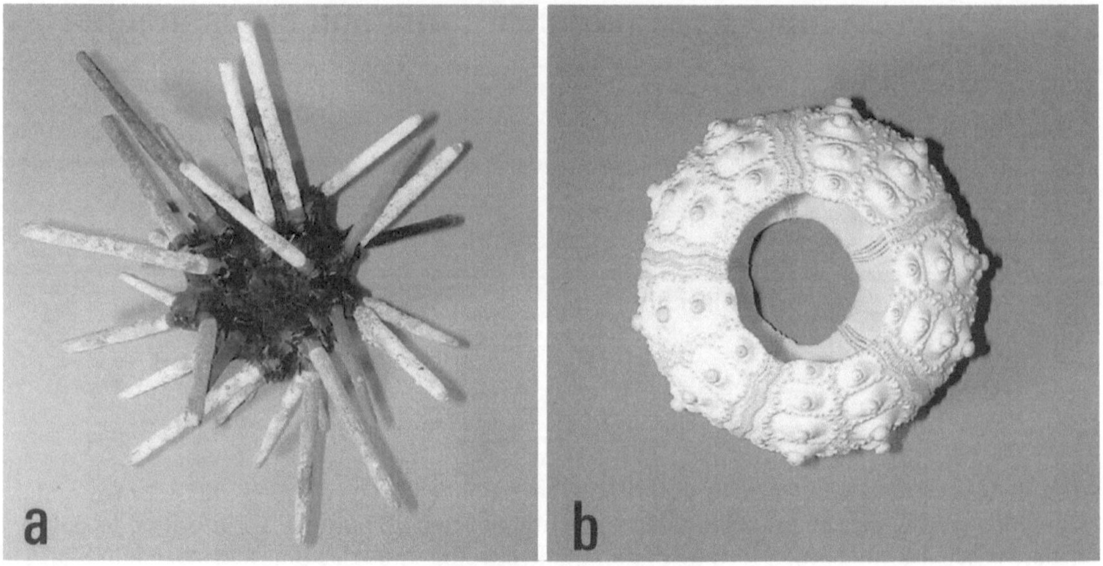

Figure 1. A cidaroid echinoid, *Phyllacanthus dubius*. (a) An intact adult specimen. (b) A denuded test of an adult.

In the first half, I will discuss the developmental mode in which the larvae metamorphose into juveniles with pentaradiate symmetry by passing through a feeding pluteus stage, which is regarded as indirect development. However, there is another mode of development, direct development, in which the embryos establish juveniles of penta radiate symmetry without passing through feeding larval stages. In the last half of this paper, the development of the direct developing species is discussed.

2. EARLY DEVELOPMENT IN TYPICAL SEA URCHINS

The embryos of typical sea urchins have radial symmetry with only an animal-vegetal (AV) body axis from the cleavage to early blastula stages. The larval body plan characteristic to sea urchins is established by the dependence on the signal cascade along the AV axis (Davidson 1989, Davidson et al. 1998). The signal(s) originally emanates from the micromeres or the micromere-descendants (Hörstadius 1973, Amemiya 1996) at the early cleavage stages (16-64-cell stages) (Ransick & Davidson 1992), and/or the early blastula stage (Minokawa & Amemiya 1999, Ishizuka et al. 2001). Another axis, the anterior-posterior (oral-aboral) axis is established by distribution of primary mesenchyme cells at the late blastula stage. The anterior-posterior axis is also established conditionally depending on the inductive signal(s) along the AV axis (Wikramanayake and Klein 1997). This axis becomes much evident by the archenteron tip to bend toward the anterior direction, and by the mesenchymal aggregates to locate a little to anterior at the late gastrula stage. On the outer surface of the embryos, many moving cilia are formed. A tuft of cilia formed on the animal pole is long and immotile (Fig. 2). A pair of coelomic sacs is formed on both sides of the archenteron tip at the period from late gastrula to prism stage (Fig. 3). A pair of pluteus arms containing skeletal arm rods are formed in the 2-armed pluteus (Fig. 4), and another pair of arms in the 4- armed pluteus. The embryos and the larvae have LR symmetry throughout the stages from late mesenchyme blastula to 4-armed pluteus. The larvae develop without taking food throughout the 4-armed pluteus stage, depending on egg yolk.

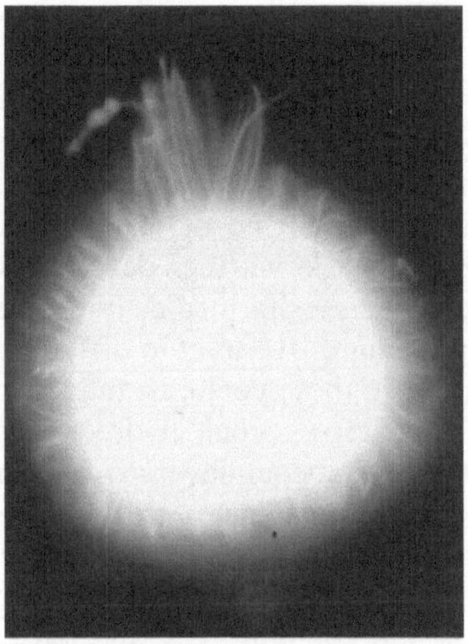

Figure 2. Dark field micrograph of a gastrula embryo of the sea urchin, *Hemicentrotus pulcherrimus*.

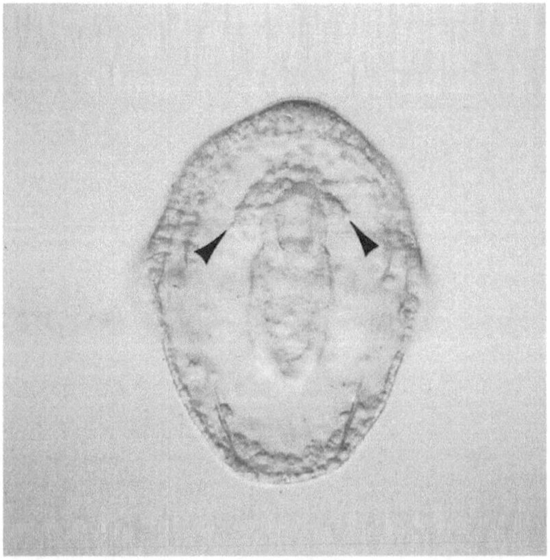

Figure 3. A prism larva of *Hemicentrotus pulcherrimus* with a pair of coelomic sacs (arrowheads) formed on the both sides of an archenteron tip.

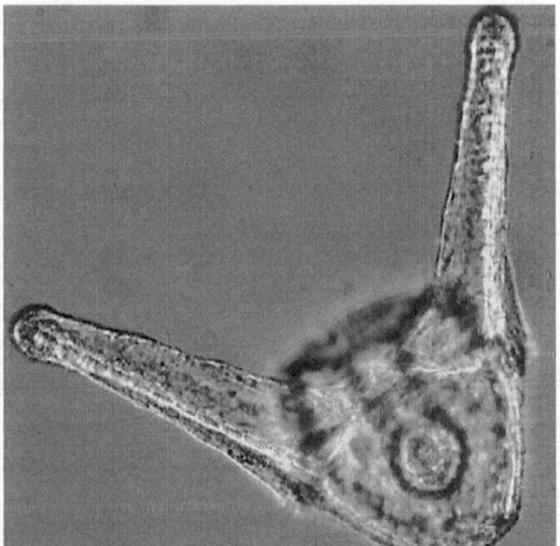

Figure 4. Polarizing micrograph of an early pluteus larva of *Hemicentrotus pulcherrimus* with a pair of pluteus arms containing skeletal arm rods.

3. LARVAL DEVELOPMENT THROUGHOUT METAMORPHOSIS IN TYPICAL SEA URCHINS

The larvae must be fed to be cultured in the laboratory after the 4-armed pluteus stage. Some diatoms, such as *Chaetoceros gracilis* are a suitable source of food for most Japanese species. However, the diatom might not be fit for some American species (Emlet, personal communication). The cultured pluteus larvae then develop into 6-armed pluteus, and finally into 8-armed pluteus. A pair of coelomic sacs formed on both sides of the archenteron tip at the late-gastrula stage elongate along the digestive tract to differentiate into three coeloms; protocoel, mesocoel and metacoel (Fig. 5). In these coeloms, the left protocoel, mesocoel, metacoel and right metacoel differentiate into left axocoel, hydrocoel, somatocoel and right somatocoel, respectively. Among them, the left axocoel develops later into a hydropore canal. Meanwhile, the ectoderm facing the hydrocoel invaginates to form the vestibule (amniotic cavity). The hydrocoel develops 5 protrusions radially (Fig. 6) which extend from the primary podia by pushing up the ectoderm in the vestibule covering the hydrocoel. Thus, the hydrocoel and the vestibule are derived from the left coelomic sac, and left ectoderm, respectively (Runnström 1917, Gustafson and Wolpert 1963, Czihak 1965, Hörstadius 1973, Pehrson and Cohen 1986).

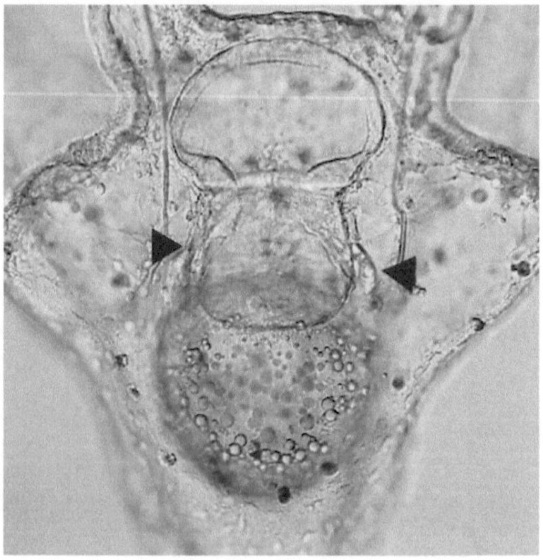

Figure 5. A pair of elongated coelomic sacs (arrowheads) located on the both sides of digestive tract in a four-armed pluteus larva of *Hemicentrotus pulcherrimus*.

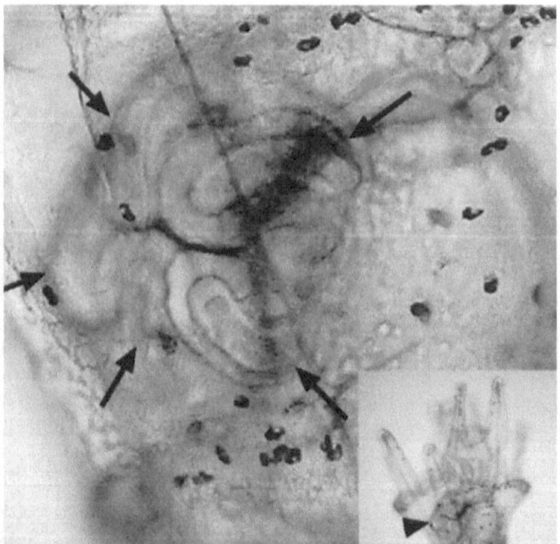

Figure 6. An echinus rudiment showing five radially formed podia (arrows) differentiated from the hydrocoel in an eight-armed pluteus larva of *Hemicentrotus pulcherrimus*. (Inset) The whole eight-armed pluteus larva with an echinus rudiment (arrowhead).

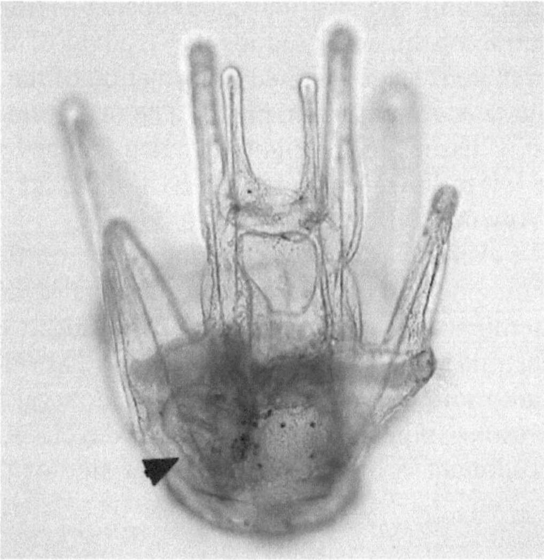

Figure 7. An eight-armed pluteus larva of the sea urchin, *Hemicentrotus pulcherrimus* with an echinus rudiment (arrowhead) formed on the left side of the body.

The corresponding parts on the right side of the larva do not develop these organs, so that the hydropore canal, the hydrocoel, and the vestibule are left-side-specific traits in normal development. Therefore, the adult rudiment of echinoids is normally formed only in the left side of the pluteus larvae (Fig. 7) (Amemiya 1996).

4. EXPERIMENTAL PERTURBATION OF NORMAL ESTABLISHMENT OF LEFT-RIGHT ASYMMETRY IN FORMATION OF ADULT RUDIMENT

As described above, the adult rudiment is normally formed in the left side of the larval body. The normal establishment of LR asymmetry in formation of the adult rudiment can experimentally be perturbed. Previous studies in which the sea urchin embryos were exposed to abnormal conditions showed that LR placement of the adult rudiment can be disrupted only in a small fraction of the larvae exposed (MacBride 1918, Ohshima 1922, Runnström 1925, Hörstadius 1973). Pronase treatment effectively impaired the LR polarity in formation of the adult rudiment among the survivors, but the survival rate of the embryos was significantly reduced (Marcus 1981).

It has been reported that the polarity for differential LR development in formation of the adult rudiment is reestablished in half embryos dissected meridionally to the AV axis. The halves separated along the first and second cleavage planes meridional to the AV axis have been shown to develop into the pluteus. The meridional halves separated in early development up to the blastula stage also develop into the pluteus (Hörstadius & Wolsky 1936, Hörstadius 1973, Marcus 1979, Cameron et al. 1996). The ability of the meridional halves to reestablish normal LR polarity was not completely lost as late as the mesenchyme blastula stage (McCain and McClay 1994). In these experiments, the left or the right half of the embryo was not identified, because the (LR) polarity in the embryo was not apparently determined at the stages.

A series of regional operation experiments excising all or a part of tissue on the left or right side of the embryos during the early gastrula stage and the two-armed pluteus stage was carried out (Aihara & Amemiya, 2000, in press). The LR position of the adult rudiment and, of its components, in the larvae derived from the operated embryos was examined. Removal of tissues on the left side of the embryos resulted in formation of a left adult rudiment as in normal development in all stages examined. By contrast, removal of tissues on the right side of the embryos resulted in three different types of stage-dependent impairment in the LR positioning of the adult rudiment. In the left half larvae dissected during the early gastrula stage, the proportions of larvae with a right adult rudiment were about equal to a left adult rudiment.

The left halves dissected in the mid-late gastrula stage showed a strong tendency to reverse LR polarity in regard to formation of the adult rudiment. Removal of the whole right side in the prism stage and two-armed pluteus stage resulted in formation of the adult rudiment on both of the left and right sides (Aihara & Amemiya, in press). The operation on the two-armed pluteus larvae to remove a portion of tissues in the right side also induced formation of a right adult rudiment in addition to the left rudiment in more than half of the larvae, resulting in the double adult rudiment (Aihara & Amemiya, 2000).

Thus, the experiments excising whole or part of the tissue on the left or the right side of embryos during the stages ranging from the early gastrula stage to the two-armed pluteus stage showed abnormal LR patterning only in the left halves but not in the right halves. Regional excisions in the mid-late gastrula stage revealed that a part of right side tissue was indispensable for the formation of an adult rudiment on the left side of the larval body, whereas that of left side tissue was not. The results indicate that, in indirect developing echiniods, LR polarity in formation of the adult rudiment is directed by the right side of the larval body (Aihara & Amemiya, in press).

5. METAMORPHOSIS

Other organs are arranged depending on the pentaradiate symmetric structure of the hydrocoel, eventually establishing the adult structure with its five-fold radial symmetry.

The five primary podia are arranged radially around the juvenile mouth in the adult rudiment before metamorphosis. A pair of secondary podia arise in between the mouth and the primary podia to squeeze the primary podia toward the aboral side. A pair of tertiary podia are formed in between the primary and secondary podia, and a fourth pair of podia are formed in between the primary and tertiary podia, and so on. In this way, each new pair of podia are always formed just on the oral side of the primary podia, as if they squeeze the primary podia toward the aboral side. Eventually, the primary podia reach to the terminal plates in the apical pole of the test.

6. DIRECT DEVELOPING ECHINOIDS

The echinoid species with the developmental mode described above are regarded as the indirect developing echinoids. This is the developmental mode of most sea urchins commonly found on the seashore. Direct developing echinoids in which expression of a part or most of the larval traits is abbreviated to accelerate expression of adult traits have been reported mainly in the Antarctic Ocean and from the deep sea. These echinoids generally have large eggs. There are two types, pelagic and brooded, in larvae of direct developers. The larvae of both types do not open the mouth so that the digestive tract is not completed. The direct developer species show various degrees of abbreviation of larval traits and various changes in timing of expression of these traits. Some direct developers appear on the seashore of temperate regions, although most are found in the deep sea or Antarctic Ocean. Among them, the development of three species, *Asthenosoma ijimai*, *Peronella japonica* and *Heliocidaris erythrogramma* has been well studied. *A. ijimai* and *H. erythrogramma* (Williams & Anderson 1975, Raff 1987) have markedly abbreviated larval traits, while *P. japonica* is comparatively less abbreviated so that the species metamorphose through pluteus-like larvae. In the followings, the development of *A. ijimai* and *P. japonica*, both of which are found around the Japanese archipelagos will be discussed.

7. DEVELOPMENT OF A DIRECT DEVELOPING ECHINOID *ASTHENOSOMA IJIMAI*

The echinothurioid echinoids, to which *A. ijimai* belongs, are characterised by living in the deep sea and having giant and flexible test. All echinothurioids with gametes are considered to be direct developers, as the eggs are large. The echinothurioids have radial muscle supporting the flexible test (Tsuchiya & Amemiya 1977). Among the echinothurioids common in the deep sea, three species of genus *Asthenosoma*, *A. ijimai*, *A. varium* and *A. intermedium*, appear in comparatively shallower sea of about 10 to 30 meters depth. *A. varium* is broadly distributed in the Red Sea, the Yellow Sea, the Indian Ocean and the sea around Malaysia. *A. ijimai* and *A. intermedium* are located in a limited area around the coast of the Japanese Islands facing the Pacific Ocean and the east coast of the Australian continent, respectively. Ecologically, *A. ijimai* lives on a rocky bottom (Fig. 8), and likes to take sponge (Fig. 9) (Birenheide et al. 1993).

A. ijimai is the only echinothurioid species in which development has been studied. The eggs of this species are 1.2 mm in diameter and are floating. The sperm have a markedly elongated head, like the sperm of other species with large eggs (Fig. 10) (Amemiya et al. 1980). Fertilized eggs were first obtained by artificial fertilization at September 15 of 1979 (Amemiya & Tsuchiya 1979). Since then, fertilizable eggs of this species have been obtained for several times. In all cases, the matured eggs were obtained during a limited period around the middle of September.

The fertilization envelope is slightly separated from the eggs. The first cleavage fallow arises from only one pole, probably the animal pole, of the eggs. The cleavage is essentially an equal division to produce blastomeres of equal size. Different from indirect developers, micromeres

Figure 8. An echinothurioid echinoid, *Asthenosoma ijimai*, living on a rocky bottom of 30 m depth.

Figure 9. An adult of *Asthenosoma ijimai* feeding on a sponge.

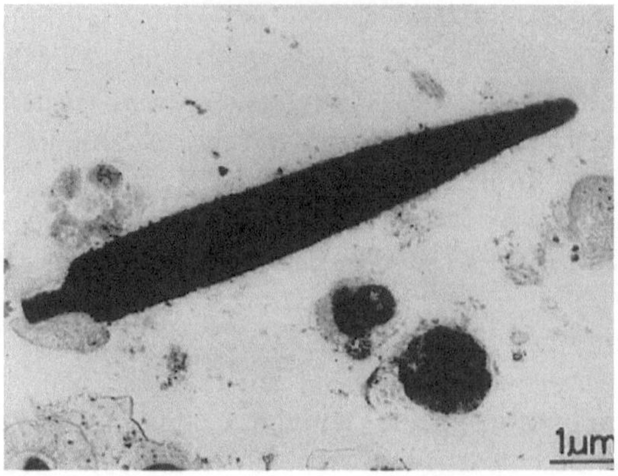

Figure 10. A transmission electron micrograph of a longitudinal section through the nuclear region of a spermatozoon of *Asthenosoma ijimai*.

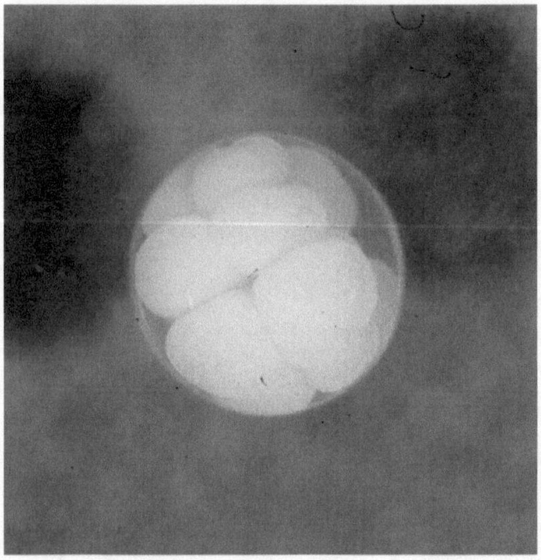

Figure 11. A 16-cell stage embryo of *Asthenosoma ijimai*.

are not produced at the fourth cleavage (Fig. 11). The wrinkled blastulae are formed at the blastula stage, the same as other blastulae that develop from large eggs. After hatching, the wrinkles are gradually canceled, to form a flattened cookie-like embryo, a novel echinoderm embryo structure. The embryos then gradually become rounded in shape, and initiate archenteron invagination from the vegetal pole. The embryos elongate along the AV axis during gastrulation. The embryos at 4 days after fertilization cultured at 20 °C give rise to a pair of bulges on the dorsal side of the body. At the same stage, both sides of the blastopore become elongated. These two pairs of bulges develop into para-arms. Five projections arranged radially arise on the left surface of the larval body at 5 days after fertilization. These are the rudiments of the primary podia. A section of the specimen shows the five projections in front of the pentaradiated hydrocoel derived from the archenteron (Amemiya & Emlet 1992). The two pairs of para-arms and five primary podia in the 6-day larvae have more developed. Five pairs of spines grow outside of the five primary podia in the 14 days larvae. The spines are increased in more developed larvae, and a mouth is formed on the central region surrounded by the 5 radially arranged podia. The larva metamorphosed into a juvenile at 55 days after fertilization. The spines in the juvenile grow further, and the test becomes rounded. The juvenile spines and adult-type spines are observed in the specimen viewed from the oral-side (Fig. 12). In more developed specimens, the juvenile spines disappear and only the adult-type spines remain (Fig. 13).

The development of *A. ijimai*, in which the larval traits are largely abbreviated, is a typical example of direct development. The development of *P. japonica* described below is an intermediate type between direct and indirect development in which the larvae develop into pluteus-like larvae and the larval traits are less abbreviated than *A. ijimai*.

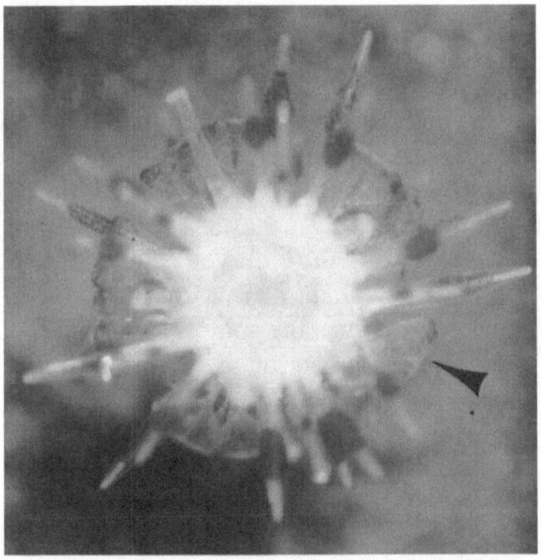

Figure 12. Aboral view of a settled juvenile *Asthenosoma ijimai* showing juvenile spines (arrowhead).

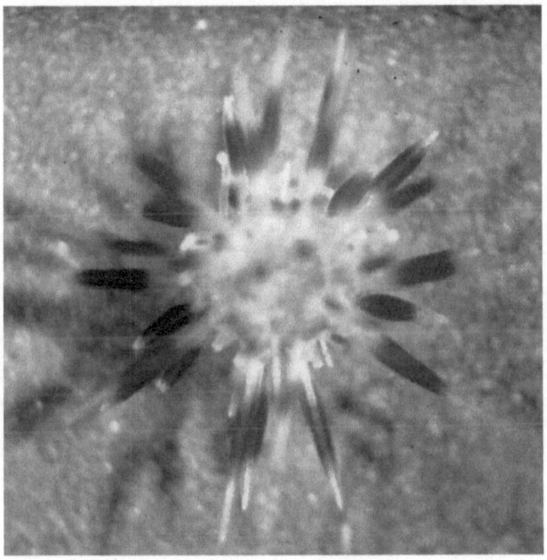

Figure 13. Oral view of another settled juvenile *Asthenosoma ijimai*.

8. DEVELOPMENT OF A DIRECT DEVELOPING ECHINOID *PERONELLA JAPONICA*

P. japonica is endemic to Japan. Distribution of this species is limited to the coast of the Japanese Islands. *P. japonica* belongs to Clypeasteroids, irregular urchins of comparatively small size with a test size of about 7 cm in long diameter (Fig. 14). The egg size, 0.3 mm in diameter, is about 5 times as large as that of indirect developers and is about one quarter that of *A. ijimai*. The fertilized eggs, different from the other direct developers, divide unequally at the fourth cleavage stage to form four micromeres in the vegetal pole (Fig. 15), although a fraction of embryos forms only one to three micromeres (Amemiya & Arakawa 1996). The larvae develop normally throughout metamorphosis regardless of the number of the micromeres. Occasionally, some embryos divide equally at the fourth cleavage to form 16 blastomeres with equal size. Most embryos without micromeres develop abnormally. The micromeres differentiate into the skeletogenic mesenchyme cells as those in the embryos of the typical sea urchins. The number of the skeletogenic mesenchyme cells derived from the micromeres in the embryos of *P. japonica* is about 240, 4-times more than that (64) in those of the typical sea urchins (Amemiya & Arakawa 1996). The micromeres of *P. japonica* might not have the potential to induce endoderm (Iijima et al. 2000). An archenteron invaginates from the vegetal pole in the same way as the indirect developers. The amniotic cavity (vestibule) precociously invaginates just after initiation of gastrulation (Fig. 16) to the region on which the stomodeum in embryos of indirect developers is formed, while the amniotic cavity in indirect developers is formed on the left side of larval body at the 6-armed pluteus stage. LiCl treatment of typical sea urchin embryos causes vegetalization and the formation of exogastrulae. The embryos of *P. japonica* show not only exogastrulation but also evagination of the amniotic cavity following treated with LiCl (Kitazawa & Amemiya 1997, 2000).

The arm rods elongate to form arms, and the embryos develop into pluteus-like larvae. Only a pair of arms corresponding to the antero-lateral arms are formed. No other arms are formed. The components of the adult skeleton, such as the spines, appear in the larvae at 50 hours after fertilization cultured at 24 °C (Fig. 17). The larvae metamorphose into juveniles as early as 3 days after fertilization (Fig. 18).

9. PERSPECTIVES

The larvae of indirect developing echinoids show distinct LR asymmetry at late larval stages by forming an adult rudiment in the left side of the body, while the young larvae demonstrate almost completely LR symmetry. The trait is extremely stable in development, so that it is suited to examine the mechanism for establishing the LR asymmetry. Among the three major phyla

Figure 14. A clypeasteroid echinoid, *Peronella japonica*, living on a sandy bottom.

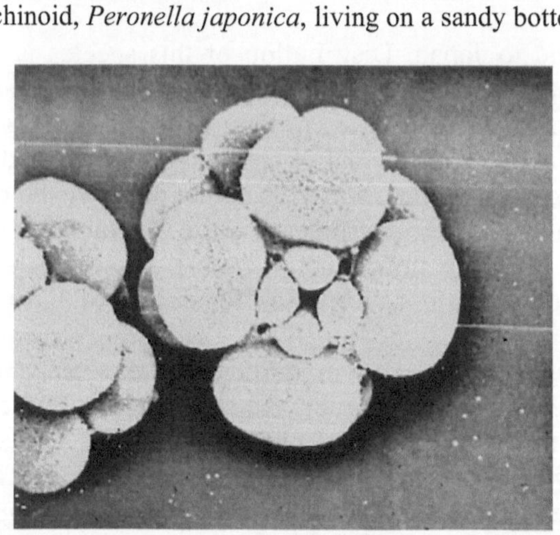

Figure 15. A 16-cell stage embryo of *Peronella japonica* with four micromeres of regular size.

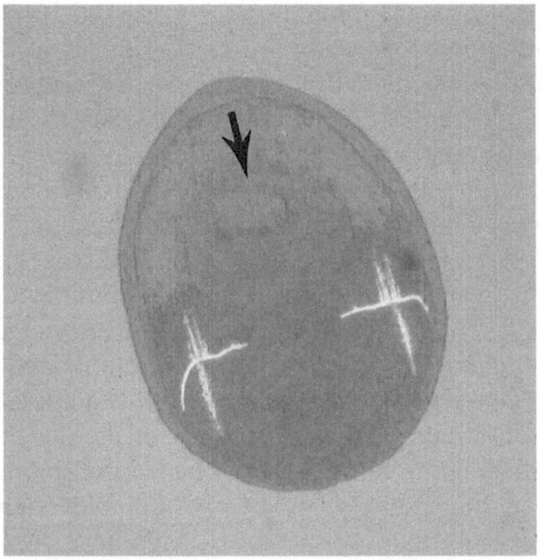

Fig. 16. A late gastrula embryo of *Peronella japonica* forming an amniotic cavity (arrow).

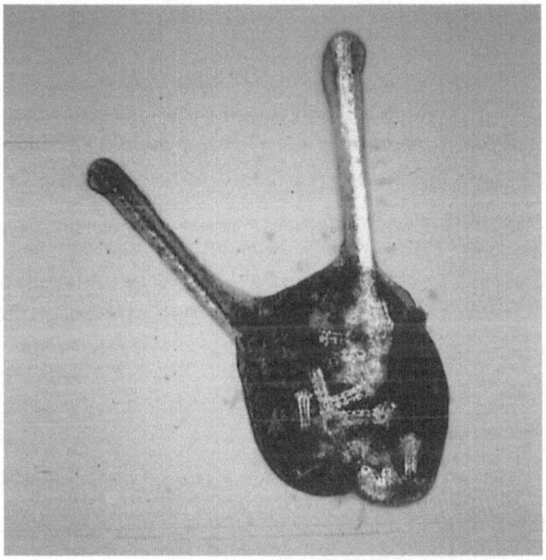

Fig. 17. A late 2-armed pluteus larva of *Peronella japonica* forming adult skeleton.

Fig. 18. A settled juvenile of *Peronella japonica* 3 days after fertilization.

in deuterostomes, the echinoderms and chordates show marked LR asymmetry in disposition of internal organs. The evolutionary processes in establishment of the LR asymmetry in the deuterostome lineages could be elucidated by comparing the mechanisms to establish the LR asymmetry between the echinoderms and chordates. The study, in turn, would be helpful for understanding the phylogenetic relation between these two phyla through estimation of the traits concerning the establishment of the LR asymmetry held by their common ancestor.

In evolutionary terms, it is thought that indirect development with feeding pluteus larvae is the more ancient type, and the direct development is derived from that by abbreviating some steps in larval development, so that some larval traits are eliminated in the direct developers (Strathmann 1978, Raff 1987). The extent of elimination is diverse among the direct developing species. In *A. ijimai* larvae, many of the larval traits are eliminated, while in *P. japonica*, a considerable part of them is conserved so that the embryos form micromeres at the 16-cell stage and the larvae develop into 2-armed pluteus. The study of the developmental mechanisms in these two direct developing echinoids would not only be helpful for elucidation of evolutionary process in developmental modes, but also be helpful for understanding the developmental mechanisms in typical indirect developing species through developmental changes caused by the elimination of various larval traits.

REFERENCES

Aihara, M. & Amemiya, S. 2000. Microsurgical operation to induce formation of double adult rudiments in sea urchin larvae. *Invertebrate Reproduction & Development* 37: 241-248.

Aihara, M. & Amemiya, S. Left-right positioning of the adult rudiment in sea urchin larvae is directed by the right side. *Development* in press.

Amemiya, S. 1996. Complete regulation of development throughout metamorphosis of sea urchin embryos devoid of macromeres. Development Growth & Differentiation 38: 465-476.

Amemiya S. and Arakawa, E. 1996. Variation of cleavage pattern permitting normal development in a sand dollar, *Peronella japonica*: comparison with other sand dollars. *Development Genes & Evolution* 206: 136-146.

Amemiya, S. & Tsuchiya T. 1979. Development of the echinothurid sea urchin *Asthenosoma ijimai. Marine Biology* 52: 93-96.

Amemiya, S., Suyemitsu T. & Uemura I. 1980. Morphological observations on the spermatozoa of echinothurid sea urchins. *Development Growth & Differentiation* 22: 327-335.

Amemiya, S. & Emlet R. B.1992. The development and larval form of an echinothurioid echinoid, *Asthenosoma ijimai*, revisited. *Biological Bulletin* 182: 15-30.

Birenheide, R., Amemiya S. & Motokawa T. 1993. Penetration and storage of sponge spicules in tissues and coelom of spongivorous echioids. *Marine Biology* 115: 677-683.

Cameron, R. A., Leahy, P. S. & Davidson, E. H. 1996. Twins raised from separated blastomeres develop into sexually mature *Strongylocentrotus purpuratus*. *Developmental Biology* 178: 514-519.

Czihak, G. 1965. Entwicklungsphysiologische untersuchungen an echiniden (experimentelle analyse der coelomentwicklung). *Wilhelm Roux' Archiv für Entwicklungsmechanics Organismen* 155: 709-729.

Davidson, E. H. 1989. Lineage-specific gene expression and the regulative capacities of the sea urchin embryo: a proposed mechanism. *Development* 105: 421-445.

Davison, E. H., Cameron, R. A., & Ransick, A. 1998. Specification of cell fate in the sea urchin embryo: summary and some proposed mechanism. *Development* 125: 3269-3290.

Gustafson, T. and Wolpert, L. 1963. Studies on the cellular basis of morphgenesis in the sea urchin embryo, formation of the coelom, the mouth, and the primary pore-canal. *Experimental Cell Research* 29: 561-581.

Hörstadius, S. 1973. Experimental embryology of echinoderms. Clarendon press: Oxford.

Hörstadius, S. & Wolsky, A. 1936. Studien über die Determimation Bilateralsymmetrie des jungen Seeigelkeimmes. *Wilhelm Roux' Archiv für Entwicklungsmechanics der Organismen* 135: 1-39.

Iijima, M., Ishizuka, Y., Minokawa, T. and Amemiya, S. 2000. Studies on the potential of micromeres to induce archenteron differentiation in embryos of a direct-developing sand dollar, *Peronella japonica. Zygote 8 Supplement*: S80.

Ishizuka, Y., Minokawa, T. and Amemiya, S. 2001. Micromere-descendants at the blastula stage are involved in normal archenteron formation in sea urchin embryos. *Development Genes & Evolution* 211: 83-88.

Kitazawa, C. and Amemiya, S. 1997 Evagination of the amniotic cavity in larvae derived from lithium-treated embryos of the sand dollar, *Peronella japonica. Journal of Experimental Zoology* 279: 309-312 (1997).

Kitazawa, C. and Amemiya, S. 2001. Regulating potential in development throughout metamorphosis of a direct developing echinoid, *Peronella japonica. Development Growth & Differentiation* 43, 73-82.

MacBride, E. W. 1918. The artificial production of echinoderm larvae with two water-vascular systems, and also of larvae devoid of a water-vascular system, *Proceedings of Royal Society of the United Kingdom Series B*, 90: 323-348.

Marcus, N. H. 1979. Developmental aberrations associated with twinning in laboratory-reared sea urchins. *Developmental Biology*. 70: 274-277.

Marcus, N. H. 1981. Effects of pronase on the development of reversed symmetry and duplications in the sea urchin, *Arbacia punctulata. Developmental Biology*. 83: 387-390.

McCain, E. R. & McClay, D. R. 1994. The establishment of bilateral asymmetry in sea urchin embryos. *Development* 120: 395-404.

Minokawa, T. and Amemiya, S. 1999. Timing of the potential of micromere-descendants in echinoid embryos to induce endoderm differentiation of mesomere-descendants. *Development Growth & Differentiation* 41: 535-547.

Mortensen T. 1921. *Studies of the Development and Larval Forms of Echinoderms*. G.E.C.Gad: Copenhagen.

Ohshima, H. 1922. The occurrence of *situs invertus* among artificially-reared echinoid larvae. *Quarterly Journal of microscopic Science* 66: 105-150.

Okazaki, K. 1975. Normal development to metamorphosis. In G. Czihak (ed.), *The sea urchin embryo*: 177-232. Berlin Heidelberg New York: Springer-Verlag.

Pehrson, J. R. and Cohen, H. H. 1986. The fate of the small micromeres in sea urchin development. *Developmental Biology* 113: 522-526.

Raff, R. A. 1987. Constraint, flexibility, and phylogenetic history in the evolution of direct development in sea urchins. *Developmental Biology* 119: 6-19.

Ransick, A. & Davidson, E. H. 1995 Micromeres are required for normal vegetal plate specification in the sea urchin embryos. *Development* 121: 3215-3222

Runnström, J. 1917. Analytische studien über die Seeigelentwicklung. . *Wilhelm Roux' Archiv für Entwicklungsmechanics Organismen* 43: 223-328.

Runnström, J. 1925. Regulatorische Bildung von Cölomanlagen bei Seeigelkeimen mit gehemmter Urdarmbildung. *Wilhelm Roux' Archiv für Entwicklungsmechanics Organismen* 105: 114-119.

Strathmann, R. R. 1978. The evolution and loss of feeding stages of marine invertebrate. *Evolution* 32: 894-906.

Tsuchiya, T. & Amemiya S. 1977. Studies on the radial muscle of an echinothurid sea-urchin, *Asthenosoma*. I. Mechanical responses to electrical stimulation and drugs. *Comparative Biochemistry & Physiology* 57C. 69-73.

Wikramanayake, A. H. & Klein, W. 1997. Multiple signaling events specify ectoderm and pattern the oral aboral axis in the sea urchin embryo. *Development* 124: 13-20.

Williams, D. H. C. and Anderson, D. T. 1975. The reproductive system, embryonic development, larval development and metamorphosis of the sea urchin Heliocidaris *erythrogramma* (Val.) (echinoidea: echinometridae). *Australian Journal of Zoology* 23: 371-403.

The Sea Urchin: From Basic Biology to Aquaculture, Yokota, Matranga & Smolenicka (eds)
© Swets & Zeitlinger, Lisse, ISBN 90 5809 379 4

Mechanisms of sea urchin metamorphosis: stimuli and responses

I. Yazaki
Tokyo Metropolitan University, Hachiohji, Tokyo, Japan

ABSTRACT: Sea urchin metamorphosis usually occurs by receiving exogenous cues followed by larval histolysis and eversion of the adult rudiment (the echinus rudiment). Many types of metamorphosis-inducing cues are known; however, the mechanisms of metamorphosis, which can be interpreted as a single pathway from the initiation to the completion of metamorphosis, still remain unclear. Metamorphosis processes, like larval histolysis and the eversion of the adult rudiment, vary according to conditions such as sea urchins species, degree of larval development, types of cues and so on. Metamorphosis should be in fact considered as a complex phenomenon, composed of multiple processes induced by different stimuli concurrently progressed. Finally, we have been able to induce the larval histolysis and the eversion of the adult rudiment separately, using different substances. In this paper, various features of metamorphosis are introduced and their possible mechanisms are discussed.

1. METAMORPHOSIS-INDUCLNG CUES

Many marine invertebrates metamorphose for changing life style from planktonic to benthic, by receiving various types of cues from the environment (Chia & Rice 1978). In sea urchins, larvae metamorphose by natural cues (bacterial films, algae, sand or seawater from adult

Table 1. Metamorphosis-inducing stimuli.

Types of stimulus	Sea urchin sp.	References
Bacterial film	*Lytechinus pictus*	Hinegardner 1969
Active seawater	*L. pictus*	Cameron & Hinegardner 1974
Sand from adult habitat	*Dendraster excentricus*	Highsmith 1982
Pheromonal peptide	*D. excentricus*	Burke 1984
Free fatty acid (red algae)	*Pseudocentrotus depressus*	Kitamura, Kitahara & Koh 1993
Dibromomethne (red algae)	*Strongylocentrotus nudus*	Taniguchi, Kurata, Maruzoi & Suzuki 1994
Floridoside-isothionic acid (red algae)	*Holopneustes purpurascens*	Williamson, Nys, Kumar, Carson & Steinberg 2000
Ulvella sp. (green algae)	*P. depressus, A. crassispina, H. pulcherrimus, S. nudus Clypeaster japonicus*	Yazaki unpublished
Electrical stimulus, Dopamine	*D. excentricus*	Burke 1983
Excess KCl	*L. variegatus*	Cameron, Tosteson & Hensley 1989
L-glutamine	*S. intermedius*	Naidenko 1990
	P. depressu	Yazaki & Harashima 1994
	H. pulcherrimus	Yazaki 1995

habitats) and chemicals (Table 1). The effective substances extracted from sand are pheromonal peptides released by adult sand dollars (Burke 1984); those from red algae are free fatty acids (Kitamura et al. 1993), dibromomethane (Taniguchi et al. 1994) and floridoside-isothionic acid (Williamson et al. 2000). Morphogenetic, histolytic and histogenic processes progressed during metamorphosis to create a juvenile, although any direct evidences for the induction mechanism of metamorphosis have not been shown yet. The facts that an electric stimulus and neurotransmitters induce rapid histolysis of larval arms (Burke 1983a) suggest an involvement of neural elements between the stimulus and the cell-response in the metamorphosis. Without natural cues from the environment, excess potassium ions can also initiate the metamorphosis, although many larvae had failed to complete it (Cameron et al. 1989).

Glutamine (Gln), an amino acid component of algae, can induce metamorphosis in larvae of many sea urchin species (Naidenko 1991, Yazaki & Harashima 1994, Yazaki 1995). In the present paper, using Gln, neurotransmitters, natural inducers (green algae) and metamorphosis processes are analyzed and possible mechanisms are shown.

2. VARIETIES OF THE METAMORPHIC PROCEDURES IN SEA URCHIN SPECIES AND CUES

In *Lytechynus pictus*, a surface covered with a mixture of algae and bacteria can induce at first larvae settling on the substrate. Within a minute, the arms flex sharply toward the right side of the larva in order to expose its left side, where the opening of the echinus rudiment is located

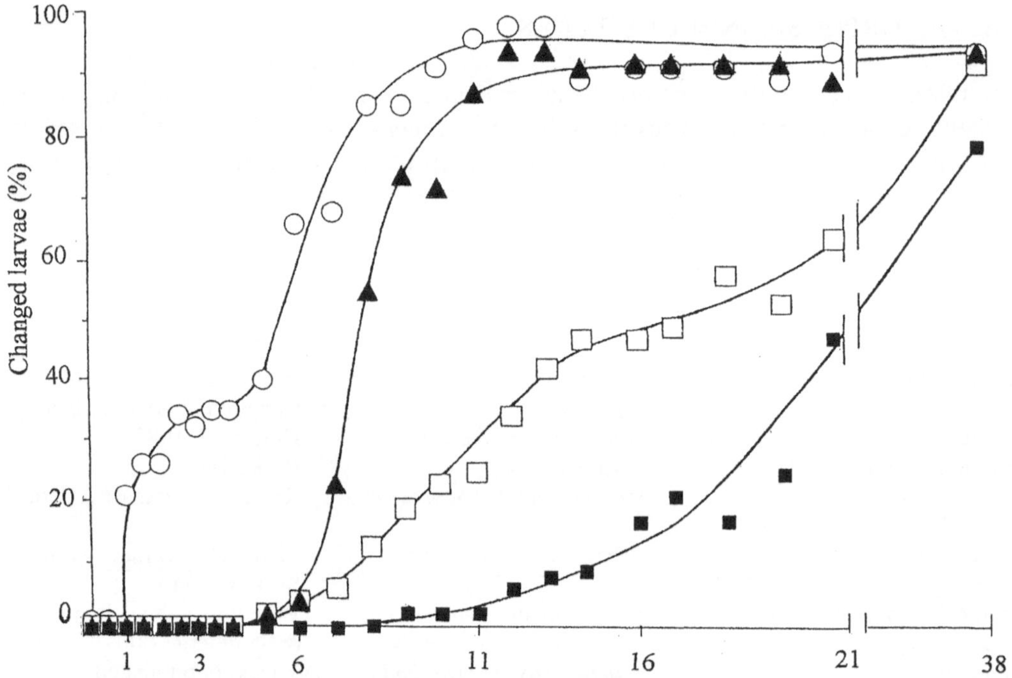

Figure 1. Time course of larval changes induced by treatment with L-Gln. Fifty-one competent larvae were kept in 6×10^{-5} M L-Gln containing the Millipore-filtered seawater (MFSW) for 21 hr and then transferred into MFSW. Observations were carried out repeatedly on four changes which occurred in each larvae; cessation of swimming (○), retraction of larval arms (▲), extrusion of the primary podia through an opening of the larval epithelium at the vestibule of the echinus rudiment (ER) (□) and eversion of ER (■). Yazaki & Harashima (1994).

Larval tissues collapse within one hour (Hinegardner 1969). Active seawater or sand from the adult habitat can also induce metamorphosis of sand dollars, *Dendraster excentricus*, within one hour (Cameron & Hinegardner, 1974). In larvae of *Anthocidaris crassispina*, *Hemicentrotus pulcherrimus*, *Pseudocentrotus depressus* and *Clypeaster japonicus*, green algae, *Ulvella sp.* can induce quick reactions in the mode similar to that mentioned above (Yazaki 2000). However, the timing of metamorphosis was varied in larvae and in *Ulvella* films employed.

Using *Ulvella* films, some larvae metamorphosed within several hours but others took two days to degenerate larval tissues and to complete the metamorphosis. When Gln was used, the metamorphosis usually progressed slowly and some processes of metamorphosis occurred inversely in their order compared to those described above. Gln-induced metamorphosis started with a cessation of swimming within one hour, then larval arm began to retract and the primary podia extruded from the vestibule 6 hr after the initiation of the Gln treatment; finally, the echinus rudiment (ER) everted (Fig. 1). Using either *Ulvella*-film or Gln, the metamorphosed juveniles could grow up by feeding *Ulvella* grown on the plastic films.

3. COMPETENCE AND STRUCTURES IN THE ECHINUS RUDIMENT OF THE 8-ARM PLUTEUS

Eight-armed plutei are conventionally divided into five stages according to the development of their echinus rudiments (ER); early 8a, middle 8a and late 8a-I, -II, -III, as shown schematically in figure 2A. The amniotic cavity (ac) is formed at the early 8a (Fig. 2B1), the primary podia (pd) at the middle 8a, and at the late stage spines (sp) are growing. The skeletal rudiments of spines (spr) are formed at the late 8a-I. Spines are growing in the late 8a-II and fully developed at the late 8a-III. The skeletal structures of the 8-arm pluteus are shown in figure 2B-D. At early 8a, no skeletal structure is present in the ER. The genital plates are originated from the meshwork skeletons formed at the end of larval arm rods (Fig. 2B1, B2). In the ER of late 8a-1, skeletal rudiments of the spines and of primary podia discs are formed (Fig. 2C). When larvae became competent (late 8a-III), two types of spines (dorsal and ventral) have elongated and discs of the primary podia have become complex. Tooth rudiments are formed (Fig 2D. Genital plates are not shown).

Scanning electron micrographs show the changes in opening of the vestibule from the early or middle stage of 8a pluteus (32-day larva, Fig. 3a) to the stage of late 8a-III (45-day larva, Fig. 3C). The small opening in the 32-day larva (arrowed in Fig 3a) becomes large and fully developed spines and podia are visible (44-day larva, Fig. 3b). During the metamorphosis, the opening of the vestibule expands extruding the primary podia and spines (45-day larva, Fig. 3c).

4. SEA URCHIN METAMORPHOSIS INDUCED BY AMINO ACIDS, NEURO-TRANSMITTERES AND GREEN ALGAE, *ULVELLA*

4.1.1 *Induction of metamorphosis by amino acids*

The effect of Gln to induce metamorphosis was initially found in *S. intermedius* (Naidenko 1983). L-Gln at the concentration of 10^{-3} - 10^{-5} M induced metamorphosis in the four Japanese sea urchins examined: *P. depressus*, *H. pulcherrimus*, *S. nudus* and *A. crassispina* (Yazaki & Harashima 1994, Yazaki 1995), although the effective concentration was different according to species (Fig. 4c, d). D-Gln was inert. Glutamic acid (Glu) was also inert at the concentration of 6×10^{-5} M for *P. depressus* (Fig. 4a) and 5×10^{-4} M for *S. nudus* (Yazaki, unpubl.), whereas Gln of each concentration was effective to induce metamorphosis. Administration of other amino acids, glycine (Gly), cystine (Cys) and asparagine (Asn), at a concentration sufficient to promote a metamorphic change, as the collapse of larval arms, resulted in larvae's death. Gly at the concentration of 5×10^{-5} M, 10^{-5} M Cys and 10^{-4} M Asn did not have any

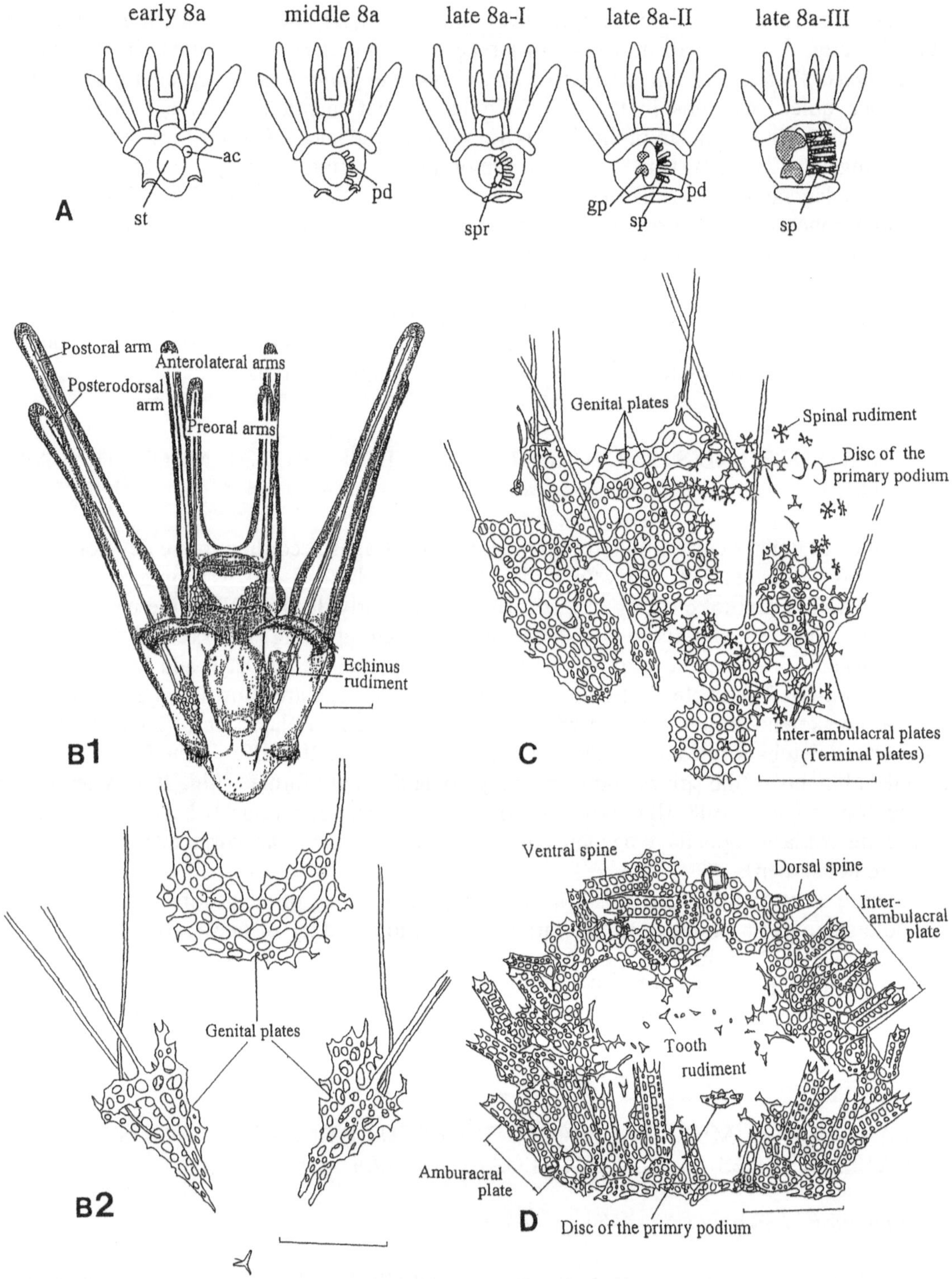

Figure 2. The 8-arm pluteus stages. A) Schematic diagrams for the developmental stages of 8-arm pluteus. In early 8a; the amniotic cavity (ac), in middle 8a; the primary podia (pd) and in late 8a-I; the spinal rudiments (spr) are formed. In late 8a-II, spines (sp) are growing and in late 8a-III, the spines have fully developed, st; stomach, gp; genital plate. B1-D) Development of the skeletal structures of early 8a (B1, B2), late 8a-I (C) and late 8a-III(D), Bars 100 mm. Sketches by M. Noguchi (Ishihara and Noguchi 1996).

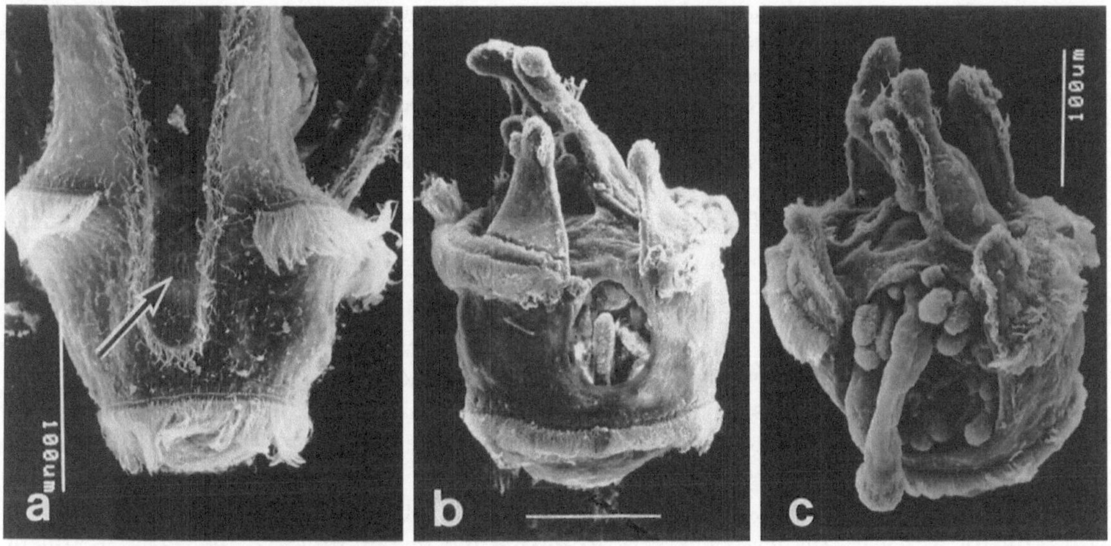

Figure 3. Scanning electron micrographs of the vestibule opening in the eight-armed plutei of *H. pulcherrimus*. Larvae on 32 day (a), 44 day (b) and 45 day (c) after fertilization. An opening of the vestibule (arrowed in a) is very small in the early or middle 8-arm pluteus. As the echinus rudiment grows, a vestibule opening enlarges (b) and widely expands at the moment of metamorphosis to extrude spines and podia (c). Bar 100 mm. Photographs by M. Noguchi.

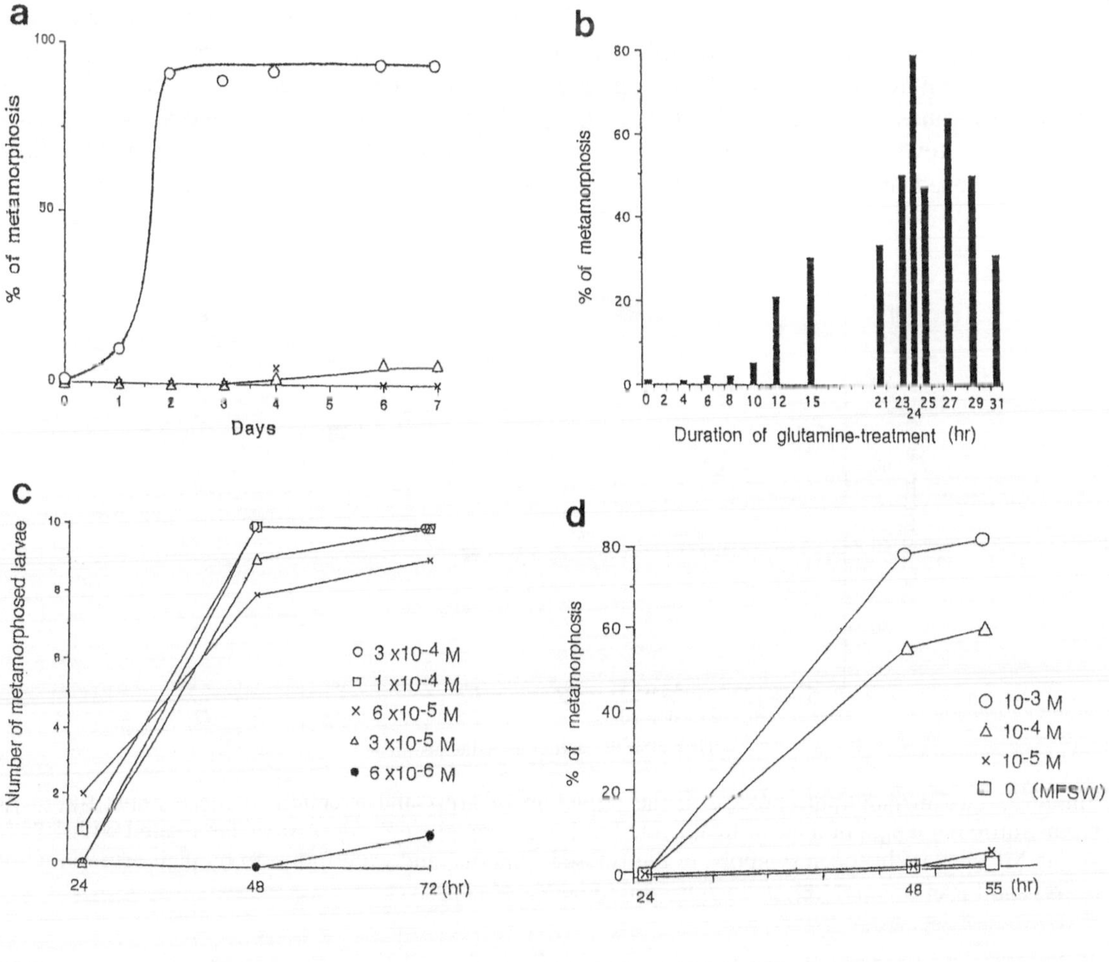

Figure 4. Effects of L-Gln on the metamorphosis. a, c) *P. depressus*. b, d) *H. pulcherrimus*. a) Larvae were treated either with 6 x 10^{-5}M Gln (○, n= 108) or glutamic acid (Δ, n=67) for 18-22hr and the metamorphosis was checked every day. 78 control larvae (×) were cultured in the Millipore-filtered seawater. b) Various steps of the Gln treatments. Each 15-20 larvae at late 8a-II, III were put into 10^{-4}M Gln containing artificial seawater and then transferred into the Gln-free seawater. Metamorphosed larvae were counted at 55hr after the start of the experiments. c,d) Effects of Gln in various concentrations. Groups of ten larvae of *P. depressus* (c) and each 30 larvae of *H. pulcherrimus* (d) were treated with Gln for 24hr. Metamorphoses were checked at the indicated times on the graphs. a,c) Yazaki & Harashima (1994) and b,d) Yazaki (1995).

metamorphic effects on larvae. Alanine (Ala) and asparate (Asp) were inert as same as Glu at the concentration of 5 × 10^{-4} M. Glutmine at the concentration of 10^{-3} M promotes both arm retraction and metamorphosis of sand dollar larvae, *C. japonicus*. In this species, 10^{-3} M Glu induces metamorphosis as Gln, while Glu at the same concentration does not cause the metamorphosis of other species.

In *H. pulcherrimus*, larvae scarcely metamorphosed by Gln-treatment for 8 hr. An incubation with Gln for at least 12hr was necessary to induce the metamorphosis; the 24-hr treatment was the most efficient, whereas the treatment longer than 25 hr resulted in a decrease of metamorphosis (Fig. 4b). In *P. depressus*, the Gln-treatment for 2 hr seemed enough to initiate the arm retraction. The 2hr-treatment never induced eversion of the ER. To make the ER evert, the treatment was required for at least 8 hr (Fig. 5). Judging from these results, two processes of the metamorphosis, the arm retraction (early process) and the ER-eversion (late process), are separable and may be induced by the different stimulus from each other.

The competence of larvae depended on their developmental stages. Total 260 larvae of *H. pulcherrimus* at the stage of 'late 8a' pluteus were subjected to 10^{-4} M Gln for 24 hr and its effect on the metamorphosis (ER eversion) were examined at one or two days after the treatment. Every larva was confirmed to be at ER stage by squashing (Fig. 6a2-d2). In 58 control larvae, none of larvae metamorphosed. In Gln-treated larvae, out of 14 larvae with 5-segmented spines (Fig. 6A, al, a2) six larvae metamorphosed (43%). Larvae in figure 4B series, having spines with 10-15 segments, metamorphosed up to 75%. In fully-developed larvae 'late 8a-III' (Fig. 6C) whose spines had 15-20 segments (Fig. 6c2), 100% of larvae accomplished metamorphosis. Figure 6D shows an over-matured larva, which failed to evert its ER. Larvae at this stage metamorphosed up to 50%, but most of metamorphosed larvae died later. It is notable that

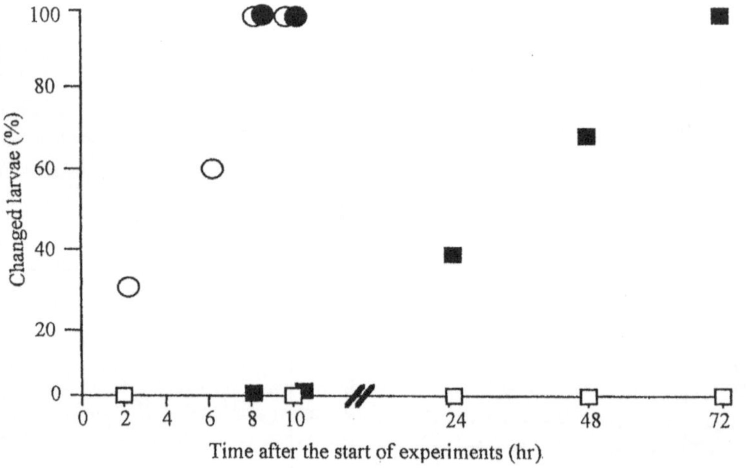

Figure 5. Two metamorphic processes, the retraction of arms and eversion of the echinus rudiment, require different length of the Gln-treatment. Ten-larvae groups of *P. depressus* were treated with 10^{-5} M Gln in MFSW for 2hr (open symbols) or 8hr (closed symbols) and were followed by their arm retraction (○,●) and ER eversion (□,■) up to 72hr. Based on Yazaki (1995).

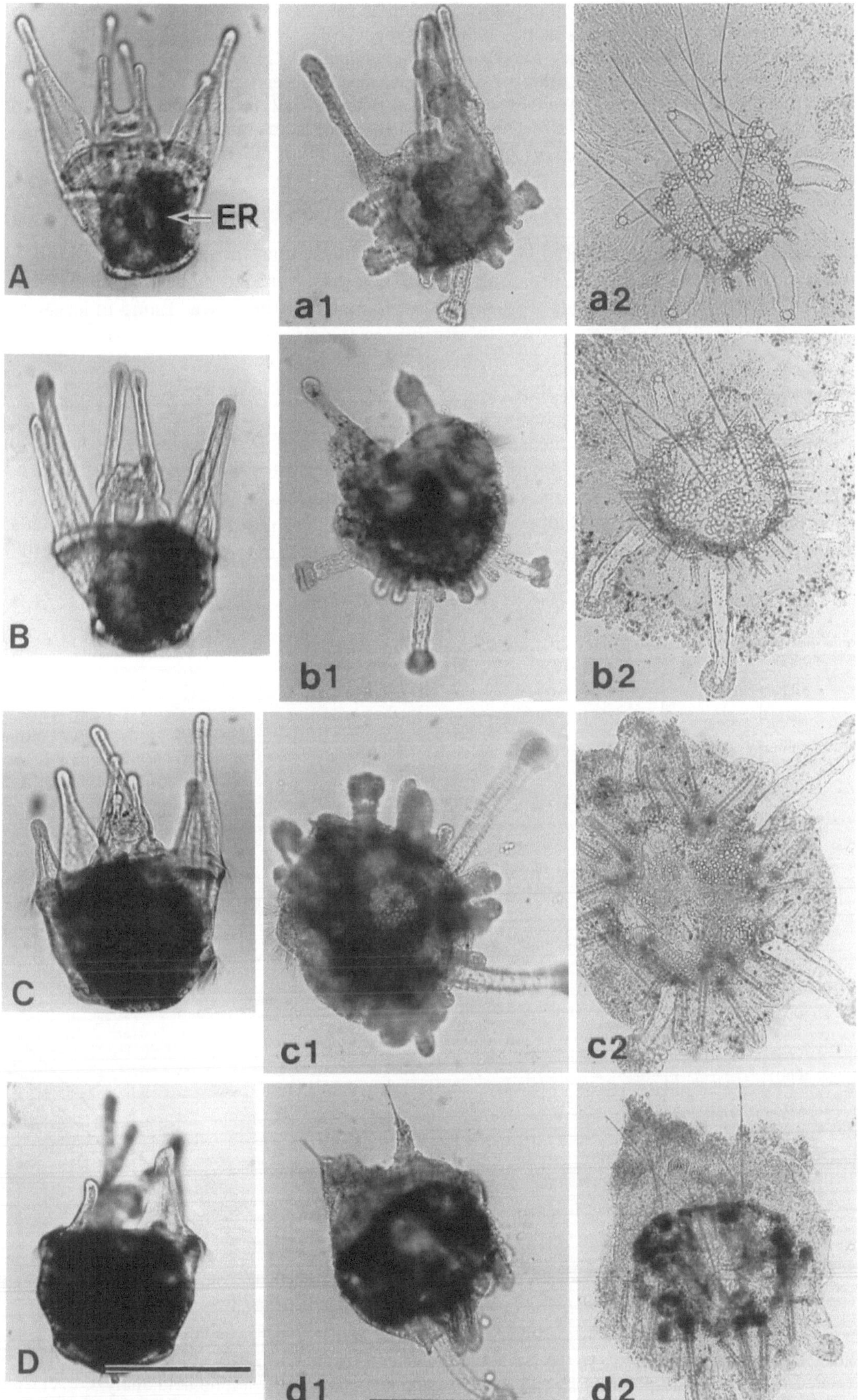
ER
A
a1
a2
B
b1
b2
C
c1
c2
D
d1
d2

Figure 6. Larval competence to metamorphose. Developmental stages of 8-arm pluteus were checked by number of steps of spinal skeletons in the echinus rudiment (ER). With the growth of ER the number of steps in the spinal skeletons increases. Larvae of *H. pulcherrimus* (A-D) were treated with 10^{-4} M Gln for 24 hr and at 48hr the larvae or juveniles (a1-d1) are squashed to observe the developmental stage of ER (a2-d2). Larvae, of which spines have five-segmented skeletons (A-a2) metamorphosed 43%, with 10-15 segments (B-b2) 75%, and with 15-20 segments (C-c2) 100%, of larvae metamorphosed, respectively. In over-matured larvae (D-d2) only 50% of larvae metamorphosed. Bars 500 mm. Yazaki (1995).

larvae at the earlier stage retracted larval arms more slowly than those at later stage (compare Fig. 6a1, b1 and c1). In the over-matured larva, skeletons were not resorbed and ER did not evert, while arms had degenerated (Fig. 6d1, d2). No metamorphosis was found in larvae at the stage earlier than 'late 8a-1'.

4.1.2 *Accumulation of glutamine and glutamic acid in sea urchin larvae*

Free amino acids in larvae were measured during development. Two species of the sea urchins, *H. pulcherrimus* and *Strongylocentrotus nudus* were cultured for 40 days at 20 °C by feeding with *Chaetoceros gracilis* up to the 8-arm plutei. The progresses of the development of these two species were similar, i.e., they grew to early pluteus, 6-arm pluteus and 8-arm pluteus, on the 3rd day, the 20th day and the 40th day, respectively. Their dry weights per larva were 0.28

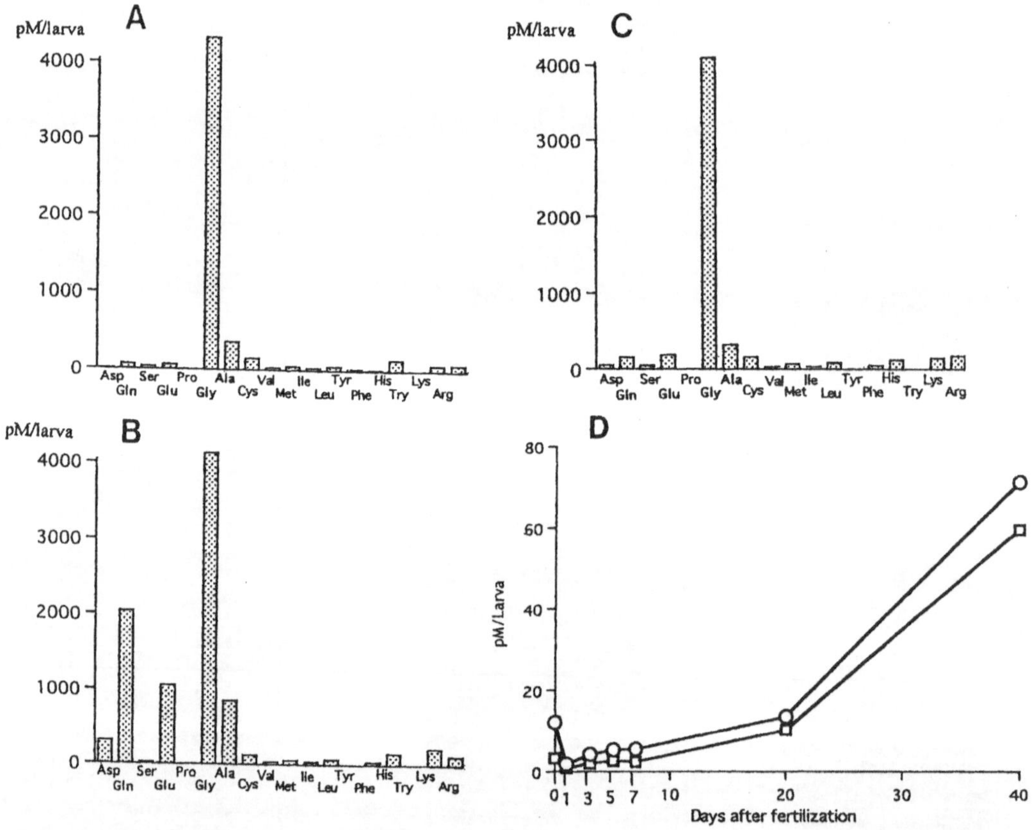

Figure 7. Amino acids contents in the larva and the effects of the Gln-treatment. Free amino acids were extracted from eight-arm plutei of *H. pulcherrimus* and subjected to the amino acid analyzer. A) Control larva. B) Larva treated with 10^{-3} M Gln and C) with 10^{-3} M glutamic acid for 24h, respectively. D) Gln and glutamic acid contents during the development of *H. pulcherrimus* (○) Gln, (□) Glu. (Yazaki, in prep.).

μg (3rd day), 1.25 μg (20th day) and 16.8 μg (40th day). Gln and Glu contents in one larva were shown in figure 7D. Unfertilized egg had 10 - 12 pM Gln and 3 pM Glu per egg. By the stage of gastrula (20 hr), Gln decreased to the level of 3.7 ± 0.3 pM/embryo (n=5 experiments) and this level was kept up to the early pluteus (3rd day). From the 3rd day, larvae were fed with *C. gracilis*. Gln contents in a larva increased from 5 pM (4-arm pluteus stage) to 68 pM (8-arm pluteus stage) after the feedings. Glu contents increase in the similar way to those of Gln in spite of fact that the amount of Glu in C. *gracilis* was much higher than that of Gln (Glu, 37.5 pM and Gln, 16.3 pM per 10^5 of *C. gracilis*). In *S. nudus* the values of Gln and Glu were quite similar to those of *H. pulcherrimus* (data not shown). Figure 7A shows the amino acid composition in the 8-arm pluteus of *H. pulcherrimus*. Out of 18 amino acids, Gly was prominent (61-81% of total free amino acids) through the developmental stages, of 4-, 6- and 8-arm pluteus examined. The second abundant amino acid was Ala (6-8%). Nagaoki and Isono (1981) compared free amino acids in unfertilized eggs of 20 species sea urchins and divided sea urchins into two groups based on the free amino acid composition, one group was overwhelmingly rich in glycine and /or alanine, and other amino acids were remarkable in the other group. The high content of glycine (70-90%) was phylogenetically specific to Echinoidea. Gln and Glu were 6% and 5 %, respectively at the 4-arm stage, both gradually decreased to about 1% at the 8-arm pluteus stage. The contents of amino acids were measured in larvae treated with 10^{-3} M Gln or Glu for 24hr. In the Gln-treated larvae, the amount of Gln increased up to 41-fold in the 8-arm larva (Fig. 7B) compared to that of the control larvae (Fig. 7A). Exposure to Gln caused the increase of Glu up to 18-fold. Ala and Asp were also increased but a little. The treatment of larvae with Glu markedly increases the larval contents of neither Glu nor Gln (Fig. 7C).

4.2.1 *Metamorphic effects of neurotransmitters*

In *D. excentricus*, 10^{-5} M dopamine induces metamorphosis with low frequency. On the other hand, dopamine at the same concentration promotes the contraction-histolyic sequence of metamorphosis with high frequency in the excised larval arms (Burke 1983b). In *P. depressus*, *C. japonicus*, *A. crassispina* and *S. nudus*, dopamine instantly induced the cessation of the ciliary movement, extrusion of the primary podia and quick degeneration of arms leaving the larval skeletons by the hour. Although cramping of arms was not so clear, ER was exposed by disintegration of the larval tissues. L-Dopa and noradrenalin showed similar effects on larvae to dopamine. Their effects, however, were not so remarkable as dopamine. Antagonists to dopamine receptors, domperidone, SKF83566 and haloperidol were examined. An antagonist to dopamine D_2-like receptor, domperidone, completely cancelled the histolyic effect of 3×10^{-5} M dopamine at the concentration of 5×10^{-5} M. In the presence of 4×10^{-6} M haloperidol, antagonist to D_2-like receptor, larval tissues degenerated either with or without dopamine. SKF83566 of 3×10^{-5} M, an antagonist to D_2-like receptor, was toxic itself and did not block the dopamine-induced histolysis. Serotonin and Glu of 10^{-3} M had no effect on sea urchin larvae (Yazaki, unpubl.).

In abalone (gastropod mollusc), g-aminobutyric acid (GABA) induced metamorphosis, while Gln was inert (Morse et al. 1979, Morse et al. 1980). In sea urchins, GABA evoked a transient eversion of the echinus rudiment but never induced metamorphosis (Yazaki & Harashima 1994). After addition of GABA in a range of 10^{-3} - 10^{-5} M, the larval arms instantly cramped and then the echinus rudiment was extruded 5-6 min later (Fig. 8b), but the retraction of arms never occurred, even after the 24hr-treatment with GABA. Usually, the cramping of arms ceased within 20-30 min, while in some larvae it continued for 3 hr and then the larva returned to the form as before the treatment (Fig. 8a). Larvae of *C. japonicus* were also induced the cramping of arms by GABA. Most larvae went back to the original larval form; however, one larva could not return (Fig. 8c). This larva moved around on the bottom of the culture plate by podia for five days more, but never retracted its arms. Figure 8d shows a 5-day juvenile metamorphosed by Gln. This fact conflicts with an idea that primary podia receive tactile stimuli and transmit them to the effectors to initiate metamorphosis (Burke 1980).

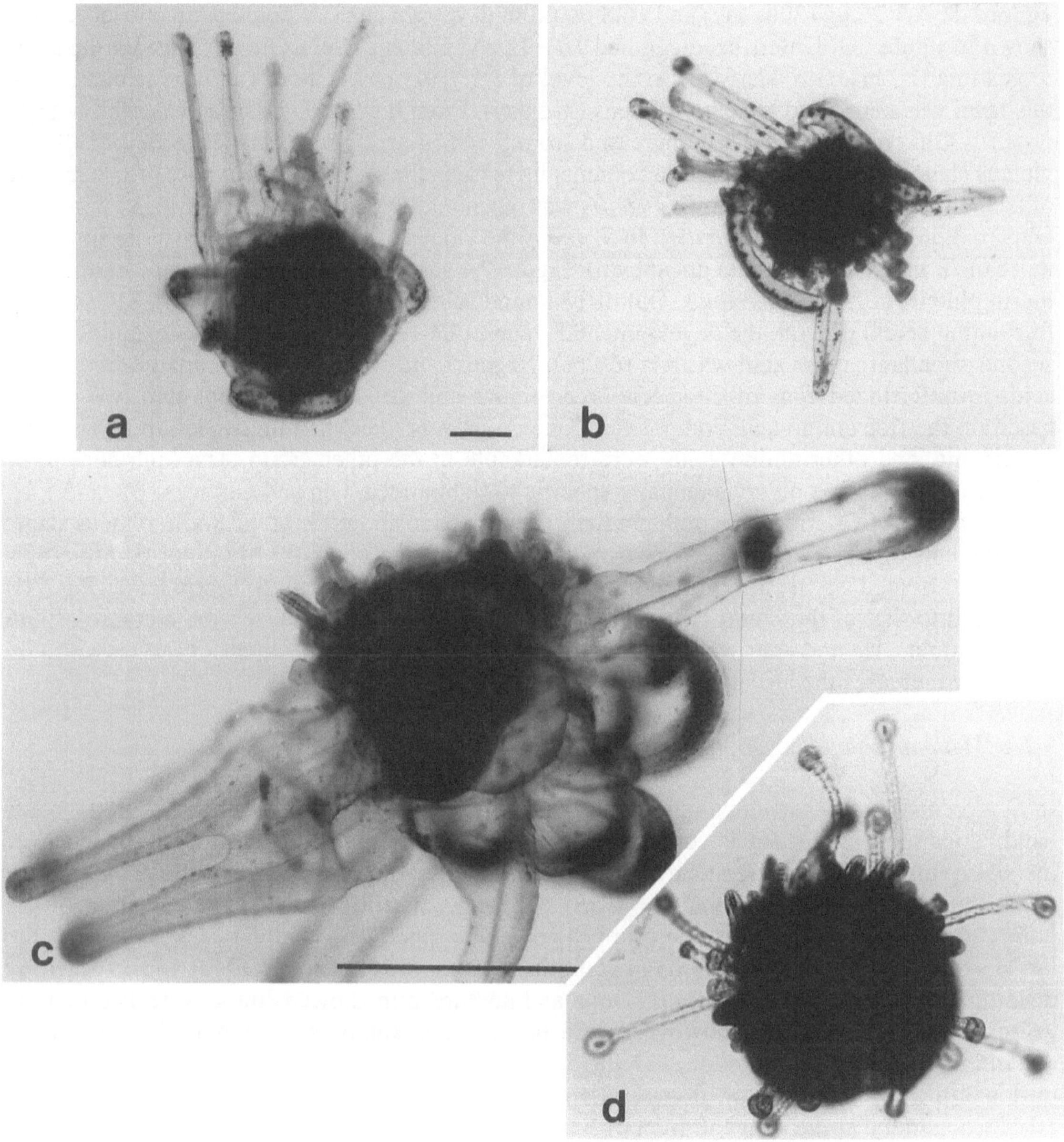

Figure 8. Spontaneous eversion of ER induced by GABA. By the addition of GABA, larval arms spread to extrude the ER, although the larva went back to the original form within 20-30 min. a,b) *P. depressus*; larva before (a) and after (b) addition of GABA. c,d) *C. japonicus*; c) Five day-larva whose ER everted with GABA but could not go back to the original form. d) A 5-day juvenile metamorphosed by Gln-treatment. Bar 100 mm. Yazaki (unpubl.).

4.2.2 *Localization of neurotransmitters in larvae*

Development of the nervous system was investigated in the early stage of the pluteus larva of *Psammechinus miliaris* (Ryberg 1977, Ryberg & Lundgren 1977) and in the 8-arm pluteus of *Strongylocentrotus droebachiensis* (Burke 1983, Bisgrove & Burke 1987) using electron microscopy, silver staining, glyoxilic acid-induced fluorescence of catecholamine and immunofluorescence. Glyoxilic acid-induced fluorescence in larvae of *H. pulcherrimus* was observed in the axon tracts of the ciliary bands and the perioral ganglions in both upper and lower lips indicated by arrowheads and arrows (Fig. 9a) similar to that in *S. droebachiensis* (Burke 1983). The fluorescence in the axon tracts of the ciliary band detected in arms of the

8-arm pluteus (Fig. 9b) disappeared from the degenerating arms after the 24 hr-Gln treatment (Fig. 9c).

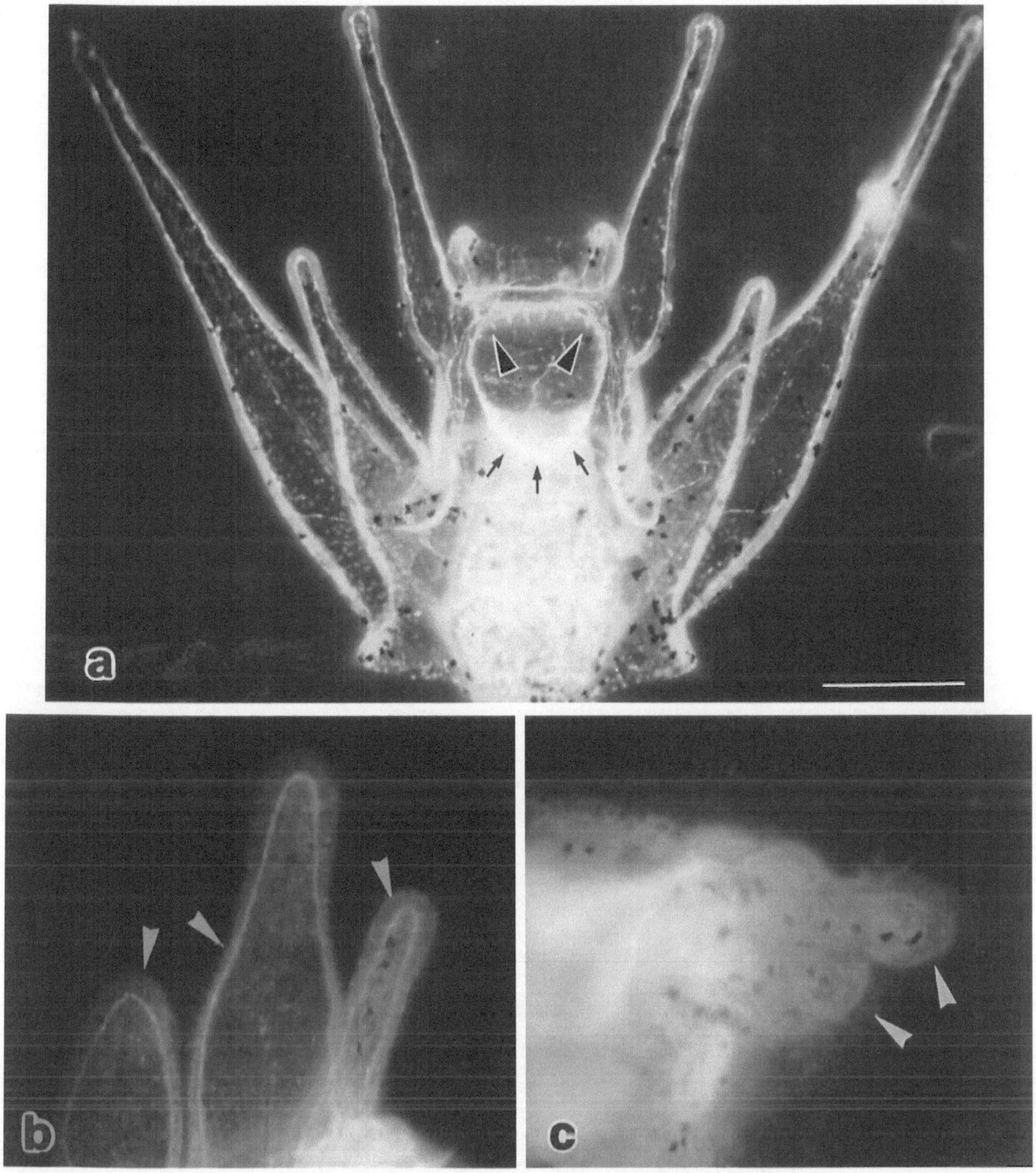

Figure 9. Nervous system of the sea urchin larva of *H. pulcherrimus* stained with glyoxilic acid. a) A larva at the late 6-arm pluteus stage. Arms of the competent larva b) before and c) after Gln-treatment. Neural axon tracts underlying the ciliary bands (arrowheads in b) had disappeared from the larval arms after the 24 hr-Gln treatment (arrowed in c). Bar 100 mm. Glyoxilic acid-staining by courtesy of T. Yazawa.

Figure 10 shows the immunostainings of 8-arm larvae of *C. japonicus* using anti-dopamine and -GABA. The dopaminergic components of the nervous system includes a ganglion in the lower lip of the mouth (arrows in Fig. 10a1) and the axon tracts along the ciliary bands of arms and epaulets or lobes (arrowheads in Fig. l0a1, a2). Bisgrove & Burke (1987) reported the first detection of 6-8 dopaminergic cells in the lower lip of 4-arm pluteus. Number of the cells increased with a progress of development to 30 at the 8-arm pluteus named the middle 8a. In the competent "late 8a-III" larva of *C. japonicus*, approximately 120 dopaminergic cells were counted each 60 cells on the left and the right sides of the oral. Dopaminergic bipolar cells, which joined to the axon tracts of the ciliary band (Bisgrove & Burke 1987), were found in the apical side of arm epithelium (data not shown).

GABAergic neural plexus were found in the larval epithelium surrounding the vestibule of the ER (arrowheads in Fig. 10b1). When larvae everted their ER by the Gln-treatment, this plexus degenerated into fragments. GABAergic neural plexus was also found at the base of postoral arms, which cramped by exposing to GABA (arrowed in Fig. 10b2). Much more GABAergic cells and plexuses were present in the preoral arms, which remained more retardingly than any other arms in the metamorphosing larva. Anti-glutamate did not stain any nerve-like structures (data not shown).

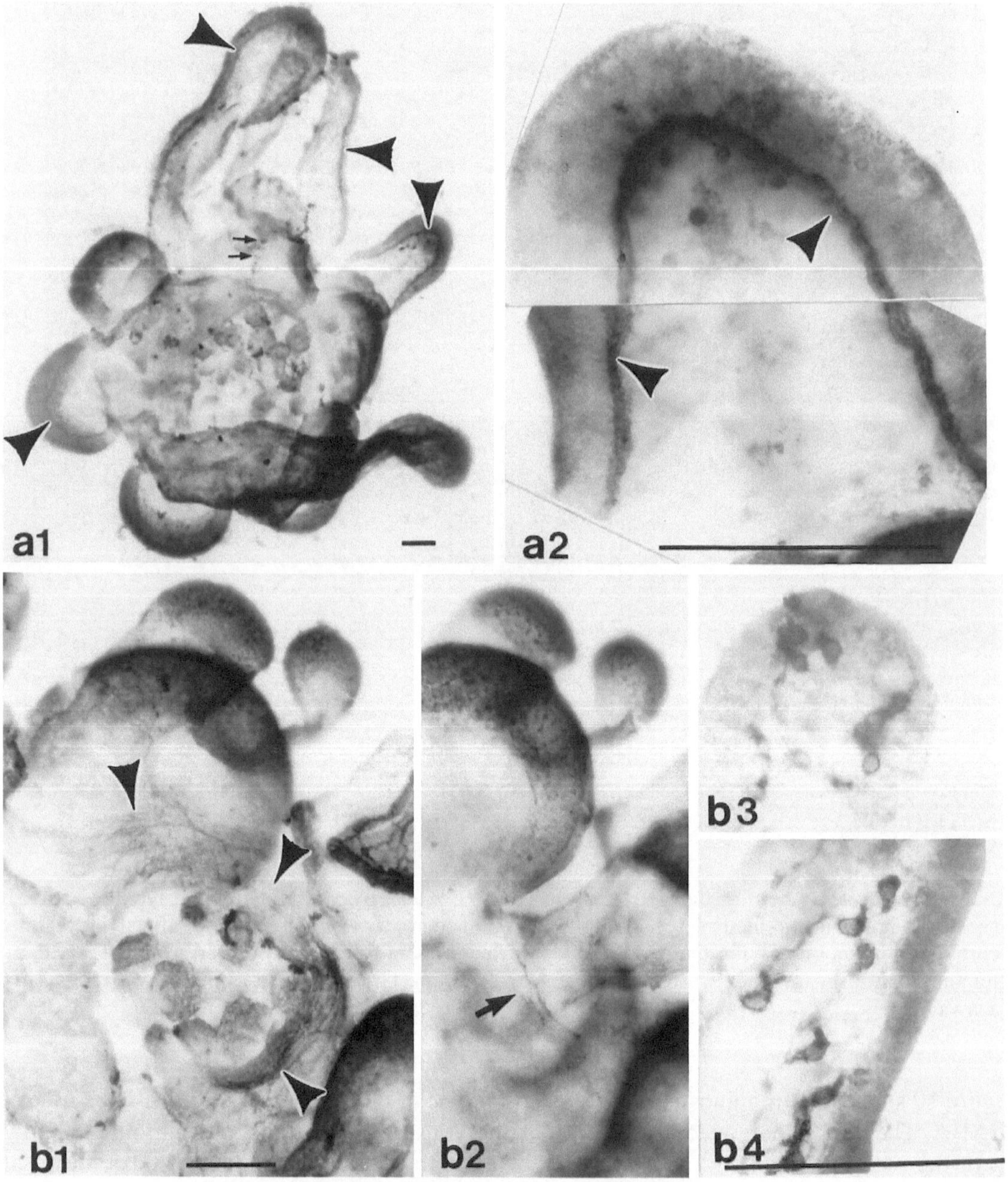

Figure 10. Nervous system in competent larvae of *C. japonicus* stained with antibodies against dopamine and GABA. a1) Whole larva stained with anti-dopamine. a2) One arm of a1. b1 -b4) A larva stained with anti-GABA. b1) Peripheral region of the vestibule opening, b2) at the root of the postoral arm and b3,b4) the preoral arm of the larva in al. Bars 100 mm. Yazaki (unpubl.).

4.3 *Metamorphosis induced by algal films*

A plastic film, on which green algae (*Ulvella* sp) were grown, were capable of inducing the metamorphosis of larvae in *P. depressus*, *H. pulcherrimus*, *A. crassispina* and *C. japonicus*. The *Ulvella* films were originally produced for the aquaculture of abalone juveniles. The processes of the metamorphosis were different in cases induced by *Ulvella*, assumed to be dependent on the growth condition of *Ulvella* and on the developmental stage of sea urchin larvae. The earliest metamorphosis completed within one hour. In another case, it took two days. In some experiments, the larval arms cramped instantly, and the arm retracted within 6 hr after the start of the experiments followed by the eversion of ER. In some cases, eversion of the ER was initiated before the retraction of arms. Since the arm retraction in the *Ulvella*-induced metamorphosis was quick, their skeletal rods were not resorbed but discarded.

Effective substances of *Ulvella* to induce metamorphosis were extractable in seawater. The metamorphosis-inducing activity was stable for heat treatment (100 °C, 5min). Larvae metamorphosed without direct contacts to *Ulvella*, separating from *Ulvella* with a meshed nylon, although it took a longer time to metamorphose than in the experiment without the mesh (Yazaki, unpubl.).

5. FATES OF LARVAL STRUCTURES IN METAMORPHOSIS

5.1 *Larval epithelia are going down to degenerate*

During the metamorphosis larval tissues undergo a series of complex changes. Figure 11 is a schematic diagram of larvae sections undergoing metamorphosis, summarized by Cameron and Hinegardner (1978) from their observations through the scanning and electron microscope transmission. The overall outline of larva is delineated by an epithelium one-cell thick, which thickens in the ciliary bands or epaulets. The larval epithelium continues to the vestibular epithelium, overlying the echinus rudiment (Fig. 11A). During the metamorphosis, the larval tissues collapse with multiple invagination along the ciliary bands and epaulets, and with retraction of larval arms. The echinus rudiment everts exposing the vestibule outside (Fig. 11B). Finally, the vestibular epithelium covers the inverted collapsed larval epithelium (Fig. 11C).

The larval epithelium was characterized by localization of the egg-derived surface substance, ES-1 (Yazaki, 1993). ES- I was exocytosed at fertilization from inside of eggs and localized at the surface plasma membrane of larva until the metamorphosis (Fig. 12a, b). In the competent

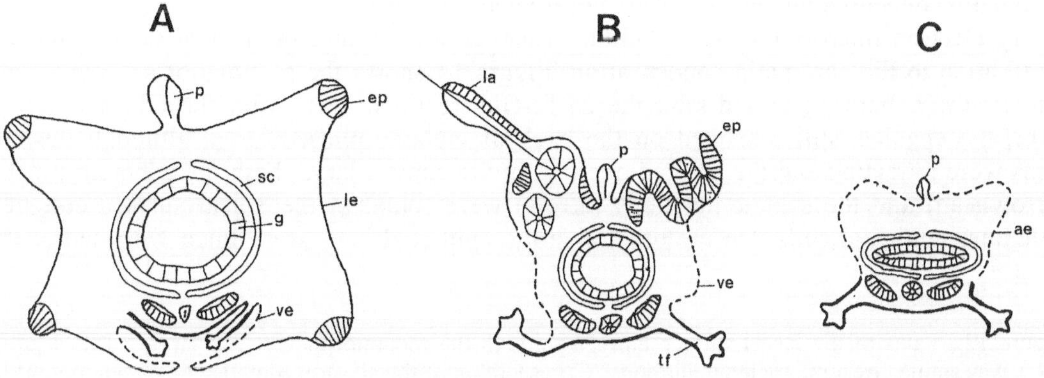

Figure 11. Morphological changes in larvae undergoing metamorphosis (a schematic diagram of sections). A competent larva (A) undergoes metamorphosis (B), and becomes a juvenile (C). The larval epithelium ( ) collapses during metamorphosis (B) and is covered by the vestibular epithelium (-----) (C). ae, adult epithelium; ep, epaulets; g gut; la, larval arm; le, larval epithelium; p, pedicellaria; sc, somatocoel epithelium; tf, tube foot; ve, vestibular epithelium, Cameron & Hinegardner (1974).

larva, ES-1 was also found in free cells (small arrows in Fig. 12b) in addition to localization on the surface membrane of the larva. The ES-l-positive cells were found in the adult epidermis and among coelomocytes taken from the adult urchin coelom (Yazaki 1993). ES-1 was not found on the surface of the newly metamorphosed juvenile any more, but present in free scattering cells (small arrows in Fig. 12c) and in the degenerating larval tissues remained on the dorsal area of juvenile (large arrows in Fig. 12C, c). The larval tissues degenerated and resorbed during metamorphosis were restricted to the ES-1 overlying epidermis; i.e. larval arms, ciliary bands, mouth and esophagus (Fig. 12a, b). ES-1 positive cells detected in the competent larva and in the newly metamorphosed juvenile were supposed to be phagocytes. A monoclonal antibody prepared against adult phagocytes stained the larval tissues remaining on the back of the newly metamorphosed juvenile (Yazaki unpubl.). The adult epidermis is formed from the vestibular epithelium as shown in figure 11. Neither epithelium of stomach nor intestine was lined with ES-1 (Fig. 12a). These tissues were reconstructed to the adult ones. The larval skeletons were discarded when the larval arm retracts so quickly, but in the case of slow metamorphosis, e.g. Gln-induced metamorphosis, the skeletons were resorbed together with retraction of the larval arms.

5.2 *Cellular changes in the metamorphosing larval tissues induced by the Gln-treatment*

In the 8-arm pluteus, high mitotic activity was found in the larval epithelium of the ciliary bands and the epaulets. Mitotic activities were detected by labeling the larvae of *H. pulcherrimus* with 5-bromodeoxyuridune (BrdU) for 4hr, i. e. either 24-28 hr or 42-46 hr from the start of the Gln treatment. The mitotic activity at either labelling times of BrdU was not changed by the Gln treatment for 6-hr, while remarkably decreased in the larvae treated with Gln for 24 hr (Yazaki 1995). Chromatin condensation in the nuclei was induced by Gln in the larval tissues of *H. pulcherrimus* after the stage of 'late 8a'. This nuclear change was never induced by the Gln-treatment less than 5 hr (Sato & Yazaki 1998).

The larval epithelium consists of the squamous epithelium with the thickness of one cell. The columnar epithelia are distributed in the ciliary band of arms and in the epaulets. Figure 13 shows sections of larva of *H. pulcherrimus* at late 8a-III stage either treated with Gln for 24 hr or not. After the 24 hr-Gln treatment, most of cells in the larval epithelium rounded and their nuclear chromatins were condensed (Fig. 13b, d, f,), while in the ER no change was found (Fig. 13g, h). TUNEL assays, by which the fragmentation of DNA is detected, showed that the fragmentation is dominant in the squamous epithelial cells of the larval body but that is not observed in ER of the Gln-treated larvae. Although the quantitative estimation was not carried out yet, it seemed that the DNA fragmentation proceeded the chromatin condensation because many TUNEL-positive cells were found in the Gln-treated larvae for 6 hr in which the chromatin condensed nuclei were scarce (Sato pers. comm.).

By electron microscopy, some cellular changes were found in cells of Gln-treated larva in addition to the chromatin condensation. Figure 14 shows the sections of an epaulet of *H. pulcherrimus*, before (A) and after the 24 hr-Gln-treatment (B). After the Gln-treatment for 24hr, the epaulets still remained on the back of metamorphosed larva, although the larval arms were retracting and the ER had everted. In the control larva, both the adherent junctions (arrowhead) and the septate-junctions (arrow) were found at the apical surface end of the columnar epithelial cells. On the basal side the epithelial cells were lined with neural tracts

Figure 12. Fates of the larval surface antigen through the metamorphosis. The larval surface antigen, ES-1 was stained by a monoclonal antibody. Cryosections of the six-arm pluteus (A, a), competent larva (late8a-III; B, b) and a newly metamorphosed juvenile (C, c). (A, B, C) Light micrographs and (a, b, c) fluorescent micrographs of A, B, C. es, esophagus; i, intestine; m, mouth; s, stomach; sp, spine; ER, echinus rudiment. The areas surrounded by a broken line are ER (B) and the mouth organ (C). Arrowheads point epaulets. The stomach is autofluorescent (indicated by flat arrows in b. Small arrows indicate the ES-1 including cells and big ones show the degenerating larval tissues. Bars 100 mm. Yazaki (1993).

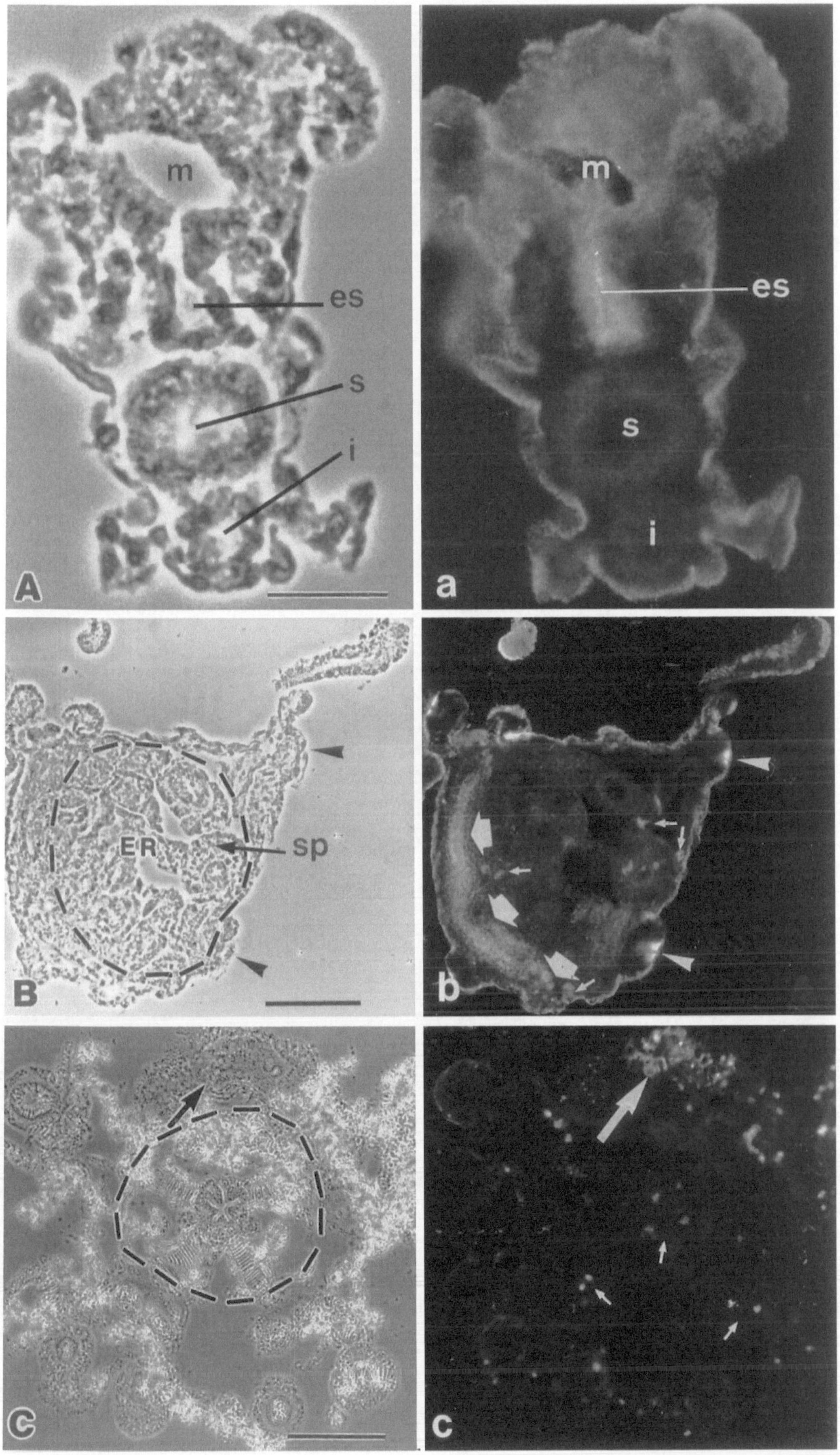
m
es
s
i
A
m
es
s
i
a
ER
sp
B
b
C
c

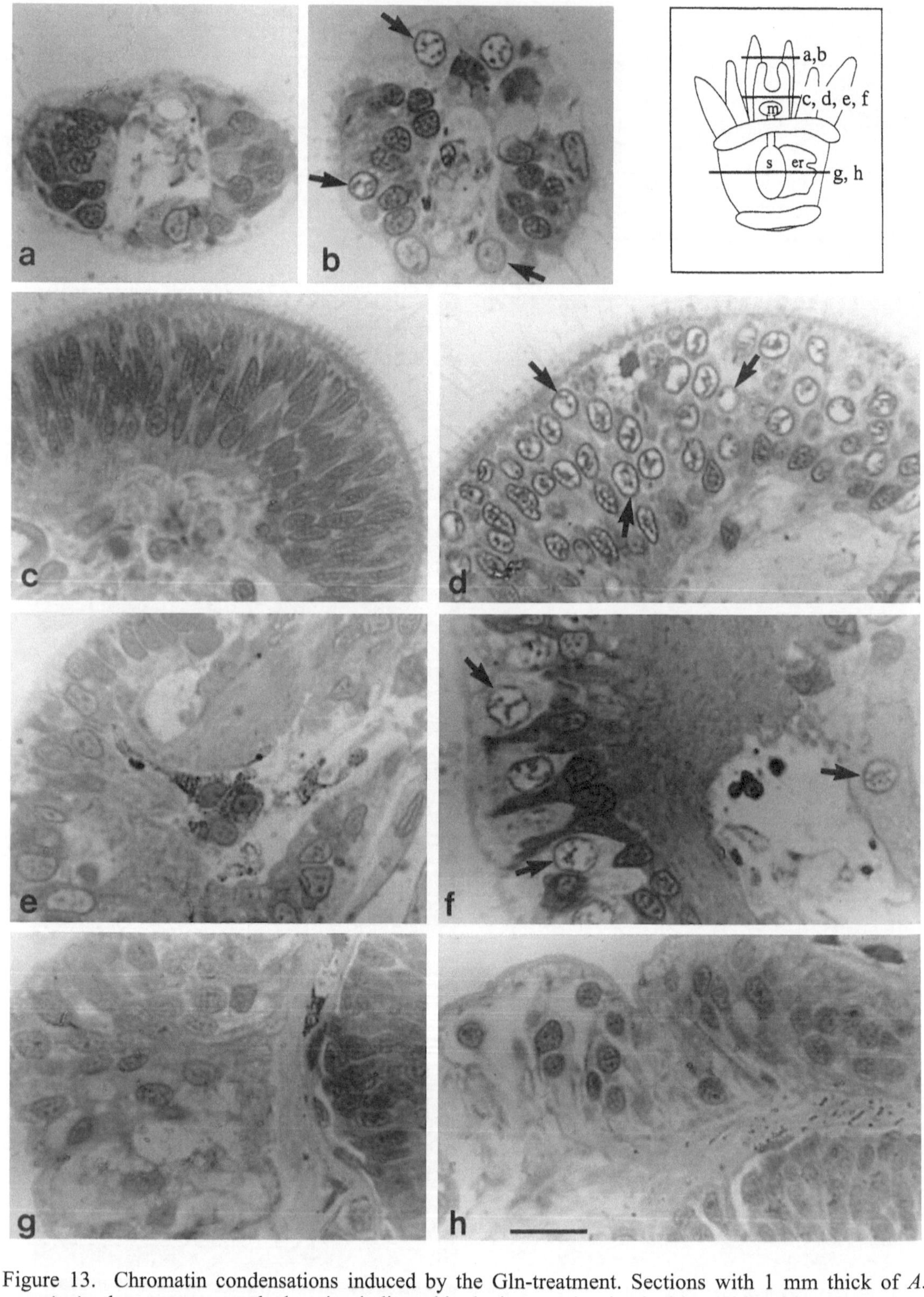

Figure 13. Chromatin condensations induced by the Gln-treatment. Sections with 1 mm thick of *A. crassispina* larvae are cut at the location indicated in the inset and stained with toluidine blue. Control (a, c, e, g) and 24hr-Gln-treated larvae (b, d, f, h); (a, b) larval arms, (c, d) oral lobes, (e, f) aboral epithelia, (g, h) echinus rudiments. er, echinus rudiment; m, mouth; s, stomach. Bars 10 mm. Sato & Yazaki (1998).

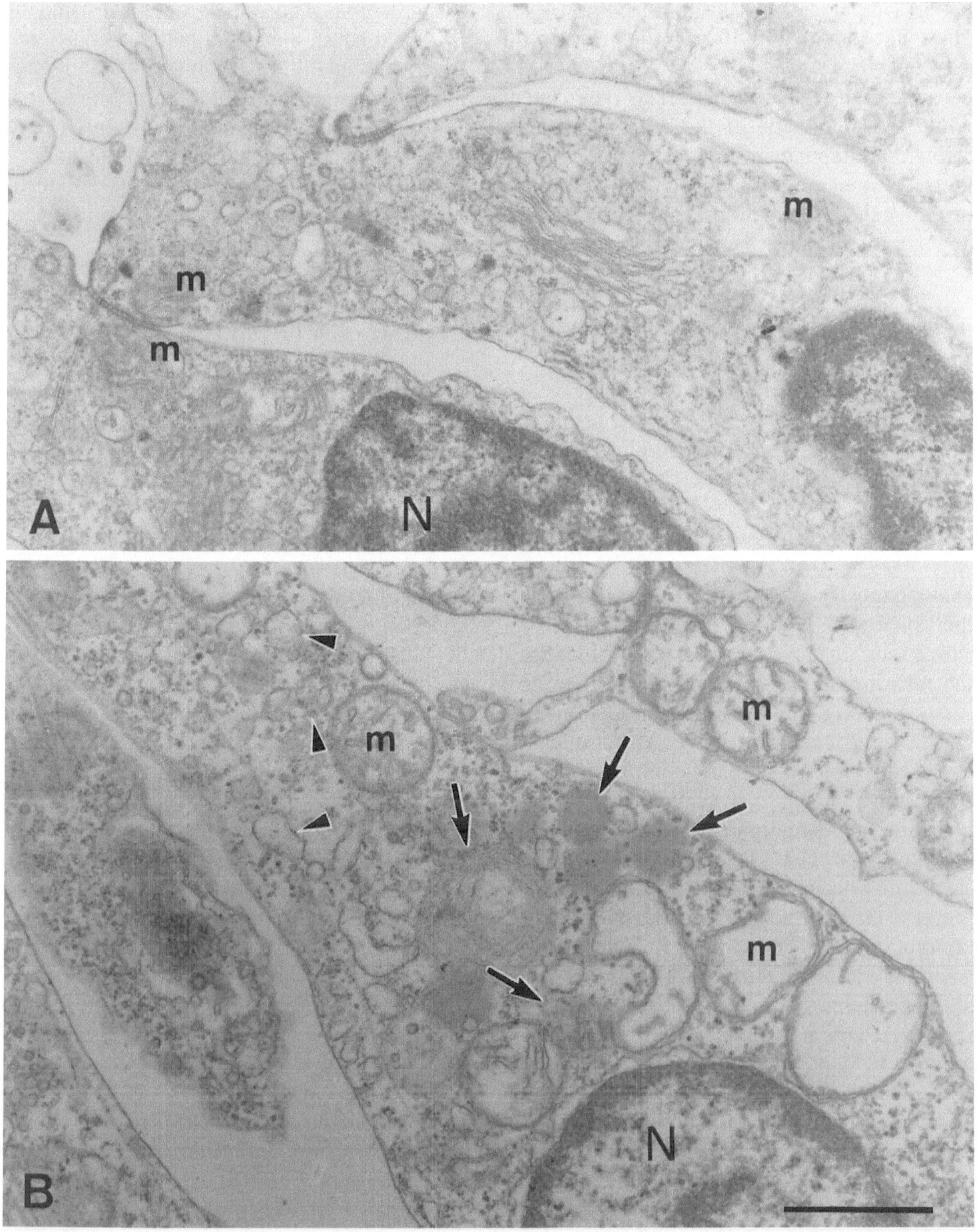

Figure 14. Effects of Gln on the larval epithelial cells of *H. pulcherrimus*. Sections of the epaulet before (A) and after (B) the 24 hr-Gln-treatment. m, mitochondria; N, nuclei, Details are explained in the text. Bar 1 mm. Sato & Yazaki (unpubl.).

being separated from the mesenchyme by the basal lamina. The neural tract consisted of 20-30 axons with 200-400 nm diameter. In the Gln-treated larva, the apical cell junctions were recognized but the locations were irregular (small arrows in Fig. 14B). Intercellular spaces became large. The mitochondria (m) expanded and those cristae partially or almost disappeared, while the membrane of the mitochondria seemed to be intact. Lysosomal structures (large arrows) and abundant autophagic vacuoles (arrowheads) were distinct (Fig. 14B) not only in these columnar cells but also in the neural axons. These cell features suggest that the cell death in Gln-induced metamorphosis is neither a typical apoptosis nor necrosis but an autophagic degeneration (Sato &Yazaki, unpubl.).

6. PROPOSED MECHANISMS OF SEA URCHIN METAMORPHOSIS

Stimuli, receptors, effectors and responses in metamorphosis of marine invertebrate larvae were reviewed by Burke (1983b). He suggested three potential means to control the metamorphosis: nervous and endocrinal control plus a control of independent effectors, in which effector cells receive and respond to the cues for metamorphosis derived from the environment.

6.1 *Neural control*

Histochemistry using glyoxilic acid showed the neural cells containing catecholamine in the competent larva (Fig. 9). Larvae of *D. excentricus* quickly metamorphosed by 10^{-5} M exogenous dopamine but relatively low incidence. Meanwhile the larval arms excised from the neural centers, the oral ganglion in the lower lip and the apical neuropile in the upper lip, frequently undergo the contraction-histolysis sequence of metamorphosis in response to the same concentration of dopamine, but not to the natural cue (Burke 1983a). Since electrical stimulation on the nervous centers frequently induced metamorphosis, Burke (1983a) suggested that nervous centers mediate the perception of the natural cue and control the initiation of metamorphosis, probably by release from the suppressive action of oral ganglion on dopaminergic cells. The effective substance of natural cue for *D. excentricus* is extracted from a sand dollar bed and supposed as a peptide, since its activities are sensitive to heat-treatment and protease (Highsmith 1982, Burke 1984). In Japanese sand dollar *C. japonicus* and other regular urchins, dopamine was effective in degeneration of the larval tissues with high incidence but not so quickly as *D. excentricus* (Yazaki 2000). In *P. depressus*, when dopamine was administrated at the concentration of 5 ¥ 10^{-5} M, the primary podia extruded instantly but it took more time for the retraction of the larval arms to start, at least more than three hours. The arm retraction induced by dopamine, though, is much faster than that by Gln-treatment. Immunostaining with anti-dopamine showed similar results to those of histochemistry using glyoxilic acid except for unstainable for the apical neuropile (Fig. 10a1, a2, Bisgrove and Burke 1987). Domperidone, a dopamine-receptor antagonist completely inhibited the arm retraction caused by dopamine (Yazaki, unpubl.). These results indicate that dopamine may act through the dopamine receptors in the arms, probably on the membrane of the epithelial cells. Bisgrove & Burke (1987) reported the presence of both GABAergic cells and serotonergic cells in sea urchin larvae. GABA evoked spontaneous ER eversions by enlarging the opening of the vestibule (Yazaki & Harashima 1994), probably by suppressive control of vestibular- and arm-muscles innervated by GABAergic nerves (Fig. 10b1, b2). In abalone, GABA induced the metamorphosis by acting at receptors on the larval epithelium to initiate settlement, plantigrade attachment, and then followed by loss of the ciliated epithelial cells of the velum beginning within 15-20hr after addition of GABA (Morse et al. 1980, Trapido-Rosenthal & Morse 1986). In marine snail *Ilyanassa obsolata*, the addition of serotonin in the culture seawater induced the metamorphosis (Levantine & Bonar 1986). Since it took much more time to metamorphose by applying serotonin into bath (20 hr) than those by the natural inducer (2hr), Levantine and Bonar (1986) suggested that serotonin did not induce metamorphosis by binding to epidermal chemoreceptors and was transported across the larval epidermis. The transported serotonin may be able to mimic the release of endogenous serotonin and induce metamorphosis. By injection of serotonin and related pharmacological agents into *Ilyanassa* larvae, Couper & Leise (1996) showed that serotonergic activity in the nervous system promotes metamorphosis. Following injection of serotonin, ciliated cells were dropped from the larval swimming organ, the velum. Serotonin is completely inert in sea urchins metamorphosis. It remains to be confirmed that the lack of sensibility to serotonin depends on the impermeability of serotonin into larvae as Glu in the sea urchin metamorphosis.

6.2 *Endocrine control or metabolic control*

Endocrine control of metamorphosis, known in insects and amphibians, has not been identified in sea urchins yet. Thyroid hormones in sea urchin larvae are supplied from the food, *C. gracilis*, and exogenously applied thyroxine promotes the development of the echinus rudiments to help facilitating their metamorphosis (Chino et al. 1994, Saito et al. 1998). However, the functions of thyroxine in sea urchin metamorphosis, such as stimuli, receptors, effectors and responses, are unknown. Gln in sea urchin larvae increases by feeding of algae, C. *gracilis* (Fig. 7) and the larval facility for histolysis in Gln-induced metamorphosis increases with the larval development (Fig. 6). The metamorphic changes induced by Gln proceed in due succession and on time. This may indicate that Gln is a substrate for the signal substances to progress the metamorphosis. Gln was incorporated into larvae and was metabolized to Glu, Ala and Asp (Fig. 7). Bath-applied Glu could not work to initiate the larval cell death or metamorphosis (Yazaki and Harashima 1994), probably due to its impermeability as shown in Fig. 7C. In cultured cells, Glu is known as an inducing factor for neuronal apoptosis. Glu within the larval cells is metabolized from Gln incorporated during the Gln-treatment and may induce the neural cell death in the larvae.

Bath-applied dopamine caused histolysis in larval arm epithelium. Is there a release of dopamine in the larva in the metamorphosis induced by Gln? D2-like Glu-receptor, which was inhibited by domperidone, may function in the larval epithelium. In either metamorphosis, induced by Gln or by *Ulvella*, it remains to be clarified whether the degenerative changes in the larval tissues involve dopamine release. Since L-DOPA, a precursor to dopamine, is also effective in histolysis of larval tissues, it is possible that dopamine synthetic pathway is present in sea urchin larvae and dopamine is also metabolized from Gln. GABA instantly causes a movement of arms to extrude the ER. When larvae are treated with Gln, the ER eversion occurs later than the larval arm degeneration. GABA may also be metabolized from Gln via Glu.

6.3 *The prospective conclusions*

In sea urchin metamorphosis, the metamorphic processes are different in time and in order according to sea urchin species and the metamorphosis-inducing cues employed. In *D. excentricus*, a quick histolysis is accompanied with actin-mediated shrinkage (Burke 1982) but slow histolysis caused by Gln, in some cases by *Ulvella*, may occur by autophagic cell death (Figs. 13, 14). When larvae are treated with dopamine, cells of the degenerating larval arms show a tendency to dissociate and are frequently discarded together with the arm skeletons. Similar histolysis is seen when *Ulvella* works on larvae strongly. However, these larvae can survive as an apparently healthy juvenile by closing larval tissues finally remained on their back. Accordingly, a perspective is assumable that sea urchins are able to metamorphose in various ways in accordance with various circumferences. We have found some substances, which are effective to evoke the metamorphic changes in larvae, dopamine for histolysis, GABA for ER eversion, and Gln for all processes of metamorphosis. Glu and Gln are prominent components of algae, and abundant GABA is included in the extracts of many algae (Morse 1979). It is of extreme interest that the oral ganglions perceive the natural cue and mediate to release dopamine in the sand dollar (Burke 1983a). It is eagerly required to know the molecular structure of the peptide supposed to be the cue molecule from the adult habitat, and to identify its receptors on the ganglion. In this case, the cue molecule may be specific to the sand dollar species. The green alga, Ulvella, is able to induce the metamorphosis of many sea urchins species, not only regular urchins but also the irregular urchin, sand dollar. The effective components from *Ulvella* are heat stable in contrast to the heat-unstable cue substance from sand described above. The metamorphic changes induced by *Ulvella* or the extract of *Ulvella* is similar to those by dopamine. Is either dopamine or dopamine-like substance contained in *Ulvella*? Excess potassium ions induce the arm retraction and metamorphosis, and excess calcium ions also stimulate the metamorphosis (Cameron et al. 1989). Excess potassium may initiate the cell death or larval histolysis by depolarization of the epithelial cell membrane to open the calcium channels localized at the apical surface membrane of epithelial cells (Yazaki 1995). It is expected that the same physiological changes, e.g. a change in the intracellular calcium ion concentration, is induced by Gln, dopamine and *Ulvella*, although the timing is to be different correlating to the process of the larval cell death.

REFERENCES

Bisgrove, B.W. & Burke, R.D. 1987. Development of the nervous system of the pluteus larva of *Strongylocentrotus droebachiensis*. *Cell Tissue Research* 248: 335-343.

Burke, R.D. 1980. Podial sensory receptors and the induction of metamorphosis in echinoids. *Journal of Experimental Marine Biology and Ecology* 47: 223-234.

Burke, R.D, 1982. Echinoid metamorphosis: retraction and resorption of larval tissues. In J.M, Lawrence (ed.) Proc. Int. Echinoderms Conf. Tampa Bay: 5 1 3-5 1 8, Rotterdam: Balkema.

Burke, R.D. 1983a. Neural control of metamorphosis in *Dendraster exentricus*. *Biological Bulletin*. 164: 176-188.

Burke, R.D. 1983b. The induction of metamorphosis of marine invertebrate larvae: stimulus and response. *Canadian Journal of Zoology* 61: 1701-1719.

Burke, R.D. 1984, Pheromonal control of metamorphosis in the pacific sand dollar, *Dendraster exentricus*. *Science* 225: 42-443.

Burke, R.D. 1985. Actin-mediated retraction of the larval epidermis during metamorphosis of the sand dollar, *Dendraster excentricus*. *Cell Tissue Research* 239: 589-597.

Cameron, R.A. & Hinegardner, R.T. 1974. Initiation of metamorphosis in laboratory cultured sea urchins. *Biological Bulletin* 146: 335-342.

Cameron, R.A. & Hinegardner, R.T. 1978, Early events in sea urchin metamorphosis, description and analysis. *Journal of Morphology* 157: 21-32.

Cameron, R.A., Tosteson, T.R. & Hensley, V. 1989. The control of sea urchin metamorphosis: ionic effects. *Development Growth and Differentiation* 31 (1): 589-594.

Chia. F-S. & Rice, M.E. (eds.) 1978. Settlement and metamorphosis of marine invertebrate larvae. New York Elsevier.

Chino, Y., Saito, M., Yamasu, K., Suyemitsu, T. & Ishihara, K. 1994. Formation of the adult rudiment of sea urchins is influenced by thyroid hormones. *Developmental Biology* 161 : 1-11.

Couper, J. M. & Leise, E.M. 1996. Serotonin injection induce metamorphosis in larvae of gastropod mollusc *Ilyanassa obsoleta*. *Biological Bulletin* 191: 178-186.

Hinegardner, R.T. 1969. Growth and development of the laboratory cultured sea urchin. *Biological Bulletin* 137: 465-475.

Highsmith, R.C. 1982. Induced settlement and metamorphosis of sand dollar (*Dendraster excentricus*) larvae in predator-free sites: adult sand dollar beds. *Ecology* 63: 329-337.

Ishihara, K. & Noguchi, M. 1996. Sea urchin. In K. Ishihara (ed.), *Developmental atlas of animals* (In Japanese), 191-220. Tokyo: Kyoritsu-Shuppan.

Kitamura, H., Kitahara, S. & Koh, H.B. 1993. The induction of larval settlement and metamorphosis of two sea urchins, *Pseudocentrotus depressus* and *Anthocidaris crassispina*, by free fatty acids extracted from the coralline red alga, *Corallina pilulifera*. *Marine Biology* 115: 387-392.

Levantine, P.L. & Bonar, D.B. 1986. Metamorphosis of *Ilyanassa obsoleta*: natural and artificial inducers. *American Zoologist* 26:14A.

Morse, D.E., Hooker, N., Duncan, H. & Jensen, L. 1979. Aminobutilic acid, a neurotransmitter, induces planktonic abalone larvae to settle and begin metamorphosis. *Science* 204: 407-410.

Morse, D.E., Duncan, H., Hooker, N., Baloun, A. & Young, G. 1980. GABA induces behavioral and developmental metamorphosis in planktonic molluscan larvae. *Fedration Proceedings* 39: 3237-3241.

Nagaoki, S. & Isono, N. 1981. Free amino acids in unfertilized sea urchin eggs. *Hiyoshi Science Review* 16: 93-110.

Naidenko,T.K. 1983. Laboratory cultivation of the sea urchin *Strongylocentrotus intermedius*. *The Soviet Journal of Marine Biology* 9 (1): 46-51.

Ishihara, K. & Noguchi, M. 1996. Sea urchin. In K. Ishihara (ed.), Developmental atlas of animals (In Japanese), Tokyo: Kyoritsu-shuppan.

Ryberg, E. 1977. The nervous system of the early echinopluteus. *Cell and Tissue Research* 179: 157-167.

Ryberg, E. & Lundgren, B. 1977. Extra-ectodermal strands in the ciliated bands of the echinopluteus. *Development, Growth and Differentiation* 19: 299-308.

Saito, M., Seki, M., Amemiya, S., Yamasu, K., Suyemitsu, T. & Ishihara, K. 1998. Induction of metamorphosis in the sand dollar *Peronella japonica* by thyroid hormones. *Development, Growth and Differentiation* 401 307-312.

Sato. Y. & Yazaki, I. 1999. A cellular analysis of sea urchin metamorphosis induced by L-glutamine. In M.C.D. Carnevali & F. Bonasoro (eds.) *Echinoderm Research 1988*: 221-226. Rotterdam: Balkema.

Taniguchi, K., Kurata, K., Maruzoi, T. & Suzuki, M. 1994. Dibromomethane, a chemical inducer of larval settlement and metamorphosis of the sea urchin *Strongylocentrotus nudus*. *Fisheries Science* 60 (6): 795-796.

Williamson, J.E., Nys, R.D., Kumar, N., Carson, D.V. & Steinberg, P.D. 2000. Induction of metamorphosis in the sea urchin *Holopneustes purpurascens* by a metabolic complex from the algal host *Delisea pulchra*. *Biological Bulletin* 198: 332-345.

Yazaki, I. 1993. A novel substance localizing on the apical surface of the ectodermal and the esophageal epithelia of sea urchin embryos. *Development, Growth and Differentiation* 35: 671-682.

Yazaki, I. 1995. Quantitative analysis of metamorphosis induced by L-glutamine in embryos of the sea urchin, *Hemicentrotus pulcherrimus*. *Zoological Science* 12: 105-112 .

Yazaki, I. 2000. An analysis of sea urchin metamorphosis. *Zygote* 8 (Supplement): S46-S47.

Yazaki, I & Harashima, H. 1994. Induction of metamorphosis in the sea urchin, *Pseudocentrotus depressus*, using L-glutamine. *Zoological Science* 11: 253-260.

Yazaki, I., Tosti, E. & Dale, B. 1995. Cytoskeletal elements link calcium channel activity and the cell cycle in early sea urchin embryos. *Development* 121: 1827-1831.

The Sea Urchin: From Basic Biology to Aquaculture, Yokota, Matranga & Smolenicka (eds)
© Swets & Zeitlinger, Lisse, ISBN 90 5809 379 4

Sea urchin development from the egg to metamorphosis: an integrated model for cell-to-cell and environment interaction

C. Falugi & C. Angelini
Dipartimento di Biologia Sperimentale, Ambientale, Applicata, (DI.BI.S.A.A.)
University of Genova. Italy

ABSTRACT: Methods and rationales to perform simple biological assays with the use of sea urchin (*Paracentrotus lividus*) gametes, early embryos and larvae are described. The usual methods have been adapted to the purpose of testing the environment potentiality in sea urchin aquaculture. The described procedures are fast, sensitive and at low cost. The tests are performed at different stages, since the effects of the xenobiotic molecules on the different developmental stages, from fertilisation to metamorphosis, seem to be different according to the biological mechanisms the pollutants interfere or compete with.

1. INTRODUCTION

Sea urchin early development as a biomarker for environmental condition

Human activities, mainly in the last decades, have caused the emission into the environment of substances that may present different degrees of toxicity. These may cause lethal as well as sub lethal damage to living organism, some of which immediately visible (acute intoxication), while others are detectable over a longer period (chronic effects). The previewing of such ecotoxicological effects is particularly complex, because, laboratory conditions aside, the action of an agent is never independent from that of others present in the environment.

A contribution to the understanding of both acute and chronic effects can be made available by the study of the molecular mechanisms that may be altered by the interference of xenobiotic substances.

Developing organisms represent the ideal material for this kind of research, as the intercellular communications taking place during the early phases are particularly delicate and sensitive to minimal interference. Embryonic development is a multiphase event, which proceeds by successive steps, each of which depends on the previous one, and all regulated by cell communication. Thus, a little alteration in signalling is successively amplified, giving relevant effects. Furthermore, the sooner the interference occurs, the higher the probability is of obtaining bad anomalies.

The embryos of aquatic invertebrates, such as sea urchins, supply further advantages: gametes and embryos are released in water, fertilisation is external, and development is generally fast. The sea urchin early developmental stages represent a good model system for a number of reasons: a) early development proceeds within a few hours, is a multi-phase process, and defects which occur during the first developmental stages are established and usually amplified during further stages; b) gametes as well as developing embryos expose cell membranes and receptors directly to the environment; c) sea urchins spawn millions of transparent eggs, which can be fertilised and developed synchronously, allowing both an immediate visual perception of anomalies and statistical elaboration of these data; d) synchronous development of a great number of embryos, moreover allows to collect an amount of samples sufficient to perform biochemical analyses at the appropriate stages; d) gametes may be easily handled in the laboratory, and no special skill is required to obtain spawning.

The analysis carried on in this work is related to the Mediterranean Sea urchin, *Paracentrotus lividus*. This species is abundant in the Mediterranean Sea. Some of the Nations facing this sea commercialise it for dietary purposes, for its orange-red roe (gonads) are appreciated in France, Greece and Southern Italy, and occasionally in other Countries, where it is consumed both cooked and raw. For the environmental bioassay, *P. lividus* is preferred to other sea urchin species present in the same waters, because the eggs are scarcely pigmented and very transparent, and therefore developmental events can be clearly observed.

For people who begin farming sea urchins for commercial purposes either in the environment or in land-based tanks, it is very important to establish whether the quality of the water is good enough for the different stages of development to take place. The bioassays carried on sea urchins represent the best procedure to assess whether the environmental state may assure perfect larval development and survival, mainly during metamorphosis, which is the crucial phase that brings about a change in the way of living of such organisms. In fact, during the planktonic larval phase, the organism feeds on micro algae, which are brought towards the mouth by the movement of ciliary bands on the perioral arms; during the benthic adult stage, it feeds on the substrate, at first by grazing the algal microfilm and then the macro algae.

An important parameter for larval survival and good development up to metamorphosis is represented by the condition of the adult specimen, that influences gamete quality, including egg yolk amount (studies from the group of Lucienne Fenaux, carried out in the bay of Villefranche sur Mer (Fenaux, 1991)). It may in turn be influenced by the quality of the environment where the adult specimens are collected. Moreover, the development of embryos obtained from adult specimens collected in areas that are thought clean from pollutants may be randomly altered by winds or currents carrying xenobiotics and changing the water quality.

A second important step is represented by gamete interaction during fertilisation, and early development: sea urchin fertilisation is particularly sensitive to environmental alterations, because they are subject to regulative and interactive processes between egg and sperm, between blastomeres, and between all of these and the environment. Male gametes are released naked into the seawater, where they will receive attractant molecules from the eggs; during fertilisation the thin egg envelope (vitelline membrane) is lysed by the hydrolytic enzymes released from the sperm acrosome; and during the entire initial phase of development. Although it takes place within the fertilisation membrane, it is sensitive to signal molecules, such as organophosphates, nicotine, etc (Falugi & Prestipino, 1987; 1988).

After cleavage, as soon as the embryos reach the blastula stage, they get rid of the fertilisation envelope, and again they expose the blastomeres directly to the environment, while gene transcription and cell-to-cell signaling begin to take place for the onset of cell and tissue differentiation. The following larval development is completely carried out in direct contact with the environment and with environmental molecules, many of which may interact and interfere with normal signaling. During subsequent differentiation and growth, the larval period lasts differently, and metamorphosis may be delayed, according to the environmental state, which may affect the ciliary movement, and consequently feeding ability (Marchi et al. 1996).

For such reasons, embryonic and larval stages of sea urchin development have often been used as biomarkers for the revelation of environmental disease, such as change in pH (Pagano et al. 1985a) salinity (Dinnel et al. 1981; Dinnel et al. 1983) as well as to detect the presence of a number of xenobiotics and metals (Pagano et al. 2000). Some of these authors, working in different laboratories, are currently collaborating to intercalibrate the method for employing the sea urchin development bioassay (Pagano et al., in progress), in order to suggest it as a standard method for pre-chemical environmental diagnosis.

1.1 *General methods*

The methods generally used consist in exposing either gametes or embryos to the water that must be tested, or to sediment elutriates, in order to record and evaluate the biological effects. In some cases, it may be also necessary to test the potential damages caused by the presence of known substances. In all these cases, the experiments may be carried out by using pelagic or artificial seawater as a control, so as to be sure that the rearing medium is not polluted.

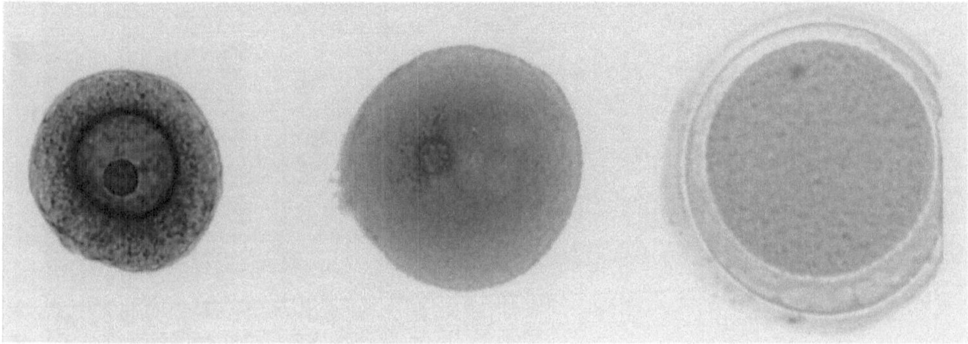

A B C

Figure 1. A) Immature oocyte. B) Mature unfertilised egg. C) Fertilised egg.

Sperm motility, fertilisation events and development are easily checked on a light microscope, with 25 to 100 magnifications. Egg activation, the first step of fertilisation, is revealed by the elevation of the fertilisation membrane (FM, due to the cortical reaction evoked by calcium release form inner storage). The FM appears as a thin circle around the egg. This endpoint is used to assess the state of gametes, by performing fertilisation in control seawater. When it takes place in at least 80% of the eggs within 40-80 seconds after the addition of sperms, the fertilisation conditions are considered good. Otherwise, the gametes are considered not good, and discharged. The egg plus presenting a high percentage of immature oocytes are also generally discharged (Fig. 1). All the chosen endpoints of exposed gametes and embryos are compared with control conditions and normal stages.

1.2 *Normal stages and timing of P. lividus development*

The listed times of the stages of development are related to a strict range of optimal temperature conditions (between 18 and 20°C). Control samples obtained by fertilising eggs in clean or artificial seawater are used step by step to assess developmental times and modes. In optimal conditions, the development must bc synchronous.

When T = 19°C:

fertilisation	within 80 sec
amphimixis	within 30 min
first cleavage	within 70 min
second cleavage	within 110 min
third cleavage	within 130 min
blastula	within 6 hours
gastrula	within 12 hours
early pluteus	within 30 hours
pluteus	within 48 hours

The aspects of the normal stages have been described and illustrated by Giudice (1973). If the plutei are properly fed, they undergo the successive larval stages, as well as metamorphosis (Fig. 3 and 4).

6 arms pluteus	within 9 days
8 arms pluteus	within 12 days
8 arms pluteus with rudiment	within 15-20 days
metamorphosis	within 18-25 days

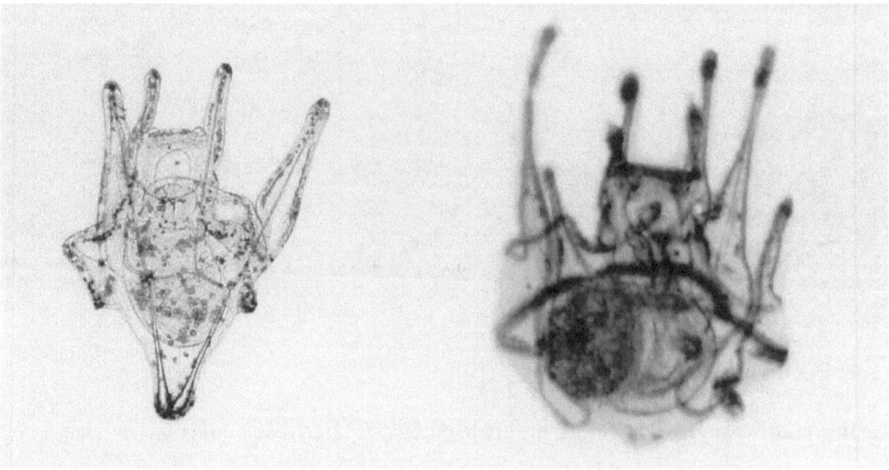

A B

Figure 2. A) 6-armed pluteus; B) 8-armed larva, with rudiment in the left side of the body.

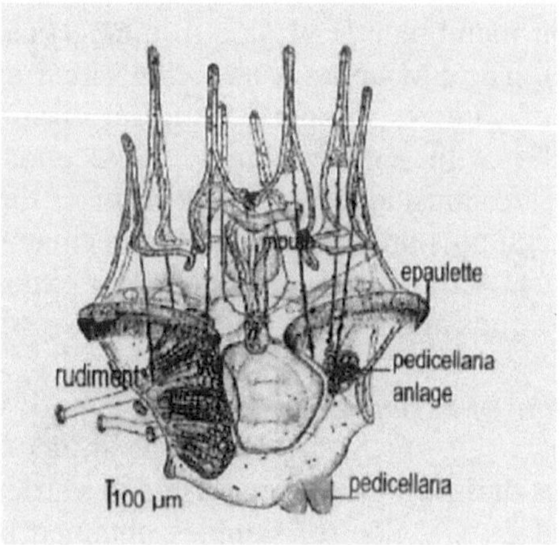

Figure 3. Schematic representation of the 8-armed pluteus (by Okazaki, 1975).

1.3 *Mechanisms representing possible toxicity targets*

1.3.1 *Sperm swimming*

Sperm swimming depends on environmental conditions, because seawater disperses acid sperm liquid containing calmodulin and raises pH levels. Basic pH levels promote sperm swimming (Falugi et al. 1993).

1.3.2 *ttraction towards the egg*

Generally, attractant substances are represented by highly dispersing peptides (e.g. resact, identified in *Arbacia punctulata,* Ward et al. (1985)). As these molecules bind membrane receptors, xenobiotic molecules may interfere with the mechanism.

1.3.3 *Sperm reception*

Sperm receptors are represented by surface glycoconjugates (Vacquier, 1977), which may be altered by bacterial contamination (Epel, 1996).

1.3.4 *Electrical change: Na^+ influx and Ca^{2+} release*

This effect may be exerted on membrane signal receptors or by membrane destabilisation by surfactants, hydrocarbons (PAH), neurotoxic pesticides etc.

The main effect is on polyspermy or nuclear aspects.

1.3.5 *Cell cycles*
Recent findings show that cell cycles may be accelerated by electromagnetic fields.

1.3.6 *Movements related to gastrulation*
Developmental cell movements and migrations are affected by anti-cholinergic substances, such as neurotoxic pesticides (Morale, 1997; Morale et al. 1997; Morale et al. 1998). Therefore, the relevance of cholinesterase activity present in and characteristic of moving structures as primary mesenchyme cells of sea urchins (as well as gastrulation of vertebrate embryos) was highlighted by Drews (1975).

1.3.7 *Ciliary movement for swimming and feeding of larvae, up to metamorphosis*
Ciliary movement is affected, as well as sperm tail movement, by substances able to affect ionic content of the cells, such as metals (e.g. Hg^{2+}, as shown by Walter et al. 1989) pesticides, etc. Recent findings indicate that different neurotransmitters are involved in regulating ciliary movements, since the ciliate epaulettes, responsible for swimming in the water column have biogenic amines as signal molecules (Nakajima, 1986; 1995). On the other hand the ciliary bands on the perioral arms, responsible for both swimming direction and feeding, are regulated by cholinergic molecules (Fig. 4) (Falugi et al. 2001, in press).

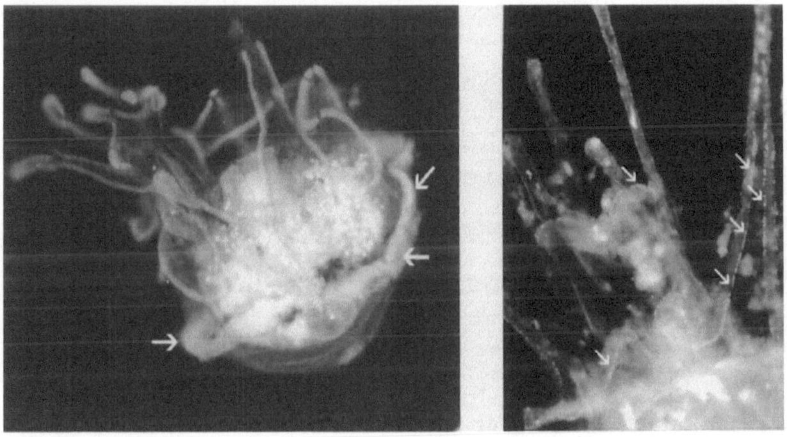

A B

Figure 4. A) formaldehyde induced fluorescence, showing biogenic amines in the ciliate epaulettes around the body; B) immunofluorescent localization of acetylcholine receptors along the perioral arms; bar equals 500 μm.

1.3.8 *Interaction among larval structures to form the rudiment*
The rudiment is formed by interaction of larval structures, carried out by gene expression (Lowe & Wray, 1997), and probably by signal exchange. An ideal development of the rudiment is largely dependant on chemical-physical parameters, such as temperature, salinity, quality of amino acid content of the microalgae used for feeding (Ikuko Yazaki, communication in Oristano, 2000), and signals due to the presence of co-specific adults in the environment (Strathman, 1978). In the laboratory rearing of larvae, the presence of protozoans, competing for food is to be considered a parameter of disturbance (personal observation).

1.3.9 *Larva-environment interaction for food availability and substrate recognition.*
Larvae which are competent for metamorphosis, may delay settlement for days and weeks, undergoing an exploratory period for substrate choice (Strathmann, 1978). When the rudiment is heavy, and the larval arms begin regressing, the larvae sink to the bottom. If the substrate is optimal, the tube feet come into contact with it, and the rudiment releases the larval structures, by apoptosis. If the substrate is not appropriate, the ciliary epaulettes are still capable

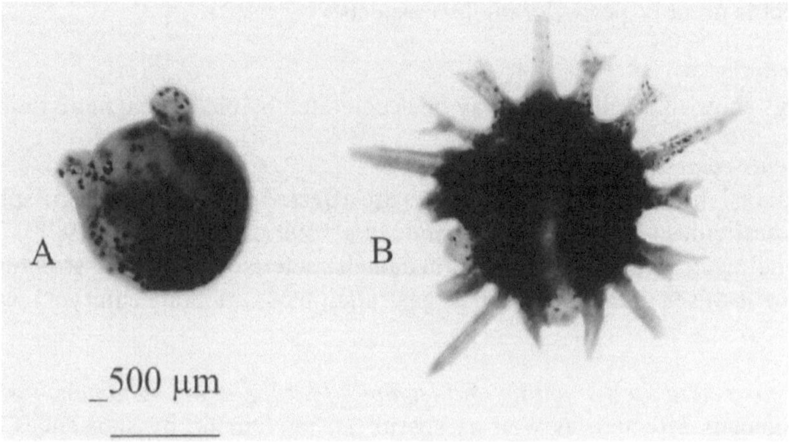

Figure 5. A : stressed juvenile; B : control juvenile, from the same pool of fertilised eggs.

of swimming, and the larvae can move to other places (personal communication by Maria Luiza Pedrotti, Observatoire Oceanologique, Villefranche- sur- Mer). This phase is critical for survival, for if the exploratory period becomes too long, the rudiment may undergo regression, and consequently die. On the other hand, under high stress environmental conditions (such as the presence of protozoa, high temperature, or with the interference of xenobiotics with chemical signalling), the rudiment release takes place before the completion of metamorphosis, and the resulting juvenile will lack of skeletal structures (in a more or less evident way) and will present a reduced number of tube feet (Fig. 5).

After metamorphosis, larvae reared in tanks undergo another critical period, lasting 1-2 weeks, in order to adapt to the new kind of diet.

2. BIOLOGICAL ASSAY

2.1 *Organisms for the assay*

Adult specimens are collected from rocky bottoms, notoriously unpolluted. The choice of the sampling areas is important, since water quality and food availability can affect reproductive success and larval health (Fenaux et al. 1988). The depth of *P. lividus* ranges between few centimetres and few meters, so that the collection of adult specimens is easy. Adult specimens, wrapped in wet paper, are carried to the Laboratory in a thermostatic bag (about 15°C). Moving adults in tanks with seawater is not suitable, or easy, because it is difficult to maintain a low water temperature during the transfer. In fact, temperature increase, even slight, could induce spawning before the arrival to the laboratory. If a thermostatic aquarium at 15 °C is available, adult sea urchins can be maintained in the Laboratory, fed with algae or boiled maize, carrots and other vegetables easily purchased.

2.2 *How to obtain spawning*:

Different "non-destructive" methods to obtain spawning are illustrated by some authors, such as the intraoral injection of acetylcholine (ACh, shown by Prof. Shonan Amemiya during the "Sea Urchin Aquaculture" workshop 2000), or electric stimulation. In the Laboratory of Genova, the method commonly employed is by intraoral injection of 0.5 M KCl, because we mainly study the biological effects of neurotoxic pesticides, which affect acetylcholinesterase (AChE) activity, and consequently the other molecules belonging to the cholinergic system.

Procedure:
Adult specimens are rinsed with clean seawater, sterilised by ultra filtration with Millipore filters (0.22 μm). In the natural position, sea urchins present ventral mouth, with 5 radial teeth, in the middle of an area lacking spines, and dorsal anus, surrounded by 5 genital pores (Fig. 6).

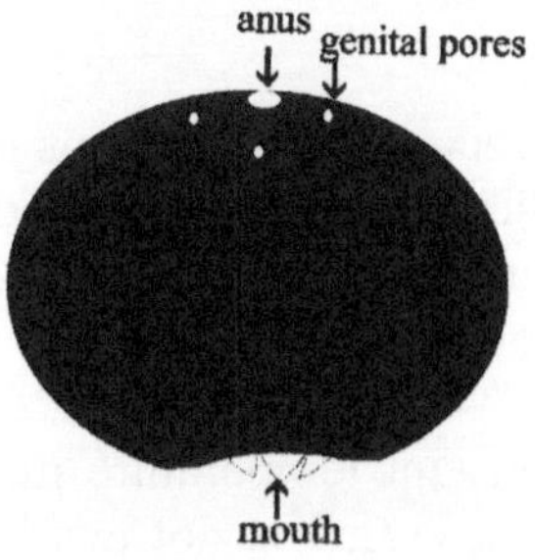

Figure 6. General organisation of the adult body (without spines and tube feet).

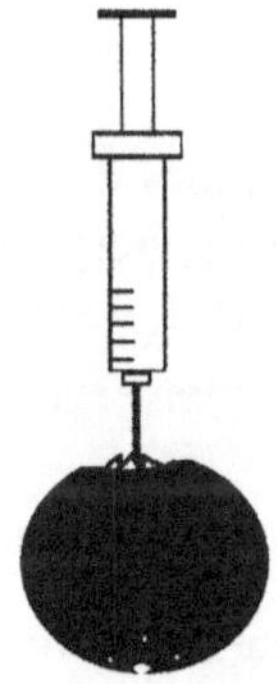

Figure 7. Mode of KCL injection.

The intraoral injection of 5 x 10^{-1} M KCl is performed in the following way: the adult specimens are taken out of the rearing seawater, and the about 0.5 mm of the syringe needle is inserted among the teeth: (Fig. 7). 0, 5 up to 1 ml (according to the size of the specimen) of KCl are injected inside the mouth. Then, the specimen is gently rotated, so that the solution can reach all the inner wall of the animal. As soon as the gonad products are seen to appear from the genital pores (a dense white liquid for males, and orange-red eggs for females) the females are put into a finger-bowl containing ultra filtered clean sea water, where the eggs sink to the bottom, forming 5 confluent masses of vivid colour orange. From here they are collected by a glass pipette, put in a Falcon and rinsed with sterile seawater. The seawater was changed 3 times, and each time the eggs are concentrated for decantation before to discharge the rinsing water.

Male gametes are collected "dry" from the genital pores, as described in the following chapters.

3 CONTROL OF LABORATORY PARAMETERS

3.1 *Light*

The quality of light is not essential for sea urchin development, as far as it is not excessive and does not alter the room temperature. It must be 18-20 °C (see also Giudice, 1973). A very recent evidence showed that also the life of adults is not affected by light duration, as moving and feeding activity is performed both during the light and the dark period, indifferently (unpublished data).

3.2 *Temperature*

The optimal temperature to obtain a correct and synchronous development in control samples is comprised between 18 and 20 °C. If the temperature is outside this range, cold or heat stress may make the organisms more sensitive to the substances present in the seawater, or added for testing the biological effects. Therefore, the temperature is important for the conduction of tests.

During the conduction of tests, a constant temperature may be obtained maintaining the test vials in a thermostatic room, or in case this is not available, by maintaining them in a thermostatic bath.

The adult specimens on the contrary have to be maintained at temperature around 15 °C, because at higher temperatures they may spawn spontaneously, emptying the gonads.

3.3 *Water oxygenation*

For an oxygenation sufficient for the first 72 hours it is necessary that the water does not fill completely the recipients (not more than 2/3 of the height), and maintained in movement. After the first stages, it is necessary to replace the rearing water every 48 h.

Water replacing is necessary also to prevent evaporation that might alter salinity and concentration of pollutants.

3.4 *Salinity*

The effects of salinity were thoroughly illustrated by Dinnel (1981).

4. METHODS FOR THE BIOASSAYS

The experiments described in the following paragraphs have been standardised by a group of researchers from different laboratories, referring to sea urchin early development for environmental analysis. The work to intercalibrate the method among the different Laboratories is still in progress (Arizzi et al. University of Venice; Giuliani & Pellegrini, ICRAM, Roma; Falugi, University of Genova; Pagano, Institute "Pascale", Napoli)

4.1 *Assay by sperm exposure*

In the first type assay, the sperms are exposed for 2 hours to the solution whose toxicity must be tested. Exposed sperms are rinsed by centrifugation to stop the toxic effect at the end of the exposure. Control sperms are maintained for the same amount of time in uncontaminated seawater, and centrifuged at the same time as test samples. Then the sperms are added to the egg cells and the subsequent success of fertilisation is evaluated by counting the eggs with an elevated fertilisation layer. A reduced fertilising success is considered an indication of toxic effects on the survival and/or fertilising efficacy of sperms. This is a short-term assay, which may last from 1 to 48 hours, and has been used in different ways since before the '50s (Tyler, 1949; Kobayashi, 1984; Dinnel et al. 1988; Bay et al. 1993) for the revelation of environmental diseases, such as changes in pH (Pagano et al. 1985b) and salinity (Dinnel et al. 1981) and for the presence of xenobiotics and metals (Pagano et al. 2000).

The sperm assay is used to evaluate acute toxicity and eventual genotoxicity, as shown by Volpi et al. (2001) and Arizzi et al. (2000).

4.1.1 *Sperm counting*

Adult specimens are sampled from rocky shores, and spawning is induced by intraoral injection of 0.5M KCl. The sperm exits from the genital pores, in the form of a dense liquid, and are collected, preventing dilution in seawater. This procedure is called "dry" collection. 0.5 μl aliquots are collected in vials and maintained for several days in the refrigerator at T=4-5 °C. In this way, sperms remain steady up to the time of use, when swimming is induced by dilution in seawater. After 5-6 days, their ability to swim rapidly decreases.

The amount of sperms contained in 1 μl of concentrated liquid may be established and standardised. From concentrated sperm, 1 μl is drawn; motility is blocked by diluting it in 1 ml 1:3 sea water/distilled water, or in 10 ml of 10% acetic acid. Other fixation could cause the cells to stick together, and counting is impossible. From this suspension, 10 μl are drawn, and cells are counted by a hemocytometer. The counted number, multiplied by 1000, represent the total amount of sperms present in 1 μl of concentrated sperm.

4.1.2 *Egg counting*
The eggs are rinsed twice with unpolluted seawater, collected in a 50 ml conic vial, and "standard" unpolluted seawater is added ("standard" sea water = pelagic water, ultra filtered through a 0, 2 μm mesh filter, or artificial sea water). Water is exchanged after egg decantation. 10 μl concentrated eggs is sampled from the vial bottom. This sample is diluted in 10 ml seawater, resuspended and 10 μl sub samples are counted with a hemocytometer.

This procedure must be repeated for each adult specimen, because different females bear different sized eggs, according to the nutritional state. This allows to establish the rate of sperms to add to subsamples, in order to perform fertilisation with not more than 20 sperms/egg, while the eggs should be about 500/ml seawater, to obtain a sensitive test. If sperms are more numerous, sensitivity to low amounts of pollutants is lost (total: 10,000 sperms for 500 eggs in 1 ml seawater).

4.1.3 *Types of experiment*
- *Testing toxic substances:* Sperms are diluted in "standard" seawater, containing different concentrations of the substances whose biological effects have to be tested.
- *Testing environmental seawater*: Sperms are diluted in seawater sampled from the monitored area, to establish the eventual presence of toxic substances, at concentration sufficient to induce acute or chronic biological effects.
- *Testing sediments*: sperms are diluted in elutriates, obtained by suspending the sediments in seawater unpolluted (or artificial) and by filtering the supernatant after a standard time of sediment decantation. (See De Rosa et al. 1997, for the elutriate preparation).

Sperms are maintained for 2 h (this time is suggested by Pagano et al. 2000) in these solutions, contained in glass vials. Then, about 20 sperms/egg are added.

For further statistical evaluation of the effects, at least 5 sperm samples are exposed to each concentration of the drugs or elutriates).

To incubate for 2 hours:

positive control =	Conc. I	Conc. II	Conc. III	Negative control =
standard sea w				standard sea w + Fe
5 vials		5 vials	5 vials	5 vials 5 vials

For these tests, Dinnel (1983) has reported that glass vials or finger bowls are preferable to either polyethylene or polystyrene recipients. Actually, after 2 h incubation in plastic recipients, a sperm/egg rate of 600/1 was needed to obtain a 100% fertilisation, and previous rinsing of recipients with seawater gave even worst results.

After sperm exposure in glass recipients, fertilisation is performed by adding 20 sperms/egg to the water containing unfertilised eggs.

After fertilisation, subsamples are collected at established times: for some experiments, the endpoint analysed is after 10 min from sperm adding (test on fertilising ability), after 30 and 70 min (test on amphimixis and block to polyspermy, or first cell cycle, respectively) 24 and 48 hours, to check the effects on gastrula and larval development, respectively.

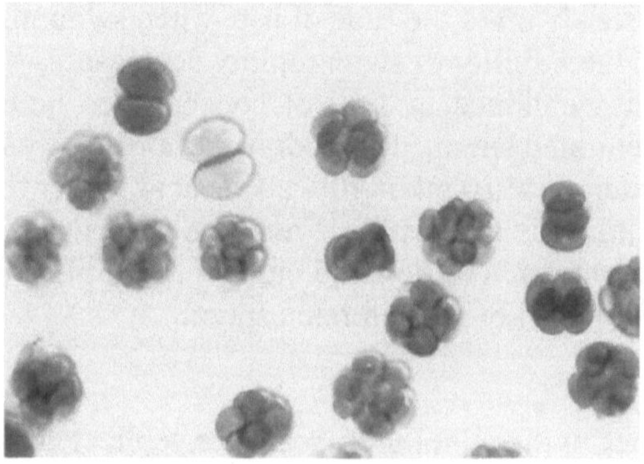

Figure 8. Example of asynchronous development.

4.2 *Assay by exposing eggs and embryos*

In the second type assay, the eggs are exposed to the test solution for 30 min prior to fertilisation that is performed by adding the sperms directly, without changing the solution where the eggs are maintained. In this type of assay, the investigated endpoints may be: fertilisation success; synchrony of development (Fig. 8), embryonic and larval state, shape, and size.

This assay can be prolonged in time, by exposing the embryos to low concentrations of the pollutant molecules, in order to obtain a long-term exposure, which will subsequently enable an evaluation of the following points: lethality caused by low doses (i.e. close to the NOEC value for man), degrees of anomalies as well as the manifestation of cell disease (apoptosis, micronuclei). All of these effects are generally more delayed in time, the more diluted the pollutants are (Trielli et al. 1995) and lesser their ability to cause biological damage is. The advantage supplied by the use of developing organisms for these tests is given by the fact that the possible biological effects are amplified as development proceeds, because embryonic development is a multi-step event, where each step arises from the previous one.

4.3 *Assay by using larval stages and metamorphosis.*

For this type of assay, the exposures are carried out from the gastrula stage onwards, to evaluate the possible interference between the pollutants presumably present in the water and the interactions which take place among larval cells and tissues to contribute to development of the rudiment. The effects of the different substances on the different developmental stages seem to differ according to the biological mechanisms with which such molecules may interfere, or compete.

5. PROCEDURE OF BIOASSAY ON TOXIC SUBSTANCES, ENVIRONMENTAL AND SEWAGE WATERS AND ELUTRIATES

5.1 *Preliminary test*

In the case of samples with unknown toxicity or suspected to be particularly toxic, it may be useful to carry out a preliminary test, to define the range of concentration for the next experiments to be performed. This experiment may be carried out quickly (a few hours will be sufficient, as explained by Viganò (1996).

In case acute and lethal effects occur, (e.g. death of all the sperms, or massive presence of blebs in the egg corteces), Viganò (1998) suggests to use the minimum concentration used in the preliminary test as the maximum concentration for the subsequent tests.

5.2 *Sediments*

When sediments is tested, it is required to obtaine elutriates routinely at three differrent sediment concentrations: 5 g, /l 50 g/l, and 500 g/l, respectively, prior to the exposure without distinction of their source. The ranges defined by these concentrations may be further divided, according to preliminary results (see the method described by De Rosa et al. 1997). Elutriates are prepared by suspending in seawater the sediments in the described proportions. The suspensions are maintained for 1 h in movement, then decantation is allowed for different times, according to the procedures of different laboratories. The supernatant is ultra filtered with 0.22 μm mesh, and diluted according to the protocol.

5.3 *Substances found in the environment, whose toxicity must be evaluated.*

These are diluted in seawater at the same concentrations as those found in the environment, or close to the NOEC for man. In this case also, at least three concentrations are preliminarily tested, then the differences among the concentrations are restricted, once the effect threshold is found. The concentration, mode and time of affecting biological targets are reported in a protocol, to be compared with the effects of environmental water samples, or elutriates.

6. TIMES OF EXPOSURE AND SAMPLING

Exposures of either of gametes or of embryos and larvae will be carried out as previously described, and the times of exposure were stopped by rinsing samples with clean seawater at the end of the exposure. Development was carried out according to the usual methods, at 19°C in covered finger bowls, and bacterial contamination was prevented by using antibiotics, according to the suggestion of D. Epel (personal communication). Embryos were sampled and fixed at regular time intervals. Fixation was performed according to the kind of biological target investigated.

6.1 *Sampling during early development*

Sampling is previewed after 10, 30 and 70 min after fertilisation, to check the effects of exposure on the first developmental stages, after 24 and 48 h, to check gastrulation and the state and dimensions at the pluteus stage. For each sampling a different test tube must be used, so that for every sampling all the embryos are used, and no errors due to different responses of embryos to the currents that are formed during sampling (e.g.: dead embryos are less heavy, or stick to the substrate, so that distribution and consequently sampling is not homogeneous). Samples are then fixed according to the subsequent preparation procedure, labeled, and then analysed at the convenient time.

6.1.1 *Parameters to be checked*

At the first sampling (10 min): fertilised eggs percentage (made evident by the presence of fertilisation layer = normality index); percentage of eggs and zygotes presenting blebs, cytoplasm fragmentation in small and numerous blastomere-like portions = stress index). At the second sampling (30 min): pronuclear fusion, amphimixis completion (= normality indexes); polyspermy, delayed nuclear fusion, anomalous nuclear shapes (= stress indexes). At the third sampling (70 min): percentage of two-blastomere stage embryos (normality index); percentage of anomalous embryos presenting an odd number of blastomeres, or unequal blastomeres (stress indexes) (Fig. 9). At the fourth sampling (24 hours): percentage of embryos with well formed archenteron (in control samples these must be < or = 80%; the transparency of embryos, the organisation of inner cells (normality indexes); the percentages of disaggregated embryos, or with big and unequal blastomeres (stress indexes). At the fifth sampling (48 hours): percentage of pluteus stage larvae, dimensions of plutei, presence of skeletal rods or dissociated blastomeres at the recipient bottom are to be observed (Fig.10 and 11).

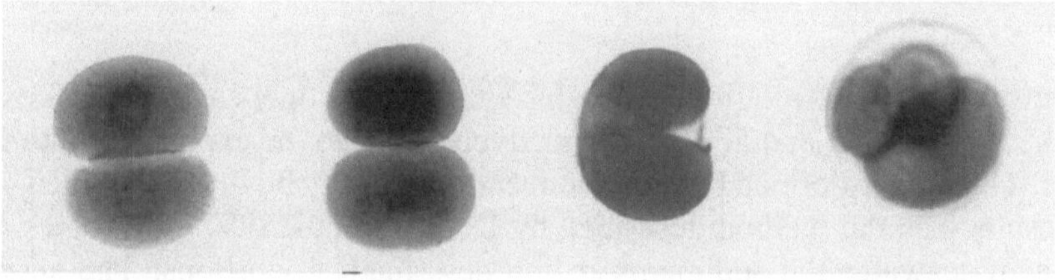

Figure 9. Some anomalies caused by xenobiotics during the first cleavage: A: normal cleavage; B: slightly asynchronous cell cycles of the 2 blastomeres; C: 2 unequal blastomeres, incomplete cleavage; D: three blastomeres, of different sizes. The dimension of the embryos is about 85 μm.

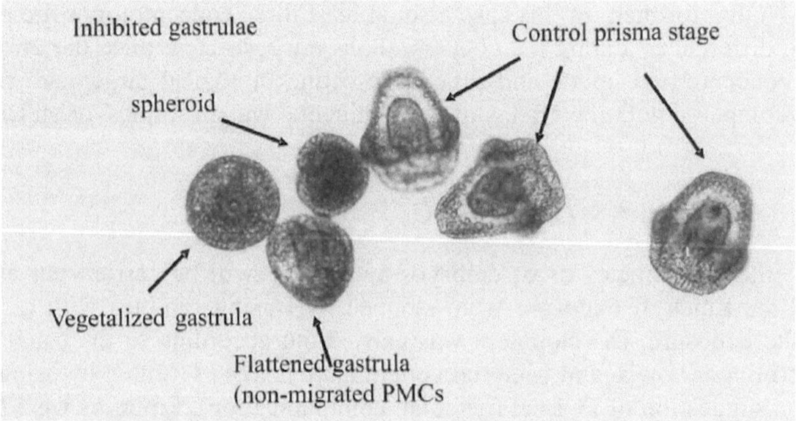

Figure 10. Different aspects of exposed larvae.

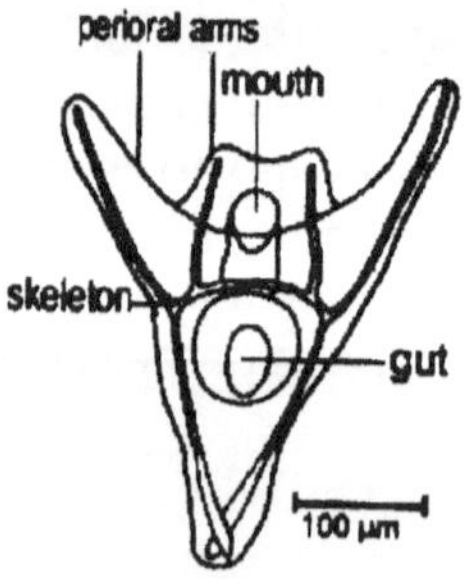

Figure 11. The dimensions of the pluteus are measured along the outer perioral arms.

Sampling during late larval development and metamorphosis: parameters to be checked: perioral arms length, appearance and diameter of the rudiment, juvenile state and survival percentage 1 week after metamorphosis (Fig. 12).

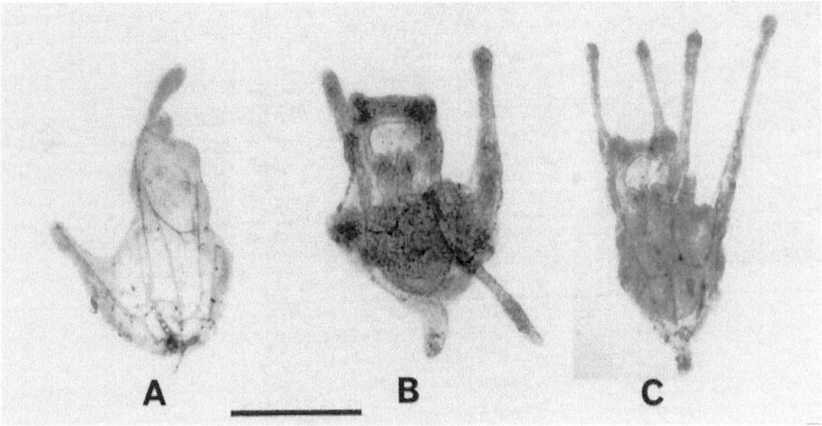

Figure 12. Various aspects of suffering late larvae (to be compared with the normal aspect represented in Figs. 2 and 3). Bar equals 500 µm.

At every sampling: synchrony of development inside the sample, and with control samples is checked. All these data must be processed with comparative statistical elaboration, on the basis of which the toxicity of the exposure water will be established.

7. SOME METHODS FOR THE ANALYSIS OF RESULTS

The outcomings of the exposures are analysed by different methods, according to the parameters and the biological targets that may clarify the action mechanisms to be understood.

7.1 *Morphological observation*

Morphological observation can be performed either on living or fixed samples, with the use of a light microscope or a stereomicroscope.

Observation of fertilisation rate: fertilisation is revealed by the elevation of the fertilisation membrane; the number of fertilised/unfertilised eggs is compared with the control samples

Observation of cleavages: this may be performed by counting the number of blastomeres of exposed samples, compared with controls. Another parameter is the shape and dimension of blastomeres. During the first 3 cleavages, the presence of an odd number of blastomeres, or of different sizes of the latter, indicate a suffering condition of the embryos. Anomalous blastulae and gastrulae do not present a wide coelomic cavity, which is moreover often full of undifferentiated opaque cells. Plutei without arms or with short arms and swollen gut, lacking inner structures are a symptom of environmental disease; 6-8 arm larvae, without rudiment, or with regressing arms and lacking inner structures represent a bad or anomalous developmental condition (see Fig. 11)

7.2 *Effects on nuclear dynamics, possible genotoxicity or polyspermy*

Fixation is performed by immersion in cold paraformaldehyde (pH 7.4) for 15 min; after decantation of samples, the fixation mixture is removed and substituted with cold methanol. Samples are then maintained for 2-3 days at T = -20 °C, or stained immediately with DAPI (4′-6-diamidine-2 phenylindole (Linn, 1977)). This staining is convenient for the earlier stages, which occur 30 min after fertilisation, and for the first cell cycles. After, the high number of cells does not allow a good observation *in toto*. At the moment of staining, rinse with 0.1 M citrate buffer, pH 7 and put 25 µl of DAPI (1:250 in buffer) solution in 100 µl of buffer, containing the embryos. Mount with Gelvatol (Lennet, 1978) as anti-fading mounting medium. Observation ia to be performed on a UV equipped microscope, wave length 360-400 nm.

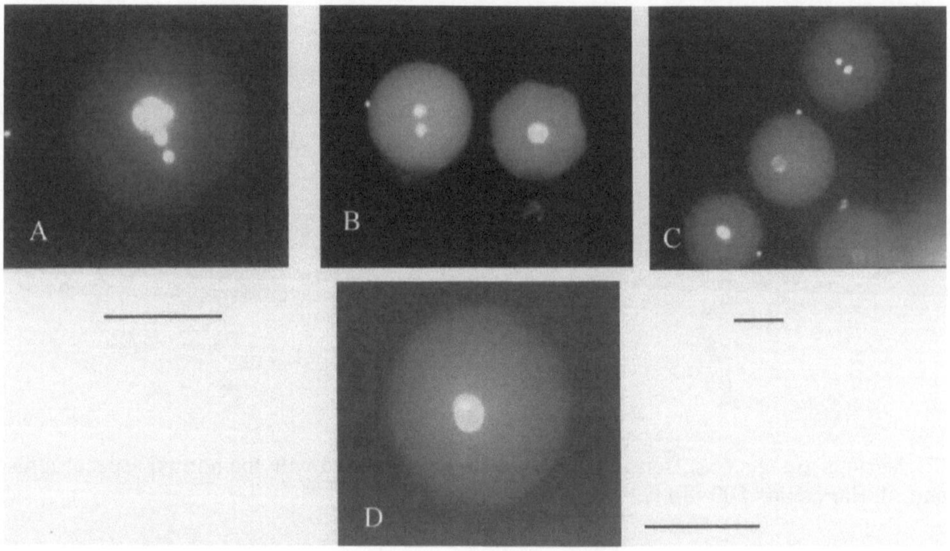

Figure 13. At 30 min development, A) Effect of a neurotoxic pesticide at concentration = NOEC for human shows polyspermy; B, D) effect of PAHs (seawater sampled from the Haven oil spill). Delayed fertilisation is shown by the fact that the two pronuclei are not yet fused; C = control: the male pronucleus is already fused to the female one, as shown by the bright spot inside the nucleus. Bars equal 70 μm.

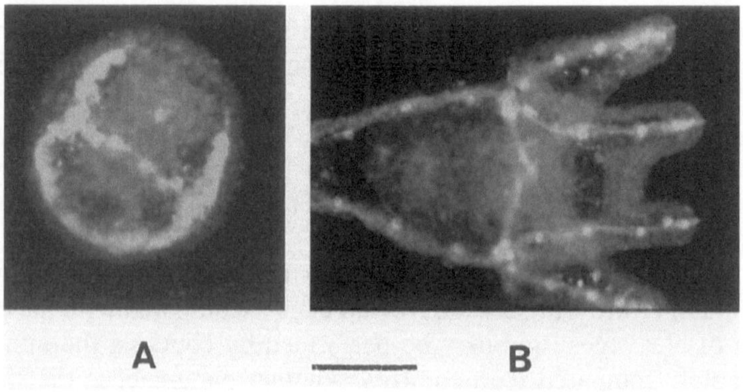

Figure 14. WGA affinity sites stained by fluorescein. A: gastrula; B pluteus (control samples). Bar equals 70 μm.

7.3 *Effects on cell migration during gastrulation*

The samples were collected at different times from 20 to 24 hours, fixed with paraformaldehyde, as above described, and maintained in 1:3 polyethylene glycol in sw at t = 5-8 °C. At the moment of staining, samples were rinsed with physiological solution and stained *in toto* with wheat germ agglutinin (WGA) conjugated with fluoresceine isothiocyanate (FITC), obtained from Sigma. FITC-WGA was diluted 25 μg/ml in physiological solution (sea water 1: 4.3 v:v distilled water), and samples were incubated for 25 min in the solution at 4 °C, in the dark. WGA is a highly specific lectin binding to primary mesenchyme cells (PMCs) (De Simone & Spiegel, 1986), which migrate inside the blastocoele to build the skeletal rods that support the pluteus structure. In sea urchin development the PMCs are strongly and selectively stained by FITC-WGA (Fig. 14). This makes it possible to measure with a micrometric ocular the migration of the PMCs and the skeletal rod elongation, since the skeletal rods are formed by inorganic salt crystallised around a matrix secreted by the PMCs (Morale et al. 1997).

7.4 *Histochemical revelation of acetylcholinesterase activity*

This enzyme is suitable for the analysis of the biological targets of neurotoxic pesticides present in the environment (when sea urchin settlement is in proximity of agricultural coasts). The histochemical revelation of ChE activity was performed by use of the Karnovsky & Roots method (1964).

In the laboratory, exposure is carried out with the use of known concentrations of the pollutants. Exposure may be performed at any stage. Samples are fixed with a short immersion (5-10 min) in cold 4% PFA (5 °C) in seawater. Rinsed in maleate buffer 0.1 M, pH 6. Incubated in the following mixture for 3h-overnight at room temperature:

Maleate buffer 0.1M, pH 6	6.5 ml
Na citrate dibasic, trihydrate 0.1 M	0.5 ml
Copper sulfate 0.03 M	1 ml
K ferricyanide 0.005 M	1 ml
H_2O	1 ml
acetyl-(or-propyonyl)-thiocholine iodide (AcTChI or PrTChI)	5 m

Embryos and larvae are mounted in DPX, and observed on a light microscope. The outcome of the reaction is seen as a dark-brownish precipitate (Fig. 15)

7.5 *Biochemical evaluation*

Unfixed samples of normal and exposed embryos are homogenised. Homogenates may be stored below 20 °C for some days. The quantitative analysis is performed with the method of Ellmann et al. (1961): in glass vials or cuvettes with the use of the following reaction mixture: 640 µl phosphate buffer, pH 8. Sixty µl homogenate and 60 µl of dithiobis nitrobenzoic acid (DTNB 0, 32 mM in 0.1 M phosphate buffer, pH 7. The absorbance at 412 nm wave length is recorded and used as a Blank. Sixty µl of substrate solution is then added and the absorbance is recorded at regular time intervals of 2 min, up to 10 min. The absorbance variation (ΔOD / min) is obtained by subtracting the blank from the maximum value and dividing by 10.

SUBSTRATE SOLUTION = 25 mM AcTChI or PrTChI in 25 ml buffer. Recordings with specific inhibitors are also necessary to establish the specificity of the enzyme reaction

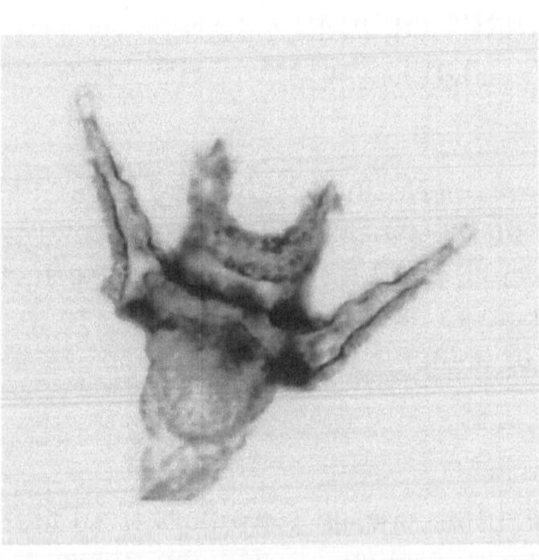

Figure 15. AChE activity in a pluteus larva, revealed by the Karnovsky & Roots histochemical method (1964). Perioral arm ciliary bands are strongly stained by the reaction products.

According to Ellman, the AChE activity expressed in Units is obtained by multiplying the obtained value $\underline{\Delta\ OD}$ x $1/1.36x10^4$

This constant was evaluated by performing the calibration curve at this wavelength, obtained by supplying known amounts of *Electrophorus electricus* AChE, diluted 5.10^{-5}U/µl,

The amount of AChE activity must be referred to the total content of protein N of the homogenates, evaluated with the Millon-Nasse reagent method, as described in the book of Oser (1965).

The difference between exposed embryos and controls represents the level of AChE activity inhibited by the drugs. This may constitute a practical and inexpensive tool to evaluate the presence of a number of pollutants, because many substances are responsible for variation in AChE activity: first of all neurotoxic pollutants, but also PAHs are known to decrease, (Jett et al. 1999) while metals have been reported to increase it, in trout brain (Dethloff et al. 1999).

Further characterisation of the molecular forms of AChE and the degree of inhibition by the different pollutants may be performed with the following methods.

7.6 *Electrophoretic characterisation of AChE activity*

Crude homogenates of the samples are electrophoresed on an 8% polyacrylamide gel and run for two hours in non-denaturing condition in order to maintain the native activity of the enzyme. After the homogenates have run in the gel, the gel is fixed with PFA 2% in PBS, rinsed with maleate buffer, and incubated in the Karnovsky and Roots (above described) incubation medium, overnight in the cold (5 °C). AChE activity is shown by bands of molecular weight around 260 kDa, corresponding to the mass of *Electrophorus electricus* AChE. If more molecular forms are present, they are evident according to the amount of enzyme isoforms and subunits.

7.7 *SDS-PAGE Electrophoresis*

Twenty – twenty-five mg of homogenates are electrophoresed on an 8% SDS-PAGE gel for two hours at 120V and transferred to a nitrocellulose membrane at 350 mA at 4 °C. After transferring, the free sites on the membrane is blocked for 1h in 3-5% milk/TBS-T (Tris 1M pH 8, NaCl, Tween 20) or 2.5% FCS, o.1% Tween 20 in PBS. The immunoreaction with an appropriately diluted primary antibody is performed in the same blocking buffer at R.T for 2h 30' or overnight. The membrane is rinsed 4 times (each in TBS-T) and reacted with the secondary antibody in blocking buffer for 1h 30'. After washing as above, the results are checked with non-radioactive methods (NBT/BCIP or other techniques).

8. EFFECTS OF SOME CLASSES OF POLLUTANTS IN THE ENVIRONMENT AND IN THE LABORATORY, STUDIED WITH THE SEA URCHIN DEVELOPMENT BIOASSAY

In the past decades, this work was performed mainly by Pagano and his group, who tested the biological effects of a number of substances and environmental parameters, including salinity and pH variations (see Pagano et al. 2000 for a review). The effects of the different substances throughout the development seem to be different according to the biological mechanisms that such molecules may interfere, or compete with.

8.1 *Neurotoxic pesticides*

Pesticides are generally released into coastal seawaters from agricultural sites, and estuaries. Their potential ability to cause biological damage depends on the physical-chemical characteristics of the different molecules. Those which are soluble in lipids may cross cell membranes and accumulate in body fats. Some of them (e.g. carbamates) do not leave residuals in the environment, but organophosphates (OP) for example, may have a longer life span than is commonly believed; as a matter of fact, they may remain for very long periods in the sediments, due to different pH and oxygen conditions. Dauberschmidt et al. (1996) reported about a disaster in the river Reno, where the lethal effects of pesticides were active for many kilometers downstream and lasted for several months. Ragnarsdottir (2000) emphasises the effect of prolonging persistence of some pesticides exerted by low pH levels. OP residuals in the coastal sediments have been evaluated in Ligurian (Italy) agricultural zones in the amount of about 10 μg/Kg (Dr. Aldo Bolla, ARPAL Savona, personal communication). Their negative effect is exerted mainly during spring-time (February-March) when periodical treatments of crops take place. This is also an ideal period for the reproduction of the sea urchin *P. lividus* that is benthic and lives in shallow coastal waters.

Generally pollutants, and in particular neurotoxic pesticides act by inhibiting enzyme activities, such as the hydrolytic enzymes of the cholinergic system: acetylcholinesterase (AChE, E.C. 3.1.1.7.); pseudocholinesterase or butyrylcholinesterase (BuChE, E.C. 3.1.1.8.) as well as propyonylcholinesterase (PrChE, linked to primitive organisms), which is studied in a lower scale and is scarcely characterised in the literature. As the main function of ChEs is to hydrolyse choline esters, such as acetylcholine (ACh), ChE inhibition will cause an overflow of ACh at its receptorial sites, and consequently the blockade of the entire system. This will be due either to a lack in receptor deactivation or to a "paradox" effect of hyper stimulation.

From our results, and from results available in the literature, dose-dependent effects were found on sperm motility (Nelson, 1990), on primary mesenchyme cells migration rate, and on the growth of pluteus perioral arms (Morale 1997, Morale et al 1998). Drews (1975) localized cholinesterase activity in primary mesenchyme cells, and hypothesized that such enzyme activity could be a marker for moving cells and tissues during differentiation. The reported experiments are consistent with this hypothesis, because AChE inhibition exerted by neurotoxic pesticides slower in dose-dependent way such a cell migration. Recently, effects on Ca^{2+} intracellular release during the cell cycle for the first cleavage have also been observed (Falugi et al. 2000).

8.2 *Mercury (Hg^{2+})*

The effects of mercurial compounds on sea urchin fertilisation and early development have been thoroughly studied by several researchers. Rünnstrom & Manelli (1964) found that exposure to mersalyl (an organic mercury compound) causes a dose-dependent inhibition of the block to polyspermy. Manelli & Ferrini (1977) forwarded the hypothesis that induced polyspermy could directly depend on the reaction between mercury and sulfhydryl groups, probably involved in binding functional membrane proteins. Moreover, recent studies have shown that the general cytotoxic effect exerted by Hg^{2+} on sea urchin fertilised eggs (zygotes) may at least in part be due to an increased cytosolic Ca^{2+} concentration (Walter et al. 1989).

Our preliminary results obtained by exposing mature unfertilised eggs to Hg^{2+} have shown a dose-dependent bioaccumulation of the metal, as evaluated on homogenates by the spectroscopic method ICP-AE (Trielli, unpublished data). Moreover, biological effects have also been caused by highly diluted metal solutions, up to 10 nM, and over: most anomalies were found on cytoskeletons, and consequently on cell cleavages, thus causing unequal or irregular chromosome distribution in the resulting blastomeres as well (Trielli et al. 1995). In this work, chromosomes of exposed samples presented a shape and distribution very similar to those observed after exposure of fertilised eggs to a-latrotoxin, the black widow venom, described by Falugi & Prestipino (1989). As such a toxin causes an anomalous influx of bivalent cations, the described results, together with those reported by Walter et al. (1989) suggest a possible interference of Hg^{2+} in the regulation of [Ca^{2+}] dynamics. The presence of micronuclei as well as apoptosis, easily shown by use of DNA staining by 4'-6-diamidine-2 phenylindole (Linn, 1977) also indicates an effect on nuclear aspects, possibly caused by cytoskeletal movements, affected in turn by intracellular anomalous modulation of Ca^{2+}. Another aspect, observed by exposing the embryos during the morula stage, was the tendency to blastomere disaggregation (Trielli et al. 1995).

8.3 *Electromagnetic fields*

The biological effects of exposure to electromagnetic fields are currently of general interest, because several areas are characterised by the presence of sources of emission. For this study, embryonic and larval development of the sea urchin have been used to perform laboratory tests with the aim of verifying the biological targets that may be affected in living cells and organisms. The main effect was found to be an enhanced speed of cell cycles, related to both cleavages and mitoses, (Falugi et al. 1987), thus underlining a possible danger concerning genetic ageing.

8.4 *Polycyclic Aromatic Hydrocarbons (PAH)*

Preliminary results on the exposure to PAHs of gametes, zygotes and early cell cycles indicate an inhibition of the block to polyspermy, possibly induced by membrane destabilisation (Thesis in preparation, unpublished data)

8.5 *Surfactant molecules.*

The main damage caused by the presence of little amounts of surfactants in the seawater (up to 0.02 ml/l) was exerted on skeletal rod structure and consequently larval shape (Bruno, 1989).

From these preliminary outcomings it appears that, at least for the classes of pollutants examined, the effects are different, in relation to the biological mechanisms that may be interfered with by the substances which are being tested. Therefore, the use of sea urchin development for the performance of biological assays could represent an ideal marker of pollution and generally of stress, because the model is simple, and the causal factors of development and metabolism are well known. For this reason, the type of effect observed by exposure to environmental waters or sediments could supply a first indication about the kind of substances present in the environment, in this way facilitating the choice of subsequent chemical analytical analyses to be made (Fig. 16).

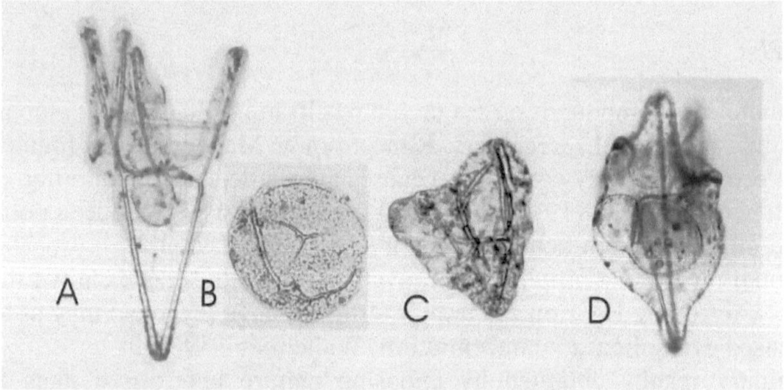

Figure 16. A control pluteus; B larva exposed to seawater containing 20 ml/l tensioactive molecules; C) larva exposed to 2 ml/l; D: larva exposed to 0.2 ml /l tensioactive molecules.

ACKNOWLEDGMENTS:

To Dr. Maria Luiza Pedrotti (Obsevatoire Oceanologique Villefranche sur Mer), for suggestions on sea urchin breeding up to metamorphosis, and for supplying information and references; to Dr. Aldo Bolla (ARPAL of Savona), for having supplied information about the pesticide residuals in the Ligurian coastal sediment; to Dr. Francesca Trielli (DIPTERIS, University of Genova) for supplying data on bioaccumulation of Hg^{2+} in unfertilised sea urchin eggs; to Prof. David Epel for suggestion how to prevent bacterial contamination and how to preserve sea urchin egg viability. To Dr. Milena Modena, for the gift of images of nuclear anomalies caused in the zygotes by PAH exposure.

To the Drs.: Alessandra Arizzi Novelli and Anna Maria Volpi Ghirardini, University of Venice; Silvia Giuliani and David Pellegrini, ICRAM, Roma; Giovanni B. Pagano, Institute "Pascale", Napoli, who have allowed me to report the first results of our cooperation in the sea urchin bioassay standardisation.

REFERENCES

Amemiya, S. 2000. Developmental Modes and Adult Rudiment Formation in Sea Urchins. In: *Sea urchin aquaculture. Textbook*. Advanced workshop in honour of Eizo Nakano: 11-12. Oristano: ICM.

Arizzi Novelli, A., Volpi Ghirardini, A. & Ghetti, P.F. 2000. Test di spermio ed embriotossicità con *P. lividus* (LMK): riproducibilità e comparazione della sensibilità per il rame con altre specie di echinoidi, XXXI Congresso SIBM, Sharm El Sheikh, 13-20 maggio *Biologia Marina Mediterranea* (in press).

Bay, S., Burgess, R., & Nacci, D. 1993. Status and applications of Echinoid (Phylum Echinodermata) toxicity test methods. In: Wayne G. Landis, Jane S. Hughes and Michael A. Lewis, (eds.), *Environmental Toxicology and Risk Assessment*, ASTM STP 1179: 281-302. Philadelphia: American Society for Testing and Materials.

Bruno, P. 1989. Lo sviluppo del riccio di mare:un modello per lo studio delle interazioni ambientali. Graduation in Biological Sciences Thesis, University of Genova. (Faculty of Sciences): 128 pp.

Dauberschmidt, C., Dietrich, D.R. & Schattler, C. 1996. Toxicity of Organophosphorus Insecticides in the Zebra Mussel, *Dreissena polymorpha* P. *Archives of Environmental Contamination and Toxicology* 30: 373-378

De Rosa, L., Marin, M.G., Fossato, V.U. & Campesan, G. 1997. Sedimenti lagunari: prove di tossicità su embrioni di riccio di mare, *Paracentrotus lividus. Biologia Marina Mediterranea* (41): 632-636

DeSimone, D.W. & Spiegel, M. 1986. Wheat germ agglutinin binds to the micromeres and primary mesenchyme cells of sea urchin embryos. *Developmental Biology* 114: 336-346.

Dethloff, G.M., Schlenk, D., Hamm, J.T. & Bailey, H.C. 1999. Alterations in physiological parameters of rainbow trout (Oncorhynchus mykiss) with exposure to copper and copper/zinc mixtures. *Ecotoxicology and Environmental Safety*. 42 (3): 253-264.

Dinnel, P.A., Pagano, G. & Oshida, P.S. 1988. A sea urchin test system for marine environmental monitoring. In: Burke, R.D., Mladenov, P.V. & Parsley, R.L. (eds.) *Echinoderm Biology*. Rotterdam: Balkema pp. 611-619.

Dinnel, P.A., Stober, Q.J. & DiJulio, D.H. 1981. Sea urchin sperm bioassay for sewage and chlorinated sea water and its relation to fish bioassay. *Marine Environmental Research* 5: 29-39.

Dinnel, P.A., Stober, Q.J., Link, J.M., Letourneau, M.W., Roberts, W.E., Felton, S.P. & Nakatani, R.E 1983. Methodology and validation of a sperm cell toxicity test for testing toxic substances in marine waters: 1-38. *Univ. Washington Sea Grant Progr. in Coop. U.S. Environmental Protection Agency*, November 1983,

Drews, U. 1975. Cholinesterase in embryonic development. *Prgress in Histochemistry and Cytochemistry* 7: 1-52.

Ellman, G.L., Courtney, K.D., Andres, V. Jr. & Featherstone, R.M. 1961. A new rapid colorimetric determination of acetylcholinesterase activity. *Biochemical Pharmacology* 7: 88-95.

Falugi, C., Delgaudio, R., Angelini, C. & Geraci, G. 2000. Calcium signalling in sea urchin zygotes. *Dev. Biol. of the Sea Urchin* XIII, Marine Biological Laboratory, Woods Hole, Massachusetts, Sept 28-Ott.2.

Falugi, C. Diaspro, A, Angelini, C., Pedrotti, M.L., Raimondo, M. & Robello, M 2001. Three dimensional mapping of cholinergic molecules by confocal laser scanning microscopy in sea urchin larvae,. Micron, in press.

Falugi, C., Grattarola, M. & Prestipino, G. 1987. Effects of low-intensity pulsed electromagnetic fields on the early development of sea urchins. *Biophysical Journal* 51: 999-1003.

Falugi, C. Pieroni, M. & Moretti, E. 1993. Cholinergic molecules and sperm functions. *Journal Submicroscopic Cytology and Pathology* 23(1): 63-69.

Falugi, C. & Prestipino, G. 1987. Effects of some inhibitors on the cholinergic system active during the sea urchin *Paracentrotus lividus* development. In: Boudouresque, C.F. (ed.) *Colloque International sur Paracentrotus lividus*: 147-155. Marseille: Gis Posidonie publ.

Falugi, C. & Prestipino, G. 1988. Localization of putative nicotinic cholinoreceptors in the early development of *Paracentrotus lividus*. *Cellular and Molecular Biology* 35(2) : 147-161.

Falugi, C. & Prestipino, G 1989. Effects of a-latrotoxin on the early developmental events of the sea urchin *Paracentrotus lividus*: a histochemical study. *Morphologisches Jahrbuch* 135: 229-239.

Fenaux, L. 1991. L'oursin pris pour modele dans l'etude de quelques facteurs du determinisme du recruitement. In: Scalera-Liaci, L. & Canicattì, C (eds.) *3rd Colloquium on Echinoderms*. Lecce.

Fenaux, L., Cellario, C. & Rassulzadegan, F. 1988. Sensitivity of different morphological stages of the larva of *Paracentrotus lividus* (Lamarck) to quality and quantity of food. In: Burke, R.D., Mladenov, P.V., Lambert, P. & Parsley, P.L. (eds.) *Echinoderm Biology*. Rotterdam: Balkema, 259-266.

Giudice, G (ed.) 1973. *Developmental Biology of the Sea Urchin Embryo*. New York: Acad. Press, 670 pages.

Jett, D.A., Navoa, R.V.& Lyons, M.A. 1999. Additive inhibitory action of chlorpyrifos and polycyclic aromatic hydrocarbons on acetylcholinesterase activity in vitro. *Toxicology* 105(3): 223-229.

Karnovsky, M.J. & Roots, L. 1964. A "direct colouring" thiocholine method for cholinesterase. *Journal of Histochemistry and Cytochemistry*. 12: 219-221.

Kobayashi, N. 1984. Marine ecotoxicological testing with echinoderms. In: Persoone G., Jaspers E., and Claus C. (eds.), *Ecotoxicological testing for the marine environment*. Bredene, Belgium: State Univ. Ghent and Inst. Mar. scient. Res. 1: 341-405.

Lennet, E.D.A. 1978. An improved mounting medium for immunofluorescence microscopy. *American Journal of Clinical Pathology* 69: 647-648.

Linn, M.S., Comings, D.E. & Alfi, O.S. 1977. Optical studies of the interactions of 4'-6-diamidino-2 phenyl indole with DNA and metaphase chromosomes. *Chromosoma* 60: 15-25.

Lowe, C.J. & Wray, G.A. 1997. Radical alterations in the role of homeobox genes during echinoderm evolution. *Nature* 389: 718-721.

Manelli, H., Ferrini, U. 1977. Ricerche biochimiche e morfologiche sul cortex della uova di riccio di mare (*Paracentrotus lividus*) rese polispermiche da un mercuriale, il Mersalyl. *Acta Embryologiae et Morphologiae Experimentalis* 3: 378.

Marchi, B., Trielli, F., Falugi, C., Corre MC. & Fenaux, L. 1996. Cholinomimetic drugs may affect growth and metamorphosis of the sea urchin larva. *Oceanologica Acta,* 19(3-4): 287-291.

Marchi, B., Trielli, F., Falugi, C., Corre, M.C. & Fenaux, L. 1994. Effects of cholinergic drugs on *Paracentrotus lividus* development up to metamorphosis. In: David, B., Guille, A., Féral, J-P. & Roux, M. *Echinoderms Through Time:* 775. Rotterdam: Balkema.

Morale, A 1997. Lo sviluppo del riccio di mare: un modello per lo studio delle interazioni cellula-cellula e cellula-ambiente. Graduation in Biological Sciences Thesis, University of Genova. (Faculty of Sciences), 153 p.

Morale, A., Coniglio, L., Angelini, C., Cimoli, G., Bolla, A., Alleteo, D., Russo, P. & Falugi, C 1998. Biological effects of a neurotoxic pesticide at low concentration early development. A Teratogenic assay. *Chemosphere* 37: 3001-3010.

Morale, A., Coniglio, L., Angelini, C., Iannone, R., Corte, G. & Falugi, C. 1997. Inhibition of acetylcholinesterase activity affects primary mesenchyme migration and morphogenesis in sea urchin. *Animal Biology* 1997: 45.

Nakajima, Y. 1986. Presence of a ciliary patch in preoral epithelium of sea urchin plutei. *Development Growth & Differentiation* 28: 243-249.

Nakajima, Y. 1995: Nervous system of echinoderm larvae: development and morphology. *2nd Biennal larval biology meeting*. Fort Pierce: Harbor Branch Oceanographic Institution. p. 28.

Nelson, L. 1990. Pesticide perturbation of sperm cell function. *Bulletin of Environmental Contamination. Toxicology*. 45: 876-882.

Okazaki, K., 1975. In: (Czihak G. Ed.) The Sea Urchin Embryo: Biochemistry and Morphogenesis. New York: Springer, 177-232.

Oser, B.L.. 1965. Hawk's Physiological chemistry. In London: McGraw-Hill Book Co., 14th ed., 179 pages.

Pagano, G., Cipollaro, M., Corsale, G., Esposito, A., Ragucci, E., Giordano, G. & Trieff, N.M. 1986. The sea urchin: bioassay for the assessment of damage from environmental contaminants. *Community Toxicity Text* 920: 66-92.

Pagano, G., Cipollaro, M., Corsale, G., Esposito, A., Ragucci, E. & Giordano, G.G. 1985a. pH-induced changes in mitotic and developmental patterns in sea urchin embryogenesis. I. Exposure of embryos. *Teratogenesis Carcinogenesis and Mutagenesis* 5: 101-112.

Pagano, G., Cipollaro, M., Corsale, G., Esposito, A., Ragucci, E. & Giordano, G.G. 1985b. pH-induced changes in mitotic and developmental patterns in sea urchin embryogenesis. II. Exposure of sperms. *Teratogen. Carcinogen. Mutagen.* 5: 113-121.

Pagano, G., Korkina, LG., Iaccarino, M., DeBiase, A., Deeva, IB., Doronin, YK., Guida, M., Melluso, G., Meriç, S., Oral, R., Trieff, N.M. & Warnau, M. 2000. Developmental, cytogenetic and biochemical effects of spiked or environmentally polluted sediments in sea urchin bioassays. In: Garrigues, P., Walker, C.H. & Barth, H. (eds.) *Biomarkers in Marine Ecosystems: a Practical Approach.* Amsterdam: Elsevier, 1-50.

Ragnarsdottir, K.V. 2000. Environmental fate and toxicology of organophosphate pesticides. *Journal of the Geological Society* 157 (Part 4): 859-876.

Runnström, J. & Manelli, H. 1964. Induction of polyspermy by treatment of sea urchin eggs with mercurials. *Experimental Cell Research* 35: 157-193.

Strathmann, R.R. 1978. Larval settlement in echinoids. In: Chia, F-S. & Rice, M.E. (eds.) *Settlement of marine invertebrate larvae:* 235-246. New York: Elsevier.

Trielli, F., Marchi, B., Falugi, C., Morale, A. & Viarengo, A. 1995. Alterations of fertilization and early development of thje sea urchin *Paracentrotus lividus* caused by Hg^{2+} treatment. In: Emson, R.H., Smith, A.B., Campbell, A.C. (eds.) *Echinoderm Biology 1995:*313-323. Rotterdam: Balkema.

Tyler, A. 1949. A simple non-injurious method for inducing repeated spawning of sea urchins and sand dollars. *Collecting Net* 19:19-20.

Vacquier, V.D. & Moy, G.V. 1977. Isolation of bindin, the protein responsible for adhesion of sperm to sea urchin eggs. *Proceedings of National Academy of Science USA* 74: 2456-2460.

Viganò L. 1996. metodo per la valutazione della tossicità acuta con *Mysidopsis bahia. Notiziario dei Metodi di Analisi delle* Acque giugno 1996: 19-31. Roma: IRSA

Viganò L. 1998. Metodo per test di tossicità cronica (7 days) con *Mysidopsis bahia. Notiziario dei Metodi di Analisi delle Acque* maggio 1998: 1-8. Roma: IRSA.

Volpi Ghirardini, A. & Arizzi Novelli, A. 2001. A sperm cell toxicity test procedure for the Mediterranean species *Paracentrotus lividus* (Echinodermata: Echinoidea). *Environmental Technology,* 22: 439-445.

Walter, P., Allemand, D., De Renzis, G., Payan, P. 1989. Mediating effect of calcium in $HgCl_2$ cytotoxicity in sea urchin egg: role of mitochondria in Ca^{2+}-mediated cell death. *Biochimica. Biophysica. Acta* 31: 15-35.

Ward, G.E., Brokaw, C.J., Garbers, D.L. & Vacquier, V.D. 1985. Chemotaxis of *Arbacia punctulata* spermatozoa to resact, a peptide from the egg jelly layer. *Journal of Cell Biology* 101: 2324-2329.

The Sea Urchin: From Basic Biology to Aquaculture, Yokota, Matranga & Smolenicka (eds)
© Swets & Zeitlinger, Lisse, ISBN 90 5809 379 4

Physiological and induced apoptosis in sea urchin embryos and larvae

M.C. Roccheri & C. Tipa
Dipartimento di Biologia Cellulare e dello Sviluppo, Università di Palermo, Palermo, Italy

ABSTRACT: Apoptosis, or programmed cell death, is a process in which cells activate intracellular death pathways to terminate themselves in a systematic way in response to a wide variety of stimuli. Within the last few years it has become apparent that apoptosis plays an important role in development besides in elimination of cells that have sustained genetic damage or that undergo uncontrolled cellular proliferation.

It has been demonstrated to be possible to induce apoptosis in sea urchin embryos and that at the pluteus stage some apoptosis occurs spontaneously especially in oral arms and intestine. Following observations of larval stages closer to metamorphosis, it was confirmed that apoptosis is a physiological and necessary event in the sea urchin development. The sea urchin embryos can represent an ideal model to study both physiological and induced apoptosis.

1. INTRODUCTION

A healthy organism is an integrated collection of differentiated cells which maintain a balance between life and death. Some cells are irreplaceable, some cells complete their functions and are then sacrificed, and some cells live a finite lifetime to be replaced by yet another generation. Cell death can occur by two distinct mechanisms: necrosis and apoptosis. What is the difference between them?

Necrosis that can be considered in "accidental cell death" is the pathological process which occurs when cells are exposed to a serious physical or chemical insult. Intracellular organelles and entire cells are lysed, membrane integrity is lost and the cytoplasmic contents including enzymes are released into the extracellular fluid. Therefore, necrotic cell death is often associated with extensive tissue damage resulting in an intense inflammatory response.

Cells undergoing apoptosis show characteristic morphological and biochemical features. These features include chromatin aggregation, nuclear and cytoplasmic condensation, partition of the cytoplasm and nucleus into membrane bound-vesicles (apoptotic bodies) which contain ribosomes and morphologically intact mitochondria. *In vivo*, these apoptotic bodies are rapidly recognised and phagocytized by either macrophages or adjacent epithelial cells. This efficient mechanism to remove the apoptotic cells *in vivo* does not cause an inflammatory response. *In vitro* this terminal phase of cell death may later become necrotic.

Apoptosis or "programmed cell death" is a genetically controlled cell death process implicated in the critical physiological mechanism of development and tissue homeostasis.

Programmed cell death (PCD) plays an essential role in sculpting structures of the body, deleting unneeded structures, controlling cell numbers and eliminating abnormal, misplaced, non-functional or harmful cells. Errors in apoptosis can have devastating consequences on development (Ellis et al. 1991, Jacobson et al. 1997).

Apoptosis is also the means for elimination of cells that have sustained irreparable levels of DNA damage, which otherwise might ultimately damage the organism. Most, if not all, physiological cell deaths occur by apoptosis, and an alteration of apoptosis may result in a variety of malignant disorders, including AIDS, cancer, Alzheimer's disease, Parkinson's disease, etc.

Numerous genes have been identified which encode products that regulate apoptosis and many of these genes are conserved throughout evolution; the complete machinery of apoptotic cell death has many similarities between phylogenetically divergent groups including invertebrates and humans (Green, 1998, 2000).

The key elements of the apoptotic pathway include: death receptors, membrane alterations, protease cascade, mitochondrial changes, and DNA fragmentation.

1.1 *Death receptors*

Apoptosis has been found to be induced via stimulation of several different cell surface receptors in association with caspase activation. Upon cross-linking by ligand, a receptor initiates a signal transduction cascade which leads to caspase-dependent programmed cell death.

1.2 *Membrane alterations*

In the early stage of apoptosis the cell surface and the plasma membrane undergo some alterations. One of the plasma membrane alterations is the translocation of phosphatidylserine (PS) from the inner side of the plasma membrane to the outer side and PS becomes exposed at the external surface of the cell.

1.3 *Protease cascade*

Signals leading to the activation of a family of intracellular cysteine proteases, the caspases (Cysteinyl-aspartate-specific proteases) play a pivotal role in the initiation and execution of apoptosis induced by various stimuli. Caspases are present in all cells as inactive pro-enzymes, most of which are activated by proteolytic cleavage. Recruitment of caspases to oligomerized receptors leads to activation via dimerization or dimerization accompanied by autoproteolytic cleavage. Active caspases generate a caspase cascade that cleaves proteins that are critical for cell survival favoring apoptosis (Enari et al. 1998).

1.4 *Mitochondrial changes*

Suppression of the anti-apoptotic members or activation of the pro-apoptotic members of the Bcl-2 family leads to altered mitochondrial membrane permeability resulting in the release of cytochrome C into the cytosol (Bradham et al. 1998). In the cytosol, or on the surface of mitochondria, cytochrome C is bound by the protein Apaf- I (apoptotic protease activating factor), which also binds caspase-9 and dATP. Binding of cytochrome C triggers activation of caspase-9, which then accelerates apoptosis by activating other caspases (Zou et al. 1999). Apoptosis inducing factor (AIF) released into the cytoplasm has proteolytic activity and is by itself sufficient to induce apoptosis. The release of AIF seems critical, since the apoptotic events become irreversible after its release. On the other hand, it has been proved that one of essential proteins in apoptosis is Bcl-2, located on the outer Mitochondria membrane that prevents the release of AIF and cytochrome C from mitochondria (Adams & Cory, 1998). The Bcl-2 expression can be down regulated by p53 that is an essential protein for the induction of apoptosis as a response to DNA damage.

1.5 *DNA fragmentation*

The biological hallmark of apoptosis is the fragmentation of the genomic DNA. It is an irreversible event that commits the cell to die and occurs before changes in the plasma membrane permeability. In many systems this DNA fragmentation has been shown to be generated by activation of an endogenous nuclear endonuclease (Fig 1).

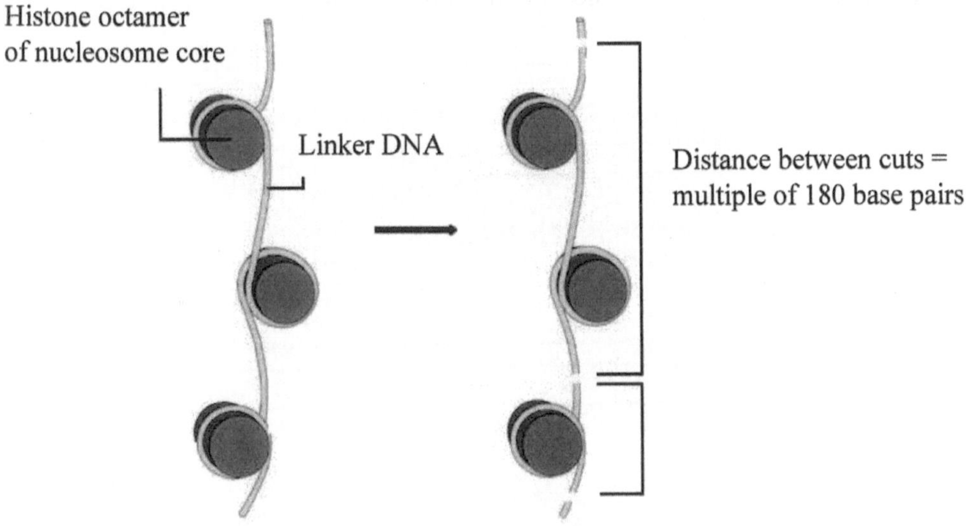

Figure 1. DNA fragmentation.
The endogenous nuclear endonuclease enzyme selectively cleaves DNA at sites located between nucleosomal units (linker DNA) generating mono- and oligo-nucleosomal DNA fragments.

1.6 *Programmed cell death in development*

Programmed cell death is particularly important during development. For example in the developing nervous system, cell death controls neuronal numbers, targeting of nerve cells and morphogenesis of the developing limb. The role of physiological cell death during the embryonic developmental stages has been explained clearly in *Caenorhabditis elegans*, where 131 of 1090 somatic cells are destined to die prior to hatching (Horvitz et al. 1982). In *Drosophila*, the mechanism of cell death is similar to that of *C. elegans*, and the Toll-Dorsal pathway demonstrates the close relationship among developmental, defensive and death signaling molecules (Grosshans et al. 1994). In *Xenopus laevis* cell death occurs during amphibian metamorphosis. Coucouvanis & Martin (1995) and Brison & Schulz (1997) have demonstrated that during mammalian development the formation of the inner and outer cell mass of the blastocyst is mediated by cell death which maintains the homeostasis of cell numbers by regulating the production of cells during mitosis.

While many molecules of the apoptotic pathways have been identified and studied, understanding of the role and control of physiological cell death during embryonic development is far from complete. The basic biochemical events involved in the control of apoptosis are similar to each other. Therefore the information obtained from one model system of apoptosis may be directly applicable to others.

2. APOPTOSIS IN SEA URCHIN EMBRYOS

Recently, we have investigated the possibility that sea urchin embryos are able to respond with apoptosis to appropriate external stimuli and that physiological cell death is implicated in development of sea urchin embryos and their metamorphosis.

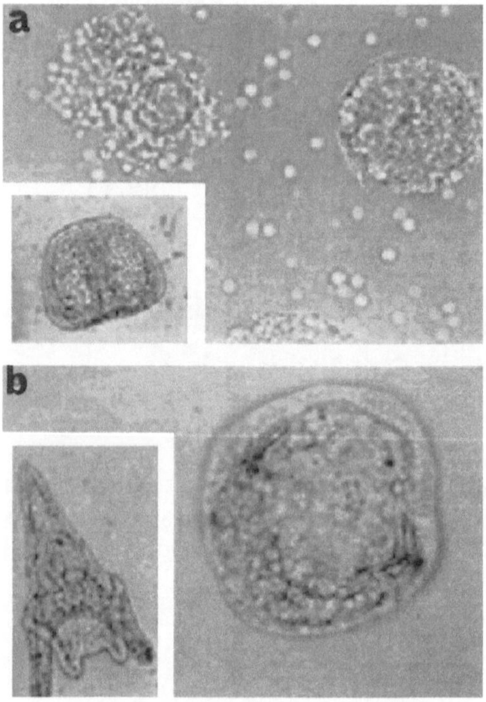

Figure 2. Morphological observation of swimming gastrulae immediately after TPA+13°C treatment (a) and 24 hours later (b). The inserts show the control embryos at the corresponding stages, at a lower magnification.

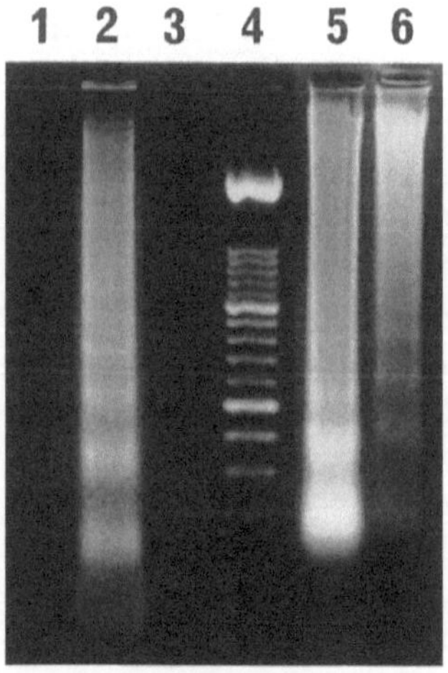

Figure 3. DNA electrophoresis analysis. I: control blastula embryos; 2: TPA+31°C gastrula embryos; 3: control gastrula embryos; 4: standard DNA ladder; 5: TPA+31°C pluteus embryos; 6: control pluteus embryos.

At first, we tested a variety of stimuli known to elicit apoptosis in other systems, and analysed the embryonic DNA for internucleosomal breakages, in *Paracentrotus lividus* embryos at the gastrula stage. Different concentrations of Staurosporin, okadaic acid, EGTA, TPA were found to have little or no effect as results from DNA fragmentation analysis; higher concentrations were lethal and caused cell lysis without apoptosis. We therefore examined a combination of TPA treatment and heat shock. The morphological observation of swimming gastrulae immediately after TPA and heat shock treatment shows that the embryos present slight cell dissociation with some variation dependent on each individual embryo. When the control embryos became plutei, the embryos treated 24 hours before showed some sign of recovery with disappearance of the de-segregated cells (Fig, 2).

We found that treatment with 25 nM TPA for 3 hours, and an increase of the temperature to 31 °C during the last hour had a clear apoptotic effect as shown by the DNA electrophoretic analysis. It has to be noted that the heat treatment prior to TPA addition results in no apoptosis observable. This fact may lead to the explanation that the heat shock reaction protected the embryos. The electrophoretic analysis indicates DNA apoptotic breakage in the treated plutei and also in the control plutei (Fig. 3).

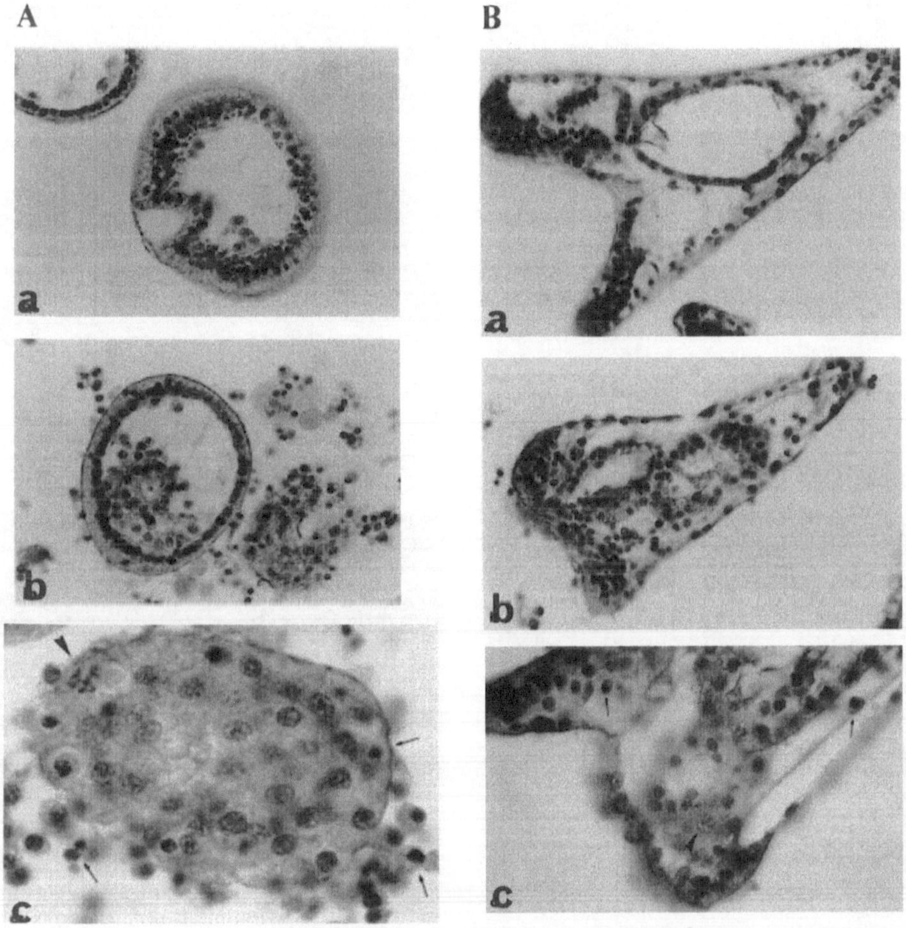

Figure 4. A: Hematoxylin and eosin stained sections of gastrula embryos; a = control; b = TPA+31°C; c = higher magnification of a tangential section of a TPA+31 °C treated gastrula. B: Hematoxylin and eosin stained sections of pluteus embryos a = control; b = TPA+31 °C; c = higher magnification of a tangential section of a TPA+31 °C treated pluteus. Picnotic nuclei and nuclei with apoptotic bodies, especially in some dissociated cells, of treated embryos were observed (see arrows).

If one looks at hematoxylin and eosin stained sections of gastrula and pluteus, one observes picnotic nuclei, especially in some dissociated cells of treated embryos (Fig. 4). In order to prove a nuclear apoptosis we examined the same sections after a specific assay: TdT reaction. This reaction is based on the recognition of the exogenous terminal deoxynucleotydil transferase (TdT) of 3' termini from DNA strand breaks. TdT can add modified, fluorescent deoxynucleotide (Cy3TM-dCTP) to the 3' termini of the cleaved fragments, which can be microscopically recognised and thus reveals apoptosis in the individual cells.

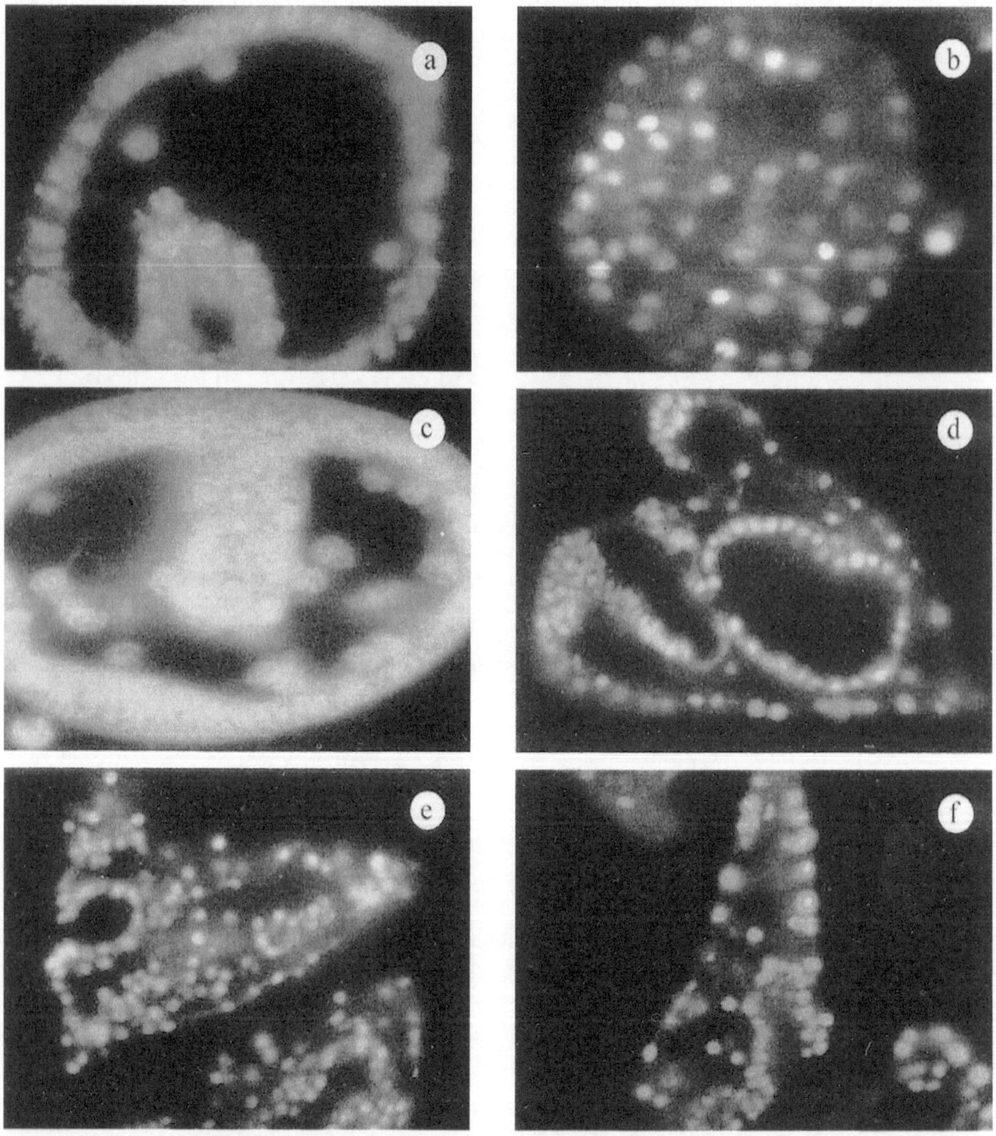

Figure 5. Apoptosis in sea urchin embryos: normal and induced. TdT assay in sections of gastrula (a-c) and of plutei embryos (d-f). a = control embryos, b = apoptotic embryos induced by TPA+31 °C, c = positive control DNase pre-treated embryos, d = control embryos, e = apoptotic embryos induced by TPA+31 °C, f = negative control embryos (reacting section without TdT enzyme). All sections were counterstained with propidium iodide. The apoptotic nuclei of treated embryos and control plutei seen as white dots in these figure.

This reaction is supposed to show fluorescent yellow nuclei when positive for apoptosis and orange nuclei when negative, due to the propidium iodide counterstaining. Treated embryos show many fluorescent yellow nuclei and the untreated ones only orange nuclei. This result was confirmed by another section but counterstaining in green using the Hoechst reactive. The positive control shows only fluorescent yellow nuclei in embryos which have been briefly treated with DNase I, in order to introduce ubiquitous DNA breaks. (Fig. 5).

Our results have in conclusion proved that it is possible to induce apoptosis in sea urchin embryos at both gastrula and pluteus stages, and that at the pluteus stage some apoptosis is spontaneous, especially in the arms and intestinal cells. No sign of apoptosis was observed in the normally developing gastrulae (Roccheri et al. 1997).

A possible explanation of that spontaneous apoptosis occurs at pluteus stage especially at the arm and intestine level, is that these embryonic structures disappear at least in part due to remodeling during metamorphosis.

The number of apoptotic cells in the arms and in intestine of control plutei has been calculated dissociating the pluteus embryos and separating the ectoderm cells from intestine cells, and after carrying out *in situ* TdT reaction. Analysis of the apoptotic cell frequency in plutei indicated that about 12% of ectoderm cells and 8% of intestinal ones underwent spontaneous apoptosis. Besides, the examination of some serial optical sections of pluteus embryos under a confocal laser microscope allowed us to observe a different distribution of the apoptotic cells in different fields of focus, which appeared concentrated especially in two districts: oral arms and intestine (Roccheri et al. in preparation).

Then we extended our observations of larval stages closer to metamorphosis to confirm if sea urchin apoptosis is a physiological and necessary event for the development to the adult.

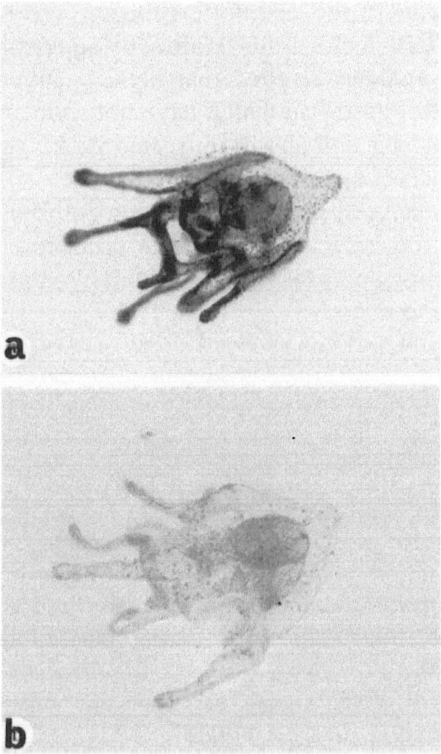

Figure 6. TUNEL in whole mount of larvae: a= control larvae, b= negative control (reacting larvae without TdT enzyme). Control larvae show apoptotic cells seen as black dots in that figure.

Plutei and larvae, tested with other TUNEL (TdT-mediated dUTP Nick-End Labeling) assay, showed apoptotic cells in the same districts. Apoptotic cells were concentrated in the oral arms, intestine, ciliary band and in the apical and oral ganglia (Fig. 6).

Moreover, the number of apoptotic cells decreases in later larval stages. This observation agrees with the fact that fewer cells are eliminated along the maturation of organisms, in the light of metamorphosis.

It can be recalled here that Sato & Yazaki, (1999) have provided some examples of apoptosis during metamorphosis of *H. pulcherrimus* plutei.

As previously discussed, it will be of extreme interest to compare the apoptotic pathways utilized during the development and those as the defensive response to stress, in order to investigate whether developmental and defensive apoptotic processes have a common origin. Vaux & Korsmeyer (1999) have hypothesized that apoptosis may have initially emerged as a defense mechanism and was later adapted to developmental phenomena with the evolution of multicellularity.

The sea urchin embryo can represent an ideal model to study both physiological and induced apoptosis in order to reveal whether cellular death during development regulates modeling and uses the same modality and molecules responsible for the defensive apoptotic response to stress.

For all these reasons, involvement of p53, pro-apoptotic molecules, and bcl-2, anti-apoptotic molecules, were tested (O'Connor et al. 2000).

In order to examine whether apoptosis in the sea urchin embryo development is p53 related, we reacted protein extracts from control and treated embryos with monoclonal human antibody anti-p53 by western blot. The immunoelectrophoretic analysis showed the presence of p53 in both control gastrulae and plutei, and that it undergoes a sharp increase in treated plutei. Experiments of immunolocalization with the same antibody anti-p53 in whole mount show that this protein is constitutive and increases in treated embryos and larvae, suggesting that p53 is involved in stress-induced apoptosis.

At the same time, we used the antibody against the anti-apoptotic protein bcl-2, known to have a contrasting function in the apoptotic process. We reacted protein extracts from control and treated embryos with a monoclonal antibody against human bcl-2, by western blot. The immunoelectrophoretic analysis showed that bcl-2 is inversely correlated to apoptosis, being more abundant in control gastrulae which have not spontaneous apoptosis, less in plutei undergoing spontaneous apoptosis and absent in treated gastrulae and plutei, which are almost completely apoptotic (Roccheri et al. in preparation).

These preliminary results suggest that the sea urchin embryo can represent an ideal model to study both physiological and induced apoptosis in order to clarify if cellular death during development regulates modeling and uses the same modality and molecules responsible for the defensive apoptotic response to stress.

REFERENCES

Adams, J.M. & Cory S. 1998. The Bcl-2 protein family: arbiters of cell survival. *Science* 281: 1322-1326.

Bradham, C.A., Qian, T., Streetz, K., Trautwein, C., Brenner, D.A & Lemasters, J.J. 1998. The mitochondrial permeability transition is required for tumor necrosis factor alpha-mediated apoptosis and cytochrome c release. *Molecular and Cellular Biology.* 18: 6353-6364.

Brison, D.R. & Schultz, R.M. 1997. Apoptosis during mouse blastocyst formation. *Biology of Reproduction.* 56: 1088-1096.

Coucouvanis, E. & Martin, G.R. 1995. Signals for death and survival: a two-step mechanism for cavitation in the vertebrate embryo. *Cell* 83: 279-287.

Ellis, R.E., Yuan, J.Y. & Horvitz, H.R. 1991. Mechanisms and functions of cell death *Annual Review of Cell Biology* 7: 663-698.

Enari, M., Sakahira, H., Yokoyama, H., Okawa, K., Iwamatsu, A, & Nagata, S. 1998. A caspase-activated DNase that degrades DNA during apoptosis, and its inhibitor ICAD. *Nature* 391: 43-50.

Green, D.R. 1998. Apoptotic pathway: the roads to ruin. *Cell* 94: 695-698.

Green, D.R. 2000. Apoptotic pathway: paper wraps stone blunts scissors. *Cell* 102: 1-4.

Grosshans, J., Bergmann, A., Haffter, P. & Nusslein-Volhard, C. 1994. Activation of the kinase pelle by tube in the dorsoventral signal transduction pathway of *Drosophila* embryo. *Nature* 372: 563-566.

Horvitz, H.R., Ellis, H.M. & Sternberg, P. W. 1982. Programmed cell death in nematode development. *Neuroscience. Commentaries* 1 : 56-65.

Jacobson, M.D., Weil, M. & Raff, M.C. 1997. Programmed cell death in animal development. *Cell* 88: 347-354.

O'Connor, L., Huang, D.C., O'Reilly, L.A. & Strasser, A. 2000. Apoptosis and cell division. *Current Opinion in Cell Biology* 12: 257-263.

Porter, A.G. 1999. Protein translocation in apoptosis. *Trends in Cell Biology* 9: 394-401.

Roccheri, M.C., Barbata, G., Cardinale, F., Tipa, C., Bosco, L., Oliva, O.A., Cascino, D. & Giudice G.1997. Apoptosis in sea urchin embryos. *Biochemical Biophysical. Research Communication*: 240: 359-366.

Sato Y. & Yazaki I. 1999. A cellular analysis of sea urchin metamorphosis induced by L-glutamine. In M.D.C. Carnevali & F. Bonasoro (eds.), *Echinoderm Research* 1998: 221-226. Rotterdam: Balkema..

Vaux, D. & Korsmeyer, S. (1999) Cell death in development. *Cell* 96: 245-254.

Zou, H., Li, Y., Liu, X. & Wang, X.1999. An APAF-1 cytochrome c multimeric complex is a functional apoptosome that activates procaspase-9. *The Journal of Biological Chemistry* 274: 11549-11556.

The Sea Urchin: From Basic Biology to Aquaculture, Yokota, Matranga & Smolenicka (eds)
© Swets & Zeitlinger, Lisse, ISBN 90 5809 379 4

Sea urchin larval ecology: food rations, development, and size at metamorphosis

R.B. Emlet
Oregon Institute of Marine Biology, University of Oregon, Charleston, Oregon, USA

ABSTRACT: Many species of sea urchin larvae can be readily cultured through metamorphosis in large numbers and this makes them amenable to aquaculture. This chapter summarizes information on larval form and function and on factors that can influence larval quality and early juvenile survival. Though the information covered here is based on ecological research on larvae and its consequences for processes in natural populations, aquaculturists can utilize the information to better understand biological issues associated with this fragile stage of the sea urchin life cycle. The most common larvae of sea urchins may be viewed as a feeding machine designed to efficiently ingest algal cells and build a juvenile sea urchin. The condition of larvae depends on many physical and biological factors including water quality and its properties, maternal condition, fertilization, larval density, and food quality and quantity. Larvae respond to certain physical and biological conditions by changing shape, and this can be used to infer condition of larvae in culture. Suitable settlement conditions and adequate food and habitat for newly metamorphosed juveniles are also essential components of successfully raising sea urchins.

Approximately 70% of sea urchin species produce relatively small eggs (65 to 200 mm diameter) that develop into feeding larvae. These larvae must feed in order to acquire energy and materials to produce a juvenile. The remaining species produce larger, yolk-rich eggs (350 mm to >1 mm diameter) that develop into pelagic, nonfeeding larvae (ca. 23%), or brooded young (7%). Species with feeding larvae are particularly common in tropical and temperate, shallow water environments of both hemispheres. Species with nonfeeding larvae are found in southern Australia, Japan, and also among deeper water species of all oceans, except the Arctic ocean (see Emlet et al. 1987, Emlet 1990, 1995). Because species with feeding larvae are so common and are most likely to be used in sea urchin aquaculture, this chapter will describe the form and function of feeding larvae and factors that can influence larval quality. Larval ecology of sea urchins has been reviewed recently by McEdward & Miner (2001a), and larval ecology and life history evolution of echinoderms has been reviewed by McEdward & Miner (2001b) and Emlet et al. (1987). Other reviews on echinoid development include Pearse & Cameron (1991), Davidson (1986), and Harvey (1956). Emlet et al. (2001) provide a summary of comparative larval anatomy, and Strathmann (1987) provides information on cultivation of larvae.

1. LARVAL FEEDING, SWIMMING AND ORIENTATION

The echinopluteus larva of sea urchins serves as a feeding machine that harvests unicellular algae and converts it into cells, tissues, and energy reserves that make up the juvenile. This bilaterally symmetric microscopic animal consists of a conical body region that grows more complex in shape with the sequential addition of 4 or 5 pairs of anterior projecting epidermal extensions called larval arms. A row of densely packed cilia (the ciliated band) forms a single, continuous loop that runs along the body and out each arm to its tip and back to the body.

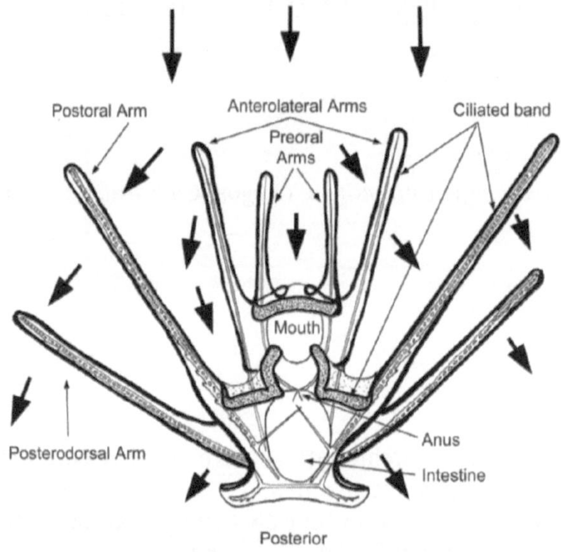

Figure 1. Ventral view of an echinopluteus larva, *Echinometra lucunter* (redrawn from Mortensen 1921, Pl. I). Heavy arrows show the direction of water current created by the ciliated band.

This ciliated band separates the anterior, oral surface (because it includes the mouth) from the posterior, aboral surface. The cilia in this band are 20-30 mm long and create water currents that move from the anterior outward through the arms and posteriorly over the body (Fig 1.). The posterior component of this current causes the larval to move, arms first, through the water. Larvae swim slowly moving approximately 0.3 mm/s. Echinoplutei are negatively buoyant and orient in the water column with their arms upward (e.g. Pennington & Emlet 1986; Pennington & Stathmann 1990). This orientation results from the different locations of the center of mass and center of buoyancy (Pennington & Stathmann 1990), and because most of the ciliated band is along the larval arms and produces a current that is outward and posterior.

Algal cells in the water current are sensed, possibly by mechanical or chemical cues, as they approach the ciliated band. Cilia in a localized region of the band react by reversing their direction of beat as a particle approaches. As a consequence of this localized reversal, the food particle is redirected along the oral surface of the arms or body toward the mouth (Strathmann 1971, 1975, Hart 1991). About 95% of the particles that go into the mouth are initially captured by a localized reversal of cilia along the band. When particles are redirected by the ciliary band they are rarely lost before they enter the mouth (Hart 1991, Appelmans 1994). The increase in numbers of larval arms is a means of increasing the length of the ciliated band, and this in turn increases the harvesting capacity of the larva (Strathmann, 1971, Hart 1991, Hart & Strathmann 1994).

2. INFLUENCES ON LARVAL QUALITY

Factors that can influence the quality of feeding larvae and their success up to and through metamorphosis include: a) maternal condition, b) spawning and fertilization, c) the larval environment - water quality including temperature and salinity, larval density, food quality and quantity, and d) adequate settlement conditions and the availability of food and suitable habitat for juveniles.

The availability and type of food in the parental environment can impact both fecundity and the quality (size and energetic content) of eggs and quality of larvae (Thompson 1983, George 1990, George et al. 1990, Meidel et al 1999). The maternal effects on larvae may be due to the energetic or nutritional content of eggs. George (1990) and George et al. (1990) reported that natural populations of the Mediterranean sea urchin *Arbacia lixula* differ in egg size, larval size and rate of larval development that are consistent with adult food supply. De Jong-Westman et al. (1995) manipulated adult diets of *Strongylocentrotus droebachiensis* by varying protein, *B*-carotene, and levels of other constituents and found that adults fed a high protein + *B*-carotene diet produced larvae with the highest rates of development, survival and metamorphosis. Tsushima et al. (1997) varied adult diets of *Pseudocentrotus depressus* and found addition of *B*-echinenone (a carotenoid) also enhanced hatching and survival of larvae. The mechanism by which these carotenoid rich diets affect larvae may be through the anti-oxidant properties of carotenoids or some other unidentified pathway.

Fertilization rates among free spawning invertebrates such as sea urchins, may be extremely sensitive to adult population density and distributions. Studies primarily on strongylocentrotid echinoids demonstrate that fertilization success drops dramatically over decimeters to meters and as a function of adult density (Pennington 1985, Levitan 1991, Levitan et al.1991, 1992). Meidel & Yund (2001) examined egg viability and fertilization in the field over a period of two days for the green urchin, *Strongylocentrotus droebachiensis*, and they found rates of fertilization were substantially higher than those reported in shorter term studies. Harvesting natural population of sea urchins may reduce densities and increase distance between remaining individuals such that they are effectively sterile or have greatly reduced fitness.

Perhaps the most important influence on larval quality and success is the larval environment. Both physical and biological factors make up the larval environment. Optimal temperature for development corresponds closely with the range of temperature during the spawning season for eight species of tropical and temperate Japanese sea urchins (Fujisawa & Shigei 1990). Fujisawa (1995) showed that the optimal temperature for development may also vary within a species, but still corresponds to the spawning temperature range for given population of *Hemicentrotus pulcherrimus*. Temperature can strongly affect rates of development. McEdward (1985) reported a Q_{10} of 3.6 for part of the developmental sequence for the temperate sand dollar *Dendraster excentricus*, and Cameron et al. (1985) reported Q_{10} of 3 and 3.6 for two tropical sea urchins. Hart & Scheibling (1988) suggested that warmer than average temperatures may have caused rapid development and intensified recruitment leading to population expansions in the green urchin *Strongylocentrotus droebachiensis*.

Most echinoderms are intolerant of low salinities. Some sea urchin larvae are also intolerant of lower salinities and show reduced developmental rates and survival at lower salinities (Roller & Stickle 1993, 1994).

Food quality and quantity are also very important for feeding echinoplutei. Larvae given natural assemblages of particulate matter can graze particles in the 2 to 30 mm range (Rassoulzadegan & Fenaux 1979). Diatoms (*Chaetoceros gracilis*), flagellated chlorophytes (*Dunaliella* spp.), gold-brown flagellates (*Cricosphaera* spp.), chrysophytes (*Rhodomonas* spp.) and mixed algal diets are particularly good foods for sea urchin larvae (e.g. Strathmann 1987, Pedrotti & Fenaux 1993). Species may vary in what is the best food for culturing larvae (Emlet, pers. obsv.). Initial starvation, low food rations or algal cells of poor quality prolong the developmental period and reduce survival (Fenaux et al. 1988, 1992, Pedrotti & Fenaux 1993). Meidel et al. (1999) showed that the amount of food (*Dunaliella*) food given to larvae of *Strongylocentrotus droebachiensis* had a greater effect on development, growth, and survival to metamorphosis than did the differing diets given to parents of these larvae during gametogenesis. Miller & Emlet (1999) demonstrated that larvae receiving higher food rations produced larger juveniles that developed more quickly than did sibling juveniles from larvae fed lower rations.

Sea urchin larvae show a phenotypic plasticity in length of larval arms in response to varying amounts of food (Boidron-Metairon 1988, Strathmann et al. 1992). Larval raised under food limited conditions have longer arms relative to their body length than to siblings fed higher rations. In contrast larvae receiving higher rations have relatively large stomachs and begin to form the echinus rudiment at an earlier developmental stage (Strathmann et al. 1992). The adaptive interpretation of this plastic response (differential allocation to arms and cilia rather than stomach and rudiment) is that it allows larvae in food-limited environments to have a longer ciliated band and capture more food particles. Hart & Strathmann (1994) demonstrated that larvae with longer ciliated bands could clear more particles. Whether this plasticity actually allows more rapid development under food limited conditions has not yet been demonstrated.

This plastic response of larvae to food has been demonstrated for temperate and tropical sea urchins and sand dollars (Boidron-Metrairon 1988, Strathmann et al. 1992, Fenaux et al. 1994). Fenaux et al. (1994) used larval form of naturally occurring larvae to infer food limitation in nature for Mediterranean echinoplutei. They reported that larvae *of Paracentrotus lividus* collected from the field had a food-limited form in both spring and fall samples. Larval form of the fall samples was even more food-limited than that in the spring samples (Fenaux et al. 1994). Other studies on food limitation (e.g. Paulay et al. 1985, Olson & Olson 1989, Boidron-Metairon 1995) indicate several regional marine environments may be naturally limited in larval food supply while others are not. Comparisons of field (N.E. Pacific) and laboratory-raised populations of larvae and juveniles of *Strongylocentrotus purpuratus*, showed field samples were more like those of laboratory larvae (and juveniles) fed high food rations (Miller & Emlet 1999). One interpretation of this is that these field-caught larvae were not food limited.

Adequate settlement conditions and the availability of food and suitable habitat for juveniles may also interact with larval quality to affect successful recruitment (Cameron & Schroeter 1980, Burke 1983). Rowley (1989, 1990) found that newly settled juveniles of *Strongylocentrotus purpuratus* feed by scraping surfaces and initially grew slightly faster and survived much better on crustose coralline algal barrens than in kelp-bed habits. After urchins reached 1 mm in test diameter they switched to macroalgal food and grew much more quickly in the kelp-bed habitat. Juveniles of *S. franciscanus* are commonly found under conspecific adults and may benefit by protection from predators (Tegner and Dayton 1977, Breen et al. 1985).

Food given to larvae can influence the test size and the rate at which newly metamorphosed juveniles develop new structures. Miller & Emlet (1999) showed that post-metamorphic juveniles of *Strongylocentrotus purpuratus,* fed a higher food ration as larvae, had faster test growth and more rapid formation of pedicellariae than those from larvae fed a lower ration. Working with *Paracentrotus lividus*, Gosselin & Jangoux (1998) described the continual development of juvenile characters during the eight or so days after metamorphosis and before the onset of juvenile feeding. More recently Vaïtilingon et al. (2001) found that delaying metamorphosis of competent larvae and feeding them for a few more days resulted in larger post-metamorphic juveniles and an earlier onset of feeding by juveniles when compared to juveniles from larvae that metamorphosed right after reaching competence. Because they also reported that the rates of metamorphosis were lower and mortality of juveniles was higher for delayed larvae and than for competent larvae, Vaïtilingon et al. (2001) recommend against delaying metamorphosis in the commercial production of sea urchins.

REFERENCES

ADDIN ENBbu Appelmans, N. 1994. Sites of particle selection determined from observations of individual feeding larvae of the sand dollar *Dendraster excentricus. Limnology and Oceanography* 39: 404-411.

Boidron-Metairon, I.F. 1988. Morphological plasticity in laboratory-reared echinoplutei of *Dendraster excentricus* (Eschscholtz) and *Lytechinus variegatus* (Lamarck) in response to food conditions. *Journal of Experimental Marine Biology and Ecology* 119: 31-41.

Boidron-Metairon, I.F. 1995. Larval nutrition. In L.R. McEdward (ed.), *Ecology of Marine Invertebrate Larvae*. Boca Raton, FL.: CRC Press.

Breen, P.A., Carolsfeld, W. & Yamanaka, K.L. 1985. Social behavior of juvenile red sea urchins, *Strongylocentrotus franciscanus* (Agassiz). *Journal of Experimental Marine Biology and Ecology* 92: 45-61.

Burke, R.D. 1983. The induction of metamorphosis of marine invertebrate larvae: stimulus and response. *Canadian Journal of Zoology* 61: 1701-1719.

Cameron, R.A. & Schroeter, S.C. 1980. Sea urchin recruitment: effect of substrate selection on juvenile distribution. *Marine Ecology Progress Series* 2: 243-247.

Cameron, R.A., Boidron-Metairon, I. & Monterrosa, O. 1985. Does the embryonic response to temperature and salinity by four species of Caribbean sea urchins parallel the reproductive synchrony? *Proc. 5th Int. Coral Reef Congr., Tahiti.* 5: 273-278.

Davidson, E.H. 1986. *Gene activity in early development, 3rd ed.* Orlando, FL: Academic Press.

de Jong-Westman, M., Qian, P.-Y., March, B.E. & Carefoot, T.H. 1995. Artificial diets in sea urchin culture: effects of dietary protein level and other additives on egg qualit, larval morphometrics, and larval survival in the green sea urchin, *Strongylocentrotus droebachiensis*. *Canadian Journal of Zoology* 73: 2080-2090.

Emlet, R.B, McEdward, L.R. & Strathmann, R.R. 1987. Echinoderm larval ecology viewed from the egg. In M. Jangoux & J.M. Lawrence (eds.), *Echinoderm Studies, Vol. II*. Rotterdam: Balkema.

Emlet, R.B. 1990. World patterns of developmental mode in echinoid echinoderms. In M. Hoshi & O. Yamashita (eds.) *Advances in Invertebrate Reproduction*. Amsterdam: Elsevier Science.

Emlet, R.B. 1995. Developmental mode and species geographic range in regular sea urchins (Echinodermata: Echinoidea). *Evolution* 49: 476-489.

Emlet, R.B., Young, C.M. and George, S.B. 2001. Phylum Echinodermata: Echinoidea. In C.M. Young, M.A. Sewell, and M.E. Rice (eds.) *Atlas of Marine Invertebrate Larvae, Chap. 28*. London: Academic Press.

Fenaux, L., Cellario, C. & Rassoulzadegan, F. 1988. Sensitivity of different morphological stages of the larva of *Paracentrotus lividus* (Lamarck) to quantity and quality of food. In R.D. Burke, P.V. Mladenov, P. Lambert & R.L. Parsley (eds). *Echinoderm Biology, Proc. 6th International Echinoderm Conference, Victoria*. Rotterdam: Balkema.

Fenaux, L., De Greef, Y. & Cellario, C. 1992 . The effects of a delay in initiation of feeding on the larval development of the sea urchin. *Internationale Revue der gesamten Hydrobiologie* 77: 651-663.

Fenaux, L., Strathmann, M.F. & Strathmann, R.R. 1994. Five tests of food-limited growth of larvae in coastal waters by comparisons of rates of development and form of echinoplutei. *Limnology and Oceanography* 39: 84-98.

Fujisawa, H. 1995. Variation in embryonic temperature sensitivity among groups of sea urchin, *Hemicentrotus pulcherrimus*, which differ in habitats. *Zoological Science* 12: 583-589.

Fujisawa, H. & Shigei, M. 1990. Correlation of embryonic temperature sensitivity of sea urchins with spawning season. *Journal of Experimental Marine Biology and Ecology* 136: 123-139.

George, S.B. 1990. Population and seasonal differences in egg quality of *Arbacia lixula* (Echinodermata: Echinoidea). *Invertebrate Reproduction and Development* 17: 111-121.

George, S.B., Cellario, C. & Fenaux, L. 1990. Population differences in egg quality of *Arbacia lixula* (Echinodermata: Echinoidea): proximate composition of eggs and larval development. *Journal of Experimental Marine Biology and Ecology* 141: 107-118.

Gosselin, P. & Jangoux, M. 1998. From competent larva to exotrophic juvenile: a morphofunctional study of the perimetamorphic period of Paracentrotus lividus (Echinodermata: Echinoida). *Zoomorphology* 118: 31-43.

Hart, M.W. 1991. Particle captures and the method suspension feeding by echinoderm larvae. *Biological Bulletin* 180: 12-27.

Hart, M.W. & Scheibling, R.E. 1988. Heat waves, baby booms, and the destruction of kelp beds by sea urchins. *Marine Biology* 99: 167-176.

Hart, M.W. & Strathmann, R.R. 1994. Functional consequences of phenotypic plasticity in echinoid larvae. *Biological Bulletin* 186: 291-299.

Harvey, E.B. 1956. *The American Arbacia. Princeton:* Princeton University Press.

Levitan, D.R. 1991. Influence of body size and population density on fertilization success and reproductive output in a free-spawning invertebrate. *Biological Bulletin* 181: 261-268.

Levitan, D.R., Sewell, M.A. & Chia, F.-S. 1991. Kinetics of fertilization in the sea urchin *Strongylocentrotus franciscanus*: interaction of gamete dilution, age, and contact time. *Biological Bulletin* 181: 371-378.

Levitan, D.R., Sewell, M.A. & Chia, F.-S. 1992. How distribution and abundance influence fertilization success in the sea urchin *Strongylocentrotus franciscanus*. *Ecology* 73: 248-254.

McEdward, L.R. 1985. Effects of temperature on the body form, growth, electron transport system activity, and development rate of an echinopluteus. *Journal of Experimental Marine Biology and Ecology* 93: 169-181.

McEdward, L.R. & Miner, B.G. 2001a. Larval ecology of echinoids. In J.M. Lawrence (ed.) *Edible sea urchins: biology and ecology* (*Developments in aquaculture and fisheries science vol. 32*) Amsterdam: Elsevier Science

McEdward, L.R. & Miner, B.G. 2001b. Larval and life-cycle patterns in echinoderms. *Canadian Journal of Zoology* 79: 1125-1170.

Meidel, S.K., Scheibling, R.E. & Metaxas, A. 1999. Relative importance of parental and larval nutrition on larval development and metamorphosis of the sea urchin *Strongylocentrotus droebachiensis*. *Journal of Experimental Marine Biology and Ecology* 240: 161-178.

Meidel, S.K. & Yund, P.O. 2001. Egg longevity and time-integrated fertilization in a temperate sea urchin (*Strongylocentrotus droebachiensis*). *Biological Bulletin* 201: 84-94

Miller, B.A. & Emlet, R.B. 1999. Development of newly-metamorphosed juvenile sea urchins (*Strongylocentrotus franciscanus* and *S. purpuratus*): morphology, the effects of temperature and larval food ration, and a method for determining age. *Journal of Experimental Marine Biology and Ecology* 235: 67-90.

Mortensen, T. 1921. *Studies of the development and larval forms of echinoderms*. Copenhagen: G.E.C. Gad.

Olson, R.R. & Olson, M.H. 1989. Food limitation of planktotrophic marine invertebrate larvae: does it control recruitment? *Annual Reviews of Ecology and Systematics* 20: 225-247.

Paulay, G., Boring, L. & Strathmann, R.R. 1985. Food limited growth and development of larvae: experiments with natural sea water. *Journal of Experimental Marine Biology and Ecology* 93: 1-10.

Pearse, J.S. & Cameron R.A. 1991. Echinodermata: Echinoidea. In A.C. Giese, J.S.Pearse, & V.B.Pearse (eds.) *Reproduction of Marine Invertebrates* Chap. 7. Pacific Grove, CA: Boxwood Press.

Pedrotti, M.L. & Fenaux, L. 1993. Effects of food diet on the survival, development and growth rates of two cultured echinoplutei (*Paracentrotus lividus* and *Arbacia lixula*). *Invertebrate Reproduction and Development* 24: 59-70.

Pennington, J.T. 1985. The ecology of fertilization of echinoid eggs: the consequences of sperm dilution, adult aggregation, and synchronous spawning. *Biological Bulletin* 169: 417-430.

Pennington, J.T & Emlet, R.B. 1986. Ontogenetic and diel vertical migration of a planktonic echinoid larva, *Dendraster excentricus* (Eschscholtz). *Journal of Experimental Marine Biology and Ecology* 104: 69-95.

Pennington, J.T & Strathmann, R.R. 1990. Consequences of the calcite skeletons of planktonic echinoderm larvae for orientation, swimming, and shape. *Biological Bulletin* 179: 121-133.

Rassoulzadegan, F. & Fenaux, L. 1979. Grazing of echinoderm larvae (*Paracentrotus lividus* and *Arbacia lixula*) on naturally occurring particulate matter. *Journal of Plankton Research* 1: 215-223.

Roller, R.A. & Stickle, W.B. 1993. Effects of temperature and salinity acclimation of adults on larval survival, physiology, and early development of *Lytechinus variegatus* (Echinodermata: Echinoidea). *Marine Biology* 116: 583-591.

Roller, R.A. & Stickle, W.B. 1994. Effects of adult salinity acclimation on larval survival and early development of *Strongylocentrotus droebachiensis* and *Strongylocentrotus pallidus* (Echinodermata: Echinoidea). *Canadian Journal of Zoology* 72: 1931-1939

Rowley, R.J. 1989. Settlement and recruitment of sea urchins (*Strongylocentrotus* spp.) in a sea-urchin barren ground and in a kelp bed: are populations regulated by settlement or post-settlement processes? *Marine Biology* 100: 485-494.

Rowley, R.J. 1990. Newly settled sea urchins in a kelp bed and urchin barren ground: a comparison of growth and mortality. *Marine Ecology Progress Series* 62: 229-240.

Strathmann, M.F. 1987. *Reproduction and development of marine invertebrates of the northern Pacific coast. Data and methods for the study of eggs, embryos and larvae.* Seattle: Univ. Washington Press.

Strathmann, R.R. 1971. The feeding behavior of planktotrophic echinoderm larvae: mechanisms, regulation and rates of suspension-feeding. *Journal of Experimental Marine Biology and Ecology* 6: 109-160.

Strathmann, R.R. 1975. Larval feeding in echinoderms. *American Zoologist* 15: 717-730.

Strathmann, R.R., Fenaux, L. & Strathmann, M.F. 1992. Heterochronic developmental plasticity in larval sea urchins and its implications for evolution of nonfeeding larvae. *Evolution* 46: 972-986.

Tegner, M.J. & Dayton, P.K. 1977. Sea urchin recruitment patterns and implications of commercial fishing. *Science* 196: 324-326

Thompson, R.J. 1983. The relationship between food ration and reproductive effort in the green sea urchin, *Strongylocentrotus droebachiensis*. *Oecologia* 56: 50-57.

Tsushima, M., Kawakami, T., Mine, M. & Matsuno, T. 1997. The role of carotenoids in the development of the sea urchin *Pseudocentrotus depressus*. *Invertebrate Reproduction and Development* 32: 149-153.

Vaïtilingon, D., Morgan, R., Grosjean, Ph., Gosselin, P. & Jangoux, M. 2001. Effects of delayed metamorphosis and food rations on the perimetamorphic events in the echinoid *Paracentrotus lividus* (Lamarck, 1816) (Echinodermata). *Journal of Experimental Marine Biology and Ecology* 262: 41-60.

The Sea Urchin: From Basic Biology to Aquaculture, Yokota, Matranga & Smolenicka (eds)
© Swets & Zeitlinger, Lisse, ISBN 90 5809 379 4

Ecology of adult sea urchins

R.B. Emlet
Institute of Marine Biology, University of Oregon, Charleston, OR. USA

There are 900+ living species of sea urchins (Kier 1977, Smith 1984). Approximately half these species are the round or regular sea urchins and the other half are the heart urchins, sand dollars and related taxa, all of which have a secondary bilateral symmetry as adults (see Smith 1984). Adult sea urchins occupy benthic habits from the intertidal to abyssal plain or (4000-5000m) or deeper, occurring in hard-bottom habitats (rocks and cobbles) and soft bottom (sand or mud) habitats. The irregular echinoids can be infaunal or epifaunal and only occur in soft bottom environments. These species are usually deposit feeders, ingesting the sediment and digesting adhering organic matter. The regular urchins are found only as epifauna in all habitat types and may be herbivores, omnivores, scavengers, or deposit feeders depending on species and habitat (e.g. Hyman 1955, Smith 1984). Moore (1966) reviewed the ecology of adult echinoids including general observations on growth rates, feeding, and responses to environmental factors. Because the regular urchins, especially those of the families Echinidae and Strongylocentrotidae, are economically important in the global industry for urchin roe (Conand & Sloan 1989, Mottet 1976), they will be the primary focus in this overview of sea urchin adult ecology.

Sea urchins are dioecious, generally long lived (several to many years) and iteroparous (Ebert 1982, Pearse & Cameron 1991). With the exception of some brooding taxa where females have indentations in the tests for retaining young, there is little obvious sexual dimorphism. In some species with small eggs and feeding larvae, genital papillae are different between the sexes (Tahara et al. 1958, 1960, Lawrence 1987, Pearse & Cameron 1991). Non-brooding taxa freely spawn gametes into the water column, with females producing thousands to millions of eggs depending on the species, adult size and egg size (Pearse & Cameron 1991). Reproduction is generally annual with distinct spawning seasons for temperate taxa and prolonged to continuous spawning seasons for tropical taxa (Pearse & Cameron 1991). The spawning season for eight species of tropical and temperate Japanese sea urchins corresponds to the time when water temperatures are optimal for rates of early development (Fujisawa & Shigei 1990). Fujisawa (1995) has reported habitat-related differences in optimal developmental temperatures for local populations of *Hemicentrotus pulcherrimus*. During the spawning season female gametes are usually stored in the ovary as mature (haploid) ova. Fertilization occurs shortly after eggs are spawned and is easily diagnosed because fertilized eggs rapidly form a fertilization envelope, which serves as protection against polyspermy. Planktonic larval development is on the order of days to weeks for lecithotrophic species and weeks to months for planktotrophic species, depending on food availability and temperature (Emlet et al. 1987).

Post-metamorphic juveniles are £ 1mm in total diameter and lack open mouths and functional guts for a week of more after metamorphosis, though they continue to develop additional podia, spines, and pedicellariae (e.g. Gosselin & Jangoux 1998, Miller & Emlet 1999). Some species can grow rapidly to a test diameter of 1-2 cm in the first year (Moore 1966, Smith 1984). Most sea urchins continue growing for several years but growth slows as animals increase in size and it may eventually stop or become immeasurably little. Many sea urchins can reproduce in their second year of life, but peak fecundity is related to body size and thus is attained later in life. Growth of the body is episodic, alternating with gametogenesis. The episodic (seasonal) growth produces rings or bands on the skeletal plates or jaws and these have been used to estimate age (see Jensen 1969, Pearse & Pearse 1975).

There is a strong positive correlation between growth rate and mortality - in wild populations, faster growing sea urchins suffer higher rates of natural mortality (Ebert 1975). Half of the 16 species included in this study are important for the global urchin roe industry, and interestingly, all but one the economically important species are among the slower growing and longer living species (Ebert 1975). The correlation between adult growth and longevity may be related to the allocation of resources to growth and reproduction and to predictable recruitment of new individuals (Ebert 1975, 1982). Species that grow rapidly and allocate large amounts to reproduction may not have resources available for maintenance. Because they are short lived as adults, these species would need to have predictable recruitment to persist. Species with unpredictable recruitment must live longer to persist, and thus may invest more resources in body construction - growing heavier skeletons, more slowly, while still allocating resources to reproduction (Ebert 1982). If these patterns are correctly interpreted (a "bet-hedging" strategy, see Stearns 1976, 1992), slower growing species of economic interest will require particular care for management of natural population harvests and perhaps have different economics than rapidly growing species in aquacultural operations. Slower growing species also have a relatively high allocation to body wall and are associated with more exposed environments with high wave energy or water movement (Ebert 1982). Given their approximately spherical shape (low surface area to volume ratio), and limited body circulation, gas exchange may be limited (Farmanfarmaian 1966, Fenner 1973). Sea urchins with heavier body walls may need relatively high degrees of water motion to eliminate boundary layers and facilitate gas exchange.

All echinoids that are commercially important or locally harvested for roe, are shallow-water grazers that consume primarily algae or sea grass (Sloan 1985). When abundant, tropical and temperate shallow-water taxa (whether of commercial value or not) usually have a strong impact the standing stock of marine plants (Ogden et al. 1973, Lawrence 1975, Harrold & Pearse 1987, Heck & Valentine 1995, Rose et al.1999). Many of the temperate taxa living on hard substrata (e.g. *Strongylocentrotus* spp., *Paracentrotus*, and *Evechinus* can either eliminate particular species of kelps or create barren grounds devoid of fleshy macroalgae and covered with crustose coralline algae (Harrold & Pearse 1987, Scheibling et al. 1999). Other echinoids (e.g. *Loxechinus*) occurring on hard substrata apparently rely heavily on drift algae and do not deplete local standing stocks. Figure 3 of Harrold & Pearse (1987) maps the geographical locations of ecological studies that examine the relative importance of echinoid herbivores in kelp forests.

Many herbivorous echinoids may also consume animal tissue such as encrusting organisms or pelagic animals that become stranded (Lawrence 1975, Duggins 1981, Briscoe & Sebens 1988). It would be interesting to know whether omnivory improves roe quality.

Because many sea urchins can strongly control macro-algal or sea grass production in their habitats, they can have a large effect on the community in general. The entire food webs and possibly other commercially important shell fish or fish species can be directly impacted by variations in sea urchin populations (Dayton 1985, Sloan 1985, Harrold & Pearse 1987, Duggins et al. 1989). Human impacts on marine communities may also modify food webs and impact sea urchin populations (Vadas and Steneck 1995). Hay (1984) demonstrated that heavily fished coral reefs had more intense urchin-algal interactions due to an increased abundance of sea urchins when compared with coral reefs not impacted have human fishing. Sea otters are known to control some populations of sea urchins along the N.E. Pacific coast, and this can lead to a mosaic of kelp forests and urchin maintained barrens grounds at the scale of otter distributions (Estes & Palmisano 1974, Duggins 1980). Storm events can also create mosaics of kelp beds and urchin barrens through destructive removal of either kelps or urchins (Ebeling et al. 1985).

There is a huge amount of literature on sea urchin-algal interactions and sea urchin recruitment. But there are few studies of fisheries ecology of echinoids. Fuji (1967) and Fuji & Kawamura (1970) have produced some of the few studies examining sea urchin biology and evaluating combined algal and urchin production in a natural setting.

REFERENCES

Briscoe, C.S & Sebens, K.P. 1988. Omnivory in *Strongylocentrotus droebachiensis* (Mueller) (Echinodermata: Echinoidea): predation on subtidal mussels. *Journal of Experimental Marine Biology and Ecology* 115: 1-24.

Conand, C. & Sloan N.A. 1989. World fisheries for echinoderms. In Caddy, J.F. (ed.), *Marine Invertebrate Fisheries and their Assessment and Management*: 647-663. John Wiley and Sons: New York

Dayton, P.K. 1985. Ecology of kelp communities. *Annual Review of Ecology and Systematics* 16: 215-245.

Duggins, D.O. 1980. Kelp beds and sea otters: an experimental approach. *Ecology* 61: 447-453.

Duggins, D.O. 1981. Sea urchins and kelp: the effects of short term changes in urchin diet. *Limnology and* 26: 391-394.

Duggins, D.O., Simenstad, C. A. & Estes, J.A. 1989. Magnification of secondary production by kelp detritus in coastal marine ecosystems. *Science* 245: 170-174.

Ebeling, A.W., Laur, D.R. & Rowley, R.J. 1985. Severe storm disturbances and reversal of community structure in a southern California kelp forest. *Marine Biology* 84: 287-294.

Ebert, T.A. 1975. Growth and mortality of post-larval echinoids. *American Zoologist* 15: 755-775.

Ebert, T.A. 1982. Longevity, life history, and relative body wall size in sea urchins. *Ecological Monographs* 52: 353-394.

Emlet, R.B., McEdward, L.R & Strathmann, R.R. 1987. Echinoderm larval ecology viewed from the egg. In M. Jangoux & J.M. Lawrence (eds.), *Echinoderm Studies II*: 55-136. Rotterdam: Balkema.

Estes, J.A.& Palmisano, J.F. 1974. Sea otters: their role in structuring nearshore communities. *Science* 185: 1058-1060.

Fenner, D.H. 1973. The respiratory adaptations of the podia and ampullae of echinoids (Echinodermata). *Biological Bulletin* 145: 323-339.

Fuji, A. 1967. Ecological studies on the growth and food consumption of Japanese common littoral sea urchin, *Strongylocentrotus intermedius* (A. Agassiz). *Memoir of Faculty of Fishery Hokkaido University* 15: 83-160.

Fuji, A. & Kawamura K. 1970. Studies on the biology of the sea urchin VII. Bio-economics of the of *Strongylocentrotus intermedius* on a rocky shore of southern Hokkaido. *Bulletin of Japanese Society of Scientific Fisheries* 36: 763-775.

Fujisawa, H. 1995. Variation in embryonic temperature sensitivity among groups of sea urchin, *Hemicentrotus pulcherrimus*, which differ in habitats. *Zoological Science* 12: 583-589.

Fujisawa, H. & Shigei, M. 1990. Correlation of embryonic temperature sensitivity of sea urchins with spawning season. *Journal of Experimental Marine Biology and Ecology* 136: 123-139.

Gosselin, P. & Jangoux, M. 1998. From competent larva to exotrophic juvenile: morphofunctional study of the perimetamorphic period of *Paracentrotus lividus* (Echinodermata: Echinoida). Zoomorphology 118: 31-43.

Harrold, C. & Pearse, J.S. 1987. The ecological role of echinoderms in kelp forest. In M. Jangoux & J.M. Lawrence (eds.), *Echinoderm Studies*: 137-233. Rotterdam: Balkema.

Hay, M.E. 1984. Pattern of fish and urchin grazing on Caribbean coral reefs: are previous results typical? *Ecology* 65: 446-454.

Heck, K.L. & Valentine, J.F. 1995. Sea urchin herbivory: evidence for long-lasting effects in subtropical sea grass meadows. *Journal of Experimental Marine Biology and Ecology* 189: 205-217.

Hyman, L.H. 1955. Echinoidea. *The Invertebrate-Echinodermata*: 413-589. New York: McGraw-Hill.

Jensen, M. 1969. Age determination of echinoids. *Sarsia* 37: 41-44.

Kier, P.M. 1977. The poor fossil record of the regular echinoid. *Paleobiology* 3: 168-174.

Lawrence, J.M. 1975. On the relationships between marine plants and sea urchins. *Oceanography and Marine Biology Annual Review* 13: 213-286.

Lawrence, J. 1987. *Functional Biology of Echinoderms*. Baltimore: John Hopkins University Press.
Miller, B.A. & Emlet. R.B. 1999. Development of newly-metamorphosed juvenile sea urchins (*Strongylocentrotus franciscanus* and *S. purpuratus*): morphology, the effects of temperature and larval food ration, and a method for determining age. *Journal of Experimental Marine Biology and Ecology* 235: 67-90.
Moore, H.B. 1966. Ecology of echinoids. In R.A. Boolootian (ed.), *Physiology of Echinodermata*: 73-85. New York: Interscience Publishers.
Mottet, M.G. 1976. The fishery biology of sea urchin in the family Strongylocentrotidae. *Washington Department of Fisheries Technical Report* 20:
Ogden, J.C., Brown, R.A. & Salesky, N. 1973. Grazing by the echinoid *Diadema antillarum* Philippi: formation of halos around West Indian patch reefs. *Science* 182: 715-716.
Pearse, J.S. & Pearse, V.B. 1975. Growth zones in echinoid skeleton. *American Zoologist* 15: 731-753.
Pearse, J.S. & Cameron, R A. 1991. Echinodermata:Echinoidea. In A.C. Giese, J.S. Pearse & V.B. Pearse (eds.), *Reproduction of Marine Invertebrates, Vol. 6*: 514-662. Pacific Grove: Boxwood Press.
Rose, C.D., Sharp, W.C., Kenworthy, W.J., Hunt, J.H., Lyons, W.G., Prager, E.J., Valentine, J.F., Hall, M.O., Whitefield, P.E. & Fourqurean, J.W. 1999. Overgrazing of a large seagrass bed by the sea urchin *Lytechinus variegatus* in outer Florid Bay. *Marine Ecology Progress Series* 190: 211-222.
Scheibling, R.E., Hennigar, A.W. & Balch, T. 1999. Destructive grazing, epiphytism, and disease: the dynamics of sea urchin – kelp interactions in Nova Scotia. *Canadian Journal of Fisheries and Aquatic Sciences* 56: 2300-2314.
Sloan, N.A. 1985. Echinoderm fisheries of the world: a review. In B.F. Keegan & B.D.S. O'Connor (eds.), *Echinodermata, Proceedings of 5th International Echinoderm Conference, Galway*: 109-124. Rotterdam: Balkema.
Smith, A.B. 1984. *Echinoid Paleobiology*. London: George Allen and Unwin Publisher.
Stearns, S.C. 1976. Life-history tactics: a review of the ideas. *The Quarterly Review of Biology* 51: 3-47.
Stearns, S.C. 1992. *The evolution of life histories*. Oxford: Oxford University Press.
Tahara Y., Okada M. & Kobayashi N. 1958. Secondary sexual characters in Japanese sea urchins. *Publications of Seto Marine Biological Laboratory* 7: 165-172.
Tahara Y., Okada M. & Kobayashi N. 1960. Further notes on the sexual dimorphism in Japanese sea urchins. *Publications of Seto Marine Biological Laboratory* 8: 183-189.
Vadas, R.L., & Steneck, R.S. 1995. Overfishing and inferences in kelp-sea urchin interactions. In H.R. Skjoldal, C. Hopkins, K.E. Erikstad & H.P. Leinaas (eds.), *Ecology of fjords and coastal waters*: 509-524. Amsterdam: Elsevier.

 ISBN 90 5809 379 4

Gonadal growth and its relationship to aquaculture in sea urchins

T. Unuma
National Research Institute of Aquaculture, Nansei, Japan

ABSTRACT: The sea urchin gonad contains two main types of the cells: nutritive phagocytes (NPs) and germ cells (GCs). NPs store nutrients necessary for gametogenesis, such as proteins, carbohydrates and lipids. The most abundant protein in NPs is a glycoprotein called the major yolk protein (MYP), originally identified as the most predominant component of yolk granules in the eggs. The proportion of NPs and GCs varies with the maturation of the gonads. Before gametogenesis, NPs fill the gonadal lumina and increase in size by accumulating nutrients derived from food. As gametogenesis proceeds, NPs supply MYP and other nutrients to the developing GCs and decrease in size. In fully mature gonads, a number of ripe ova or spermatozoa fill the gonadal lumina, whereas NPs lose their nutrients and shrink. The quality of the gonad as a food product usually deteriorates while GCs proliferate and develop. Promotion of NP growth and suppression of gametogenesis are prerequisites to achieving high profitability in sea urchin aquaculture.

1. INTRODUCTION

The sea urchin gonad is a popular seafood in Asian, Mediterranean and some other countries; it is one of the most expensive seafood in Japan where market demand for this delicacy has been growing for decades. Japan is the largest consumer of sea urchins, annually importing more than 6,000 tons of the gonads (FAO 1996). As wild stocks of sea urchins gradually decline as a result of over-fishing, an increasing number of people are interested in the sea urchin aquaculture (Keesing & Hall 1998). Any aquaculturist planning to export sea urchins to Japan should be aware of their commercial value in this country, where superior-quality gonads are usually sold in attractive packages, such as wooden trays (Fig. 1). The wholesale price of a package containing about 300 g is around 12,000YEN (about US$100.00; US$333.33/kg: US$1.00 = 120YEN in 2001). One sea urchin weighing 100 g with a gonad index (the percentage of the gonad weight relative to the body weight) of 20% would yield gonads worth US$6.67. However, this is the case for superior-quality gonads only. The price varies according to quality; that of low-quality gonads can be less than 10% of that of high-quality ones. Moreover, gonad index sometimes does not even reach 10%.

Sales of the cultured sea urchin are largely dependent on the size and quality of the gonads, the latter being determined by taste, colour, shape and firmness. Increasing the size and improving the quality of the gonads are both crucial prerequisites to successful sea urchin aquaculture. It is therefore useful for aquaculturists to understand the fundamental biology of the sea urchin gonad. In this paper, gonadal growth and maturation in relation to aquaculture are discussed, using *Pseudocentrotus depressus* as a model, the most popular edible sea urchin in the southern part of Japan.

Figure 1. Sea urchin gonads placed in a wooden tray. In Japan, superior-quality gonads are usually sold in this kind of package, whereas low-quality gonads are sold in plastic trays. (Courtesy of Murakami Shooten, Hakodate, Japan)

2. STRUCTURE OF THE GONAD

2.1 *Nutritive phagocyte and germ cell*

The sea urchin has five gonads attached to the inner wall of its test. Each gonad consists of hundreds of acini and resembles a bunch of grapes. There are two main types of cells inside the acinus: germ cells (GCs) and nutritive phagocytes (NPs). GCs include the oogonium, oocyte and ovum in females and the spermatogonium, spermatocyte, spermatid and spermatozoon in males. NPs are somatic cells that store nutrients necessary for gametogenesis (Walker 1982). The proportion of GCs and NPs in the gonadal acini varies with the maturation of gonads. Before the initiation of gametogenesis, the lumina of the gonadal acini are filled with NPs. During gametogenesis, NPs supply nutrients to the developing GCs and decrease in size. In the fully mature gonads, the lumina are filled with gametes. In most countries that consume sea urchins, immature gonads containing fewer GCs than NPs are preferred as food (Walker et al. 2001), making NPs more important than GCs as food.

2.2 *Structural change during gametogenesis*

During the annual reproductive cycle, structural changes inside the gonad are classified into several stages. Among the stage classification systems which have been described (Fuji 1960, Kobayashi & Konaka 1971, Nicotra & Serafino 1988, Yamamoto et al. 1988, Byrne 1990, Spirlet et al. 1998, Walker et al. 1998), this paper presents the system of Fuji with slight modifications for *P. depressus* (Unuma et al. 1996). The annual reproductive cycle is classified into five stages from recovering (Stage 1) to spent (Stage 5). The histological characteristics at each stage are observed by light microscopy in paraffin-embedded sections stained with haematoxylin and eosin.

Stage 1 (Recovering): Immature gonads before gametogenesis. In both sexes, each acinus is filled with NPs (eosinophilic cell populations). The determination of the boundaries of individual NPs is difficult in paraffin sections but possible in plastic semi-thin sections (Walker et al. 1998). In females, a few young oocytes are present at the periphery of the acini (Fig. 2A).

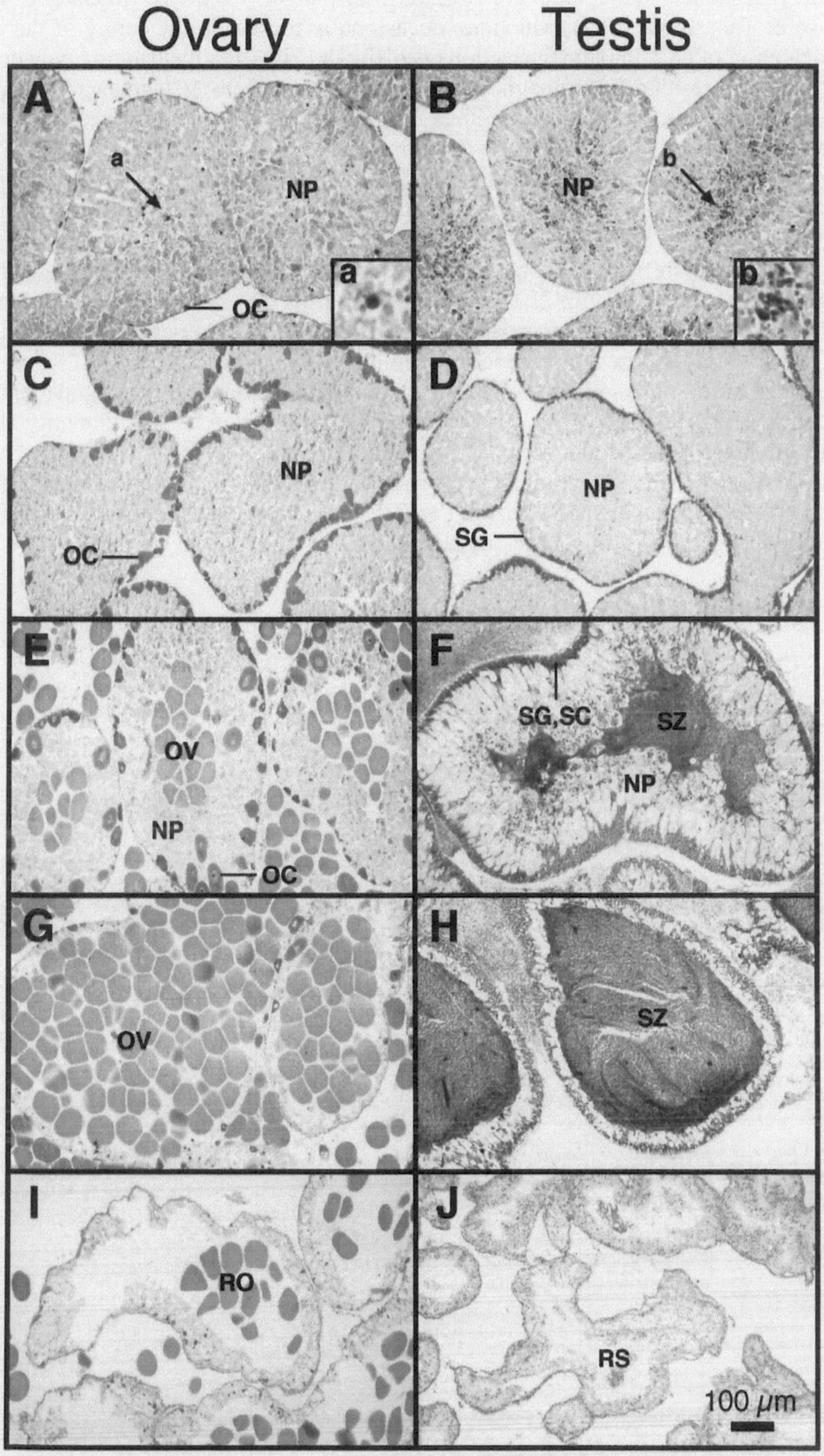

Figure 2. Histological changes in ovaries (left) and testes (right) in *Pseudocentrotus depressus*. Paraffin-embedded sections are stained with haematoxylin and eosin. A,B) Stage 1; C,D) Stage 2; E,F) Stage 3; G,H) Stage 4; I,J) Stage 5. NP) nutritive phagocyte; OC) oocyte; OV) ripe ovum; RO) residual ovum; SG) spermatogonium; SC) spermatocyte; SZ) spermatozoon; RS) residual spermatozoon; a) round spot derived from phagocytized residual ova; b) amorphous speckles derived from phagocytized residual spermatozoa.

Haematoxylin-stained round spots are residue from phagocytized ova (Masuda & Dan 1975, Tominaga & Takashima 1987), which are occasionally present in the centre of the lumina. In males, detection of GCs in paraffin sections is difficult (Fig. 2B). Instead, many haematoxylin-stained speckles, residue from phagocytized spermatozoa (Kato & Ishikawa 1982), are often present in NPs. These speckles, amorphous unlike the round spots observed in the immature ovary, are a useful key to distinguish testes from ovaries.

Stage 2 (Growing): Beginning of gametogenesis. Many developing oocytes (Fig. 2C) or clusters of spermatogonia (Fig. 2D) are present at the periphery of the acini, and the gonadal lumina are still filled with NPs.

Stage 3 (Pre-mature): Middle of gametogenesis. In the centre of the gonadal lumina, NPs are replaced with ripe ova (Fig. 2E) or spermatozoa (Fig. 2F). Numerous developing oocytes or clusters of spermatogonia and spermatocytes are present at the periphery of the acini. NPs gradually decreasing in size are present between the GCs.

Stage 4 (Mature): Fully-mature gonads at the end of gametogenesis. The gonadal lumina are filled with ripe ova (Fig. 2G) or spermatozoa (Fig. 2H). Shrunken NPs, which have already lost nutrients, are present only at the periphery of the acini.

Stage 5 (Spent): After spawning. The gonadal lumina have numerous empty spaces and a few residual ova (Fig. 2I) or spermatozoa (Fig. 2J). NPs gradually phagocytize residual gametes and re-start growing. After this stage, gonads return to Stage 1 and a new cycle starts.

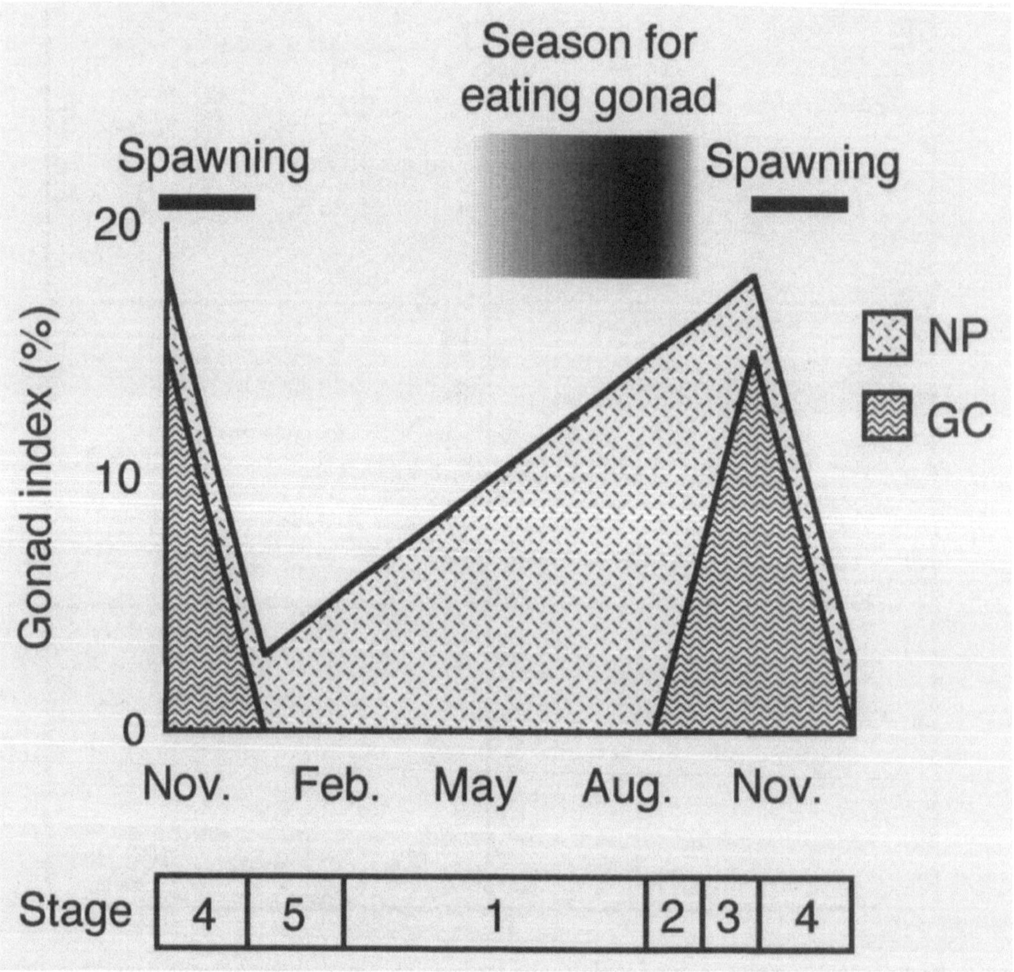

Figure 3. Diagrammatic representation of the seasonal changes in the gonad index and in the proportion of nutritive phagocytes (NPs) and germ cells (GCs) in *Pseudocentrotus depressus*. A typical stage of the gonad is shown below the graph. Spawning occurs around November. The best season to eat gonads is from June to August.

2.3 *Annual reproductive cycle*

Figure 3 shows the seasonal changes in the gonad index and in the proportion of NPs and GCs in *P. depressus*. This species spawns around November when gonad indices rapidly decrease. After spawning, an increase in the gonad index continues until the next spawning. The increase before gametogenesis is attributed to the growth of NPs (not of GCs) as they accumulate various nutrients derived from ingested food and increase in size. Proliferation and development of GCs occur for only two months before spawning. The best season to eat gonads is from June to August, when NPs have grown sufficiently but gametogenesis has not yet begun.

3. A PROTEIN STORED IN NUTRITIVE PHAGOCYTES

3.1 *MYP as a nutrient reserve*

Nutrients needed for gametogenesis, such as protein, lipid, glycogen, are stored in NPs as various kinds of granules (Takashima & Takashima 1965, Holland 1967, Chatlynne 1969). The most abundant protein contained in NPs of immature gonads is a glycoprotein with a molecular weight of 170 kDa (Fig. 4), originally identified as the most predominant component of yolk granules in the sea urchin eggs and termed the major yolk protein or major yolk glycoprotein (MYP; Harrington & Easton 1982, Kari & Rottman 1985, Yokota & Kato 1988, Scott &

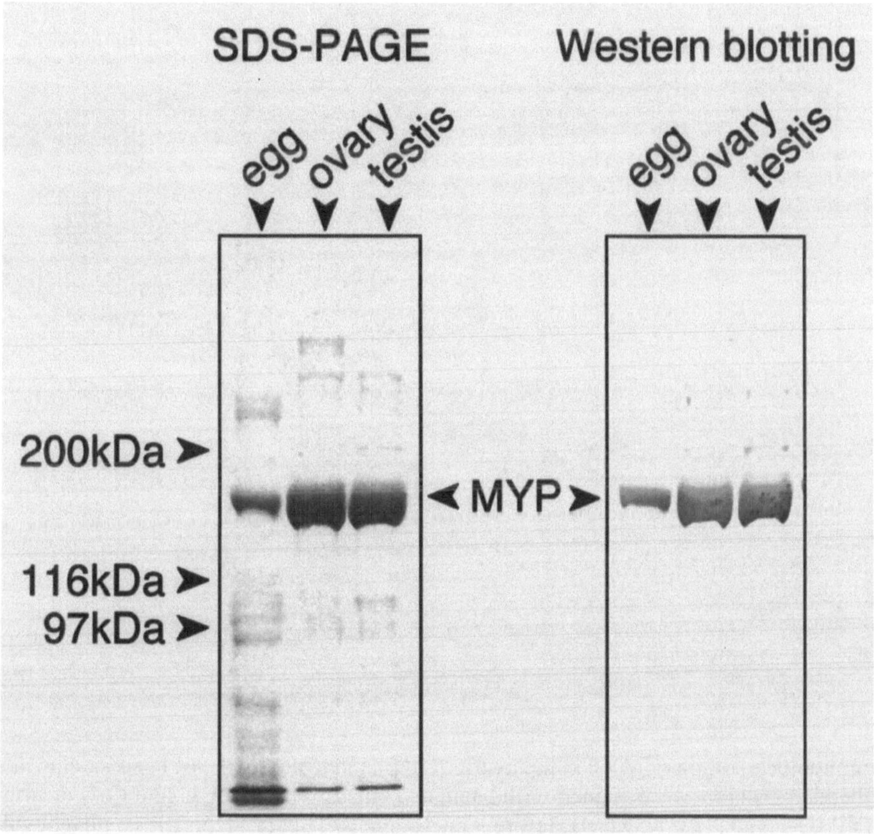

Figure 4. SDS-PAGE and Western blot analysis of MYP in extracts from egg, immature ovary and immature testis of *Pseudocentrotus depressus*. The gel was stained with Coomassie-Brilliant-Blue and the membrane immunostained with an antibody specific to MYP. The positions of the molecular weight markers are indicated on the left margin. The predominant protein in the egg extract with a molecular weight of 170 kDa is called the major yolk protein (MYP); it is also contained in the immature ovary and testis.

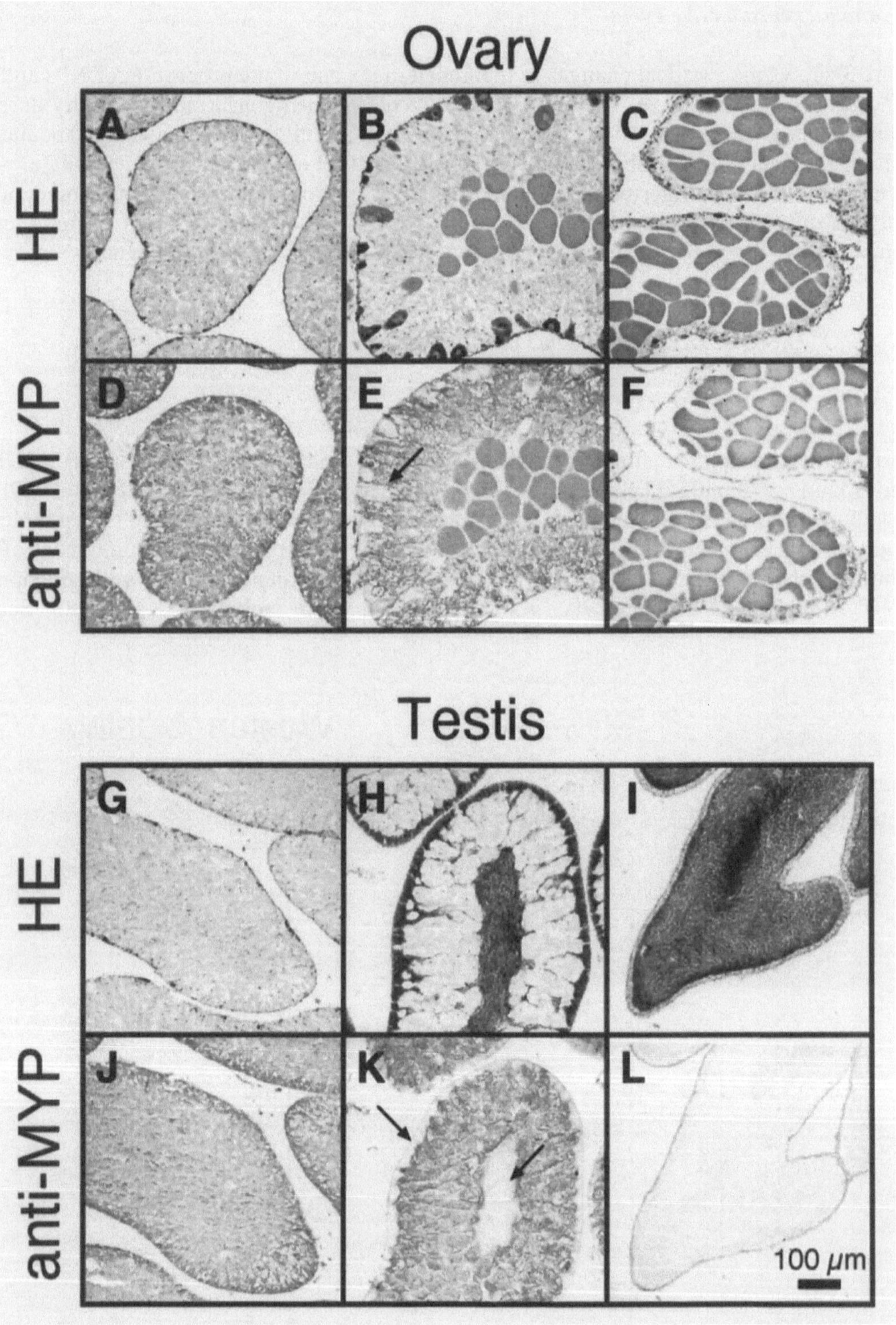

Figure 5. Immunolocalization of MYP in the ovaries (A-F) and testes (G-L) of Pseudocentrotus depressus. Paraffin-embedded sections were stained with haematoxylin and eosin (A-C and G-I) or immunostained with anti-MYP (D-F and J-L). A,D,G,J) Before gametogenesis (Stage 1). Nutritive phagocytes filling the gonadal acini are immunoreactive with anti-MYP in both sexes. B,E,H,K) The middle of gametogenesis (Stage 3). Nutritive phagocytes are still immunoreactive with anti-MYP. Ripe ova are reactive with the antibody at the centre of the ovarian acini, whereas oocytes and spermatogenic cells are not (arrows). C,F,I,L) End of gametogenesis (Stage 4). Ripe ova are reactive with anti-MYP, unlike shrunken nutritive phagocytes at the periphery of the ovarian acini. Neither nutritive phagocytes nor germ cells are reactive with anti-MYP in testes.

Lennarz 1989). In oviparous animals other than sea urchins, yolk protein, a nutrient reserve for embryogenesis, is usually female specific because the male does not need to produce the yolk. In sea urchins, however, MYP is synthesized in both females and males (Shyu et al. 1986), which store a large quantity of MYP in the NPs of their immature gonads to use during gametogenesis (Unuma et al. 1998). As oogenesis proceeds, MYP is transported from ovarian NPs to the oocytes to form yolk granules (Ozaki et al. 1986), and as spermatogenesis proceeds, it gradually disappears from the testis, suggesting that it is used as a nutrient source to produce sperm (Unuma et al. 1998).

Immunolocalization of MYP in the gonads of *P. depressus* is performed by using an antibody specific to MYP (Fig. 5). In both sexes before gametogenesis (Stage 1), NPs filling the gonadal acini accumulate MYP (Figs. 5A, D, G, J). In the middle of gametogenesis (Stage 3), NPs become small though still contain MYP (Figs. 5B, E, H, K). In the ovaries, oocytes at the periphery of the acini do not contain MYP, whereas ripe ova in the centre of the acini do, suggesting that oocytes incorporate MYP during their move from the periphery to the centre of acini. In the testes, none of the male GCs contain MYP. In mature ovaries (Stage 4), ripe ova contain MYP, but NPs observed only at the periphery of the acini do not (Figs. 5C, F). In mature testes (Stage 4), neither GCs nor NPs contain MYP (Figs. 5I, L).

3.2 *Synthesis and accumulation of MYP*

The synthesis and accumulation mechanism of MYP appears rather complicated. Vitellogenin, a putative precursor of MYP with a slightly higher molecular weight than MYP (180-200 kDa) is contained in the coelomic fluid, a body fluid which fills the inner cavity of the test (Harrington & Easton 1982). Partial genomic DNA and full-length cDNA encoding vitellogenin are cloned from *S. purpuratus* (Genbank Y00513; Shyu et al. 1986, 1987) and *P. depressus* (Genbank AF318300; Unuma et al. 2001), respectively. Vitellogenin mRNA expression is detected in NPs, digestive tracts (esophagus, stomach, intestine and rectum) and coelomocytes, cell populations in the coelomic fluid (Unuma et al. 2001). Vitellogenin synthesized in digestive tracts and coelomocytes is probably secreted into coelomic fluid, taken up by the NPs, and then processed to MYP. Before gametogenesis, the greater part of MYP accumulated in NPs is suggested to be synthesized inside the NPs, since mRNA expression is higher in NPs than in other tissues. The vitellogenin gene has a palindrome sequence similar to the estrogen responsive element, suggesting that the synthesis of vitellogenin is controlled by estrogen (Shyu et al. 1987). Although some steroids are reported to promote gonadal growth in *P. depressus* (Unuma et al. 1999), the relationship between estrogen and MYP accumulation is still unclear.

In addition to the role as a nutrient reserve for gametogenesis and embryogenesis, MYP and its related proteins have various physiological roles in sea urchins. MYP and its proteolytically processed products, called toposomes, function as cell adhesion molecules in sea urchin embryogenesis (Noll et al. 1985, Matranga et al. 1986, 1991, Cervello & Matranga 1989). Also, vitellogenin may be involved in the clotting phenomenon in coelomic fluid (Cervello et al. 1994). These physiological characteristics of MYP as well as the process from synthesis to degradation are well reviewed by Yokota and Sappington (in press). As a whole, MYP has been inspiring a lot of interests in the field of basic biology. Hereafter, however, studies on sea urchin aquaculture should focus on MYP because this protein is crucial for the gonadal growth of sea urchins.

4. RELATIONSHIP BETWEEN MATURATION AND AQUACULTURE

4.1 *Quality deterioration associated with maturation*

As mentioned before, most countries prefer consuming immature gonads containing scarce GCs, although there may be some exceptions (Walker et al. 2001). Considering sea urchin gonads as a food product, their quality deteriorates as gametogenesis proceeds. This is due to some adverse events, the most serious of which is gamete ooze. The tissues of the gonads become more fragile as gametogenesis proceeds. After the mature gonads are taken out of the test, they appear to be melting (Fig. 6). Eggs or sperm ooze from the gonoduct, and the gonads cannot maintain their consistency. This phenomenon, common among all the edible sea urchins, considerably reduces the commercial value of the gonads. To suppress this gamete ooze, alum, an aluminum compound, is usually used in Japan. After removing the gonad from the test, they are lightly washed in salt water and then immersed in a 0.5-1.5% alum solution for 10-20 minutes. In this way, the mature gonads can maintain their shape for a longer time. The alum cannot, however, prevent gamete ooze completely; besides, it alters the taste of gonads astringent and bitter.

Another drawback in quality is the bitterness of mature ovaries. In some species including *P. depressus* and *Hemicentrotus pulcherrimus* (another popular species in the southern part of Japan), mature ovaries have an extremely bitter taste, contrary to the mature testis. Recently, pulcherrimine, a novel sulfur-containing amino acid, has been isolated as the bitter principle from the mature ovary of *H. pulcherrimus* (Murata & Sata 2000, Murata et al. 2001, Murata et al. in press). Similarly, the mature ovary of *P. depressus* may also contain pulcherrimine, although it has not been investigated yet. This bitterness shortens the season of consuming these two species compared to that of other species whose mature ovaries do not have such bitterness.

In addition to the deterioration in quality with the ongoing gametogenesis, subsequent spawning causes a large year-to-year decrease in the size of the gonad. After the spawning, it takes time for the gonads to grow to the marketable size again. The best season for harvesting sea urchin gonads is rather short because of these problems. Low-quality or small gonads can be harvested for a longer period of time, but superior-quality and sizable ones can be harvested for only about two or three months. Therefore, gametogenesis and subsequent spawning are not desirable for culturing adult sea urchins, although they are necessary for the production of seedlings.

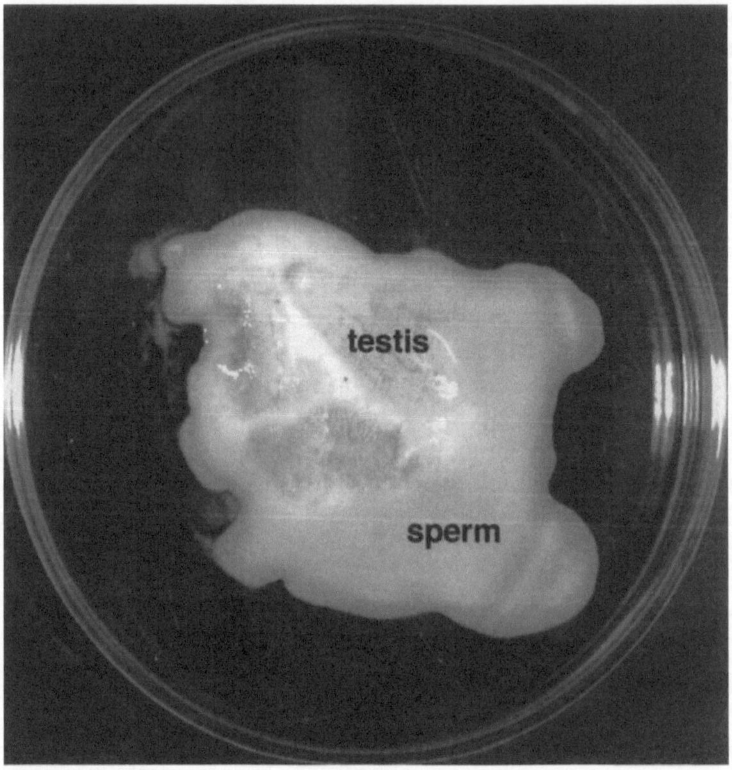

Figure 6. Quality deterioration in the gonad of *Pseudocentrotus depressus*. Mature testes were taken out of the test and placed in a petri dish for 30 minutes. The testes appeared to be melting because of oozing sperm. The same occurs in the mature ovaries also.

4.2 *Strategy to extend the season- I. Acceleration of NP growth*

One possible strategy to extend the season for harvesting high-quality gonads is to accelerate the growth of NPs as sea urchins recover from spawning (Fig. 7A). In autumn-spawning species such as *P. depressus*, *Strongylocentrotus intermedius*, *S. nudus*, the water temperature after spawning is too low for increased feeding and rapid recovery. The feeding ratio, the daily percentage of the food weight ingested relative to body weight, correlates with the water temperature (Akiyama et al., in prep.). Keeping the water warm during this period can enhance feeding of sea urchins and the growth of NPs (McBride et al. 1997). Temperature control is uneconomical for commercial-scale sea urchin aquaculture, but could be utilized if warm water such as the effluent from an electric power station is available. Temperature control should be applied carefully to obviate out-of-season maturation, since, in some species, initiation of

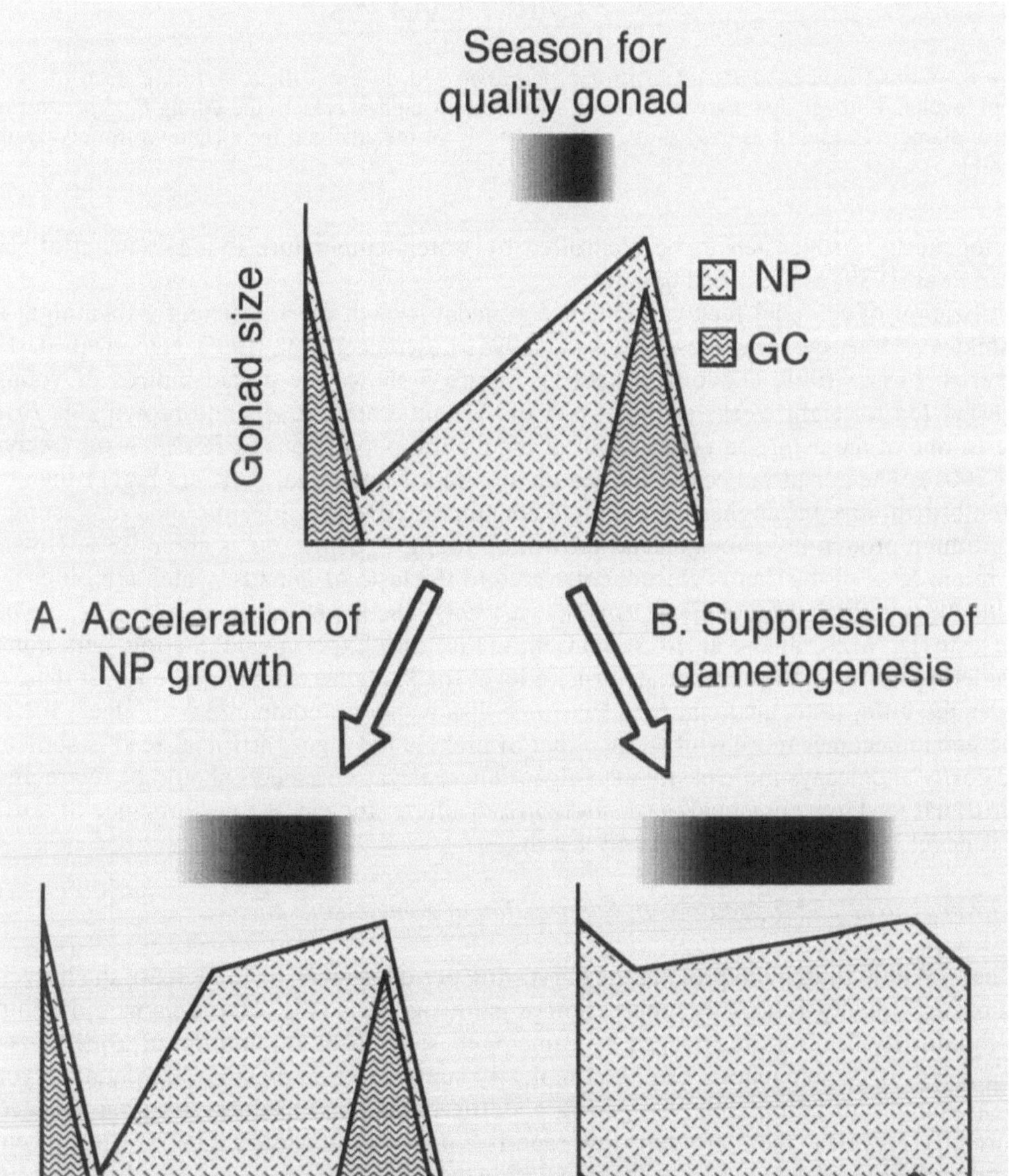

Figure 7. Strategies for extending the season of harvesting high-quality gonads. Acceleration of growth of nutritive phagocytes (A) and suppression of gametogenesis (B) are suggested.

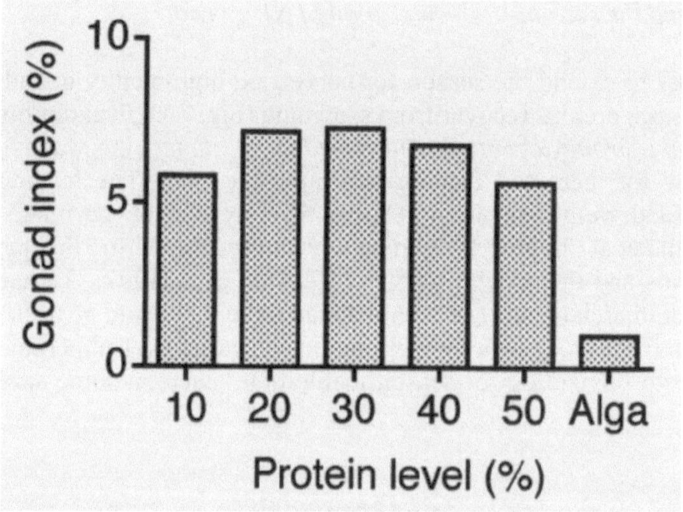

Figure 8. Gonad indices of *Pseudocentrotus depressus* fed different diets. Artificial feed of 10-50% protein level and brown alga, *Eisenia bicyclis*, were fed for eight weeks to the young *P. depressus* of 15 mm test diameter. Casein was used as the protein source for the artificial feed. (Drawn from Akiyama et al. 2001)

gametogenesis is suggested to be controlled by water temperature (Yamamoto et al. 1988, Sakairi et al. 1989) as described below.

Utilization of artificial feed can enhance gonadal growth more efficiently than algal food (Hoshikawa 1993, de Jong-Westman et al. 1995, Lawrence et al. 1997, Barker et al. 1998, Walker & Lesser 1998, Olave et al. 2001). Figure 8 shows the gonad indices of young *P. depressus* fed for eight weeks artificial diet of various protein levels and brown alga *Eisenia bicyclis*, one of the main sources of food for the natural populations of *P. depressus* (Akiyama et al. 2001). The gonad indices of the sea urchins fed the artificial diets are higher than those fed the brown alga and highest in the diets containing 20% and 30% protein, suggesting that the optimum protein level for gonadal growth of young *P. depressus* is about 25%. However, the protein level of the diet is closely correlated to the taste of gonads, which are bitter in sea urchins fed with high protein diets (Hoshikawa 1994). The lower the protein level is, the better is the taste (Dr. M. Kitagawa at Hokkaido Central Fisheries Experimental Station, pers. comm.). Considering the taste, the appropriate protein level for *P. depressus* should be lower than 25%. Besides the bitter taste, artificial feed has some disadvantages compared with algae: the color of the gonad becomes more whitish than that of urchins fed algae; artificial feed is sometimes more costly and decays more easily than algae. These negative aspects should be overcome for the artificial feed to be practical in sea urchin aquaculture. Indeed, the performance of artificial feed is developing dramatically year after year.

4.3 *Strategy to extend the season- II. Suppression of gametogenesis*

Another possible strategy to prolong the harvesting period, as well as to increase the harvesting yield of high-quality gonads, is suppression of gametogenesis (Fig. 6B). This would minimize both the quality deterioration caused by gametogenesis and the diminution of gonad size due to spawning. If gametogenesis can be completely suppressed, harvesting high-quality gonads all year round will be possible. To generate sterile organisms, triploidy is often induced. In sea urchins, however, there are very few reports concerning triploidy. One of the difficulties in generating triploid sea urchins is that sea urchin eggs complete their meiotic divisions before spawning. Therefore, retention of the polar body, a popular method of generating triploid fish and shellfish, cannot be applied to sea urchins. Production of a tetraploid sea urchin by suppressing the first cleavage and its subsequent crossing with the normal urchin may be an efficient method to generate triploid sea urchins.

Control of environmental conditions can also be applied to the suppression of gametogenesis (Walker et al. 2001). In *S. purpuratus*, *S. droebachiensis* and *Euchidaris tribuloides*, the environmental cue to initiating gametogenesis is suggested to be the photoperiod (Pearse et al. 1986, Bay-Schmith & Pearse 1987, McClintock & Watts 1990, Walker & Lesser 1998), whereas in *P. depressus*, *H. pulcherrimus* and *Anthocidaris crassispina*, it is suggested to be the water temperature (Yamamoto et al. 1988, Sakairi et al. 1989). Actually, in some hatcheries in Japan, which produce seedlings of fish and shellfish including sea urchins for restocking and aquaculture, out-of-season maturation of brood stock of *P. depressus* and *H. pulcherrimus* is induced by manipulating water temperature (Ito et al. 1989, Masaki & Kawahara 1995, Noguchi et al. 1995, Hagen 1996). For culturing adult sea urchins, however, not the induction but the supression of gametogenesis is required. In order to suppress gametogenesis by manipulating photoperiod and/or water temperature, the control regime needs to be further investigated. Needless to say, high cost of this method should be avoided to make it applicable to commercial-scale aquaculture of sea urchins.

5. ACKNOWLEDGEMENT

I am grateful to Dr. Hiromi Ohta, Dr. Kooichi Konishi, Dr. Hiroyuki Okamoto, Dr. Tohru Suzuki, Dr. Tadahide Kurokawa, Dr. Takeshi Yamamoto, Dr. Hirofumi Furuita at National Research Institute of Aquaculture and Dr. Toshio Akiyama at Fukui Prefectural Fisheries Experimental Station for their collaborative work on the reproduction of *P. depressus*. I appreciate the valuable advice from Professor Charles W. Walker at the University of New Hampshire and from Professor Katsuyoshi Mori at Tohoku University. I thank Professor Yukio Yokota at Aichi Prefectural University for his helpful discussion of my research.

6. REFERENCES

Akiyama, T., Unuma, T. & Yamamoto, T. 2001. Optimum protein level in a purified diet for young red sea urchin *Pseudocentrotus depressus*. *Fisheries Science* 67: 361-363.

Barker, M.F., Keogh, J.A., Lawrence, J.M. & Lawrence, A.L. 1998 Feeding rate, absorption efficiencies, and enhancement of gonad production in the New Zealand sea urchin *Evechinus chloroticus* Valenciennes (Echinoidea: Echinometridae) fed prepared and natural diets. *Journal of Shellfish Research* 17: 1583-1590.

Bay-Schmith, E. & Pearse, J.S. 1987. Effect of fixed daylengths on the photoperiodic regulation of gametogenesis in the sea urchin *Strongylocentrotus purpuratus*. *International Journal of Invertebrate Reproduction and development* 11: 287-294.

Byrne, M. 1990. Annual reproductive cycles of the commercial sea urchin *Paracentrotus lividus* from an exposed intertidal and a sheltered subtidal habitat on the west coast of Ireland. *Marine Biology* 104: 275-289.

Cervello, M. & Matranga, V. 1989. Evidence of a precursor -product relationship between vitellogenin and toposome, a glycoprotein complex mediating cell adhesion. *Cell Differentiation and Development* 26: 67-76.

Cervello, M., Arizza, V., Lattuca, G., Parrinello, N. & Matranga, V. 1994. Detection of vitellogenin in a subpopulation of sea urchin coelomocytes. *European Journal of Cell Biology* 64: 314-319.

Chatlynne, L.G. 1969. A histochemical study of oogenesis in the sea urchin, *Strongylocentrotus purpuratus*. *Biological Bulletin* 136: 167-184.

de Jong-Westman, M., March, B.E. & Carefoot, T.H. 1995. The effect of different nutrient formulations in artificial diets on gonadal growth in the sea urchin *Strongylocentrotus droebachiensis*. *Canadian Journal of Zoology* 73: 1495-1502.

FAO (Food and Agriculture Organization of the United Nations). 1996. *FAO yearbook of fishery statistics*. Rome: FAO.

Fuji, A. 1960. Studies on the biology of the sea urchin. I. Superficial and histological gonadal changes in gametogenic process of two sea urchins, *Strongylocentrotus nudus* and *S. intermedius*. *Bulletin of Faculty of Fishery, Hokkaido University* 11: 1-14.

Hagen, N.T. 1996. Echinoculture: from fishery enhancement to closed cycle cultivation. *World Aquaculture*. Dec.: 7-19.

Harrington, F.E. & Easton D.P. 1982. A putative precursor to the major yolk protein of the sea urchin. *Developmental Biology* 94: 505-508.

Hoshikawa, H. 1993. Uni ni daizu wo tabesaseru hanashi. *Hokusuishi Dayori* 22: 9-14. (in Japanese)

Holland, N.D. 1967. Gametogenesis during the annual reproductive cycle in a cidaroid sea urchin (*Stylocidaris affinis*). *Biological Bulletin* 133: 578-590.

Ito, S., Shibayama, M., Kobayakawa, A. & Tani, Y. 1989. Promotion of maturation and spawning of sea urchin *Hemicentrotus pulcherrimus* by regulating water temperature. *Nippon Suisan Gakkaishi* 55: 757-763. (in Japanese)

Kato, K.H. & Ishikawa, M. 1982. Flagellum formation and centriolar behavior during spermatogenesis of the sea urchin, *Hemicentrotus pulcherrimus*. *Acta Embryologia Morphologia Experimentalis n.s.* 3: 49-66.

Kari, B.E. & Rottmann, W.L. 1985. Analysis of changes in a yolk glycoprotein complex in the developing sea urchin embryo. *Developmental Biology* 108: 18-25.

Keesing, J.K. & Hall, K.C. 1998. Review of harvests and status of world sea urchin fisheries points to opportunities for aquaculture. *Journal of Shellfish Research* 17: 1597-1604.

Kobayashi, N. & Konaka, K. 1971. Studies on periodicity in oogenesis of sea urchin I. Relation between the oogenesis and the appearance and disappearance of nutritive phagocytes detected by some histochemical methods. *The Science and Engineering Review of Doshisha University* 12: 27-45. (in Japanese)

Lawrence, J.M., Olave, S., Otaiza, R., Lawrence, A.M. & Bustos, E. 1997. Enhancement of gonad production in the sea urchin *Loxechinus albus* in Chile fed extruded feeds. *Journal of World Aquaculture Society* 28: 91-96.

Masaki, K. & Kawahara, I. 1995. Promotion of gonadal maturation by regulating water temperature in the sea urchin *Pseudocentrotus depressus* -I. *Bulletin of Saga Prefectural Sea Farming Center* 4: 93-100. (in Japanese)

Masuda, R. & Dan, J.C. 1977. Studies on the annual reproductive cycle of the sea urchin and the acid phosphatase activity of relict ova. *Biological Bulletin* 153: 577-590.

Matranga, V., Kuwasaki, B. & Noll, H. 1986. Functional characterization of toposomes from sea urchin blastula embryos by a morphogenetic cell aggregation assay. *EMBO Journal* 5: 3125-3132.

Matranga, V., Ferro, D.D., Cervello, M., Zito, F. & Nakano, E. 1991. Adhesion of sea-urchin embryonic cells to substrata coated with cell adhesion molecules. *Biology of the Cell* 71: 289-291.

McBride, S.C., Pinnix, W.D., Lawrence, J.M., Lawrence, A.L. & Mulligan, T.M. 1997. The effect of temperature on production of gonads by the sea urchin *Strongylocentrotus franciscanus* fed natural and prepared diets. *Journal of World Aquaculture Society* 28: 357-365.

McClintock, J.B. & Watts, S.A. 1990. The effects of photoperiod on gametogenesis in the tropical sea urchin *Euchidaris tribuloides* (Lamarck) (Echinodermata: Echinoidea). *Journal of Experimental Marine Biology and Ecology* 139: 175-184.

Murata, Y. & Sata N.U. 2000. Isolation and structure of pulcherrimine, a novel bitter-tasting amino acid, from the sea urchin (*Hemicentrotus pulcherrimus*) ovaries. *Journal of Agricultural and Food Chemistry* 48: 5557-5560.

Murata, Y., Sata, N.U., Yokoyama, M., Kuwahara, R., Kaneniwa, M. & Oohara I. 2001. Determination of a novel amino acid, pulcherrimine, in the gonad of the green sea urchin *Hemicentrotus pulcherrimus*. *Fisheries Science* 67: 341-345.

Murata, Y., Yokoyama, M., Unuma, T., Sata, N.U., Kuwahara, R. & Kaneniwa, M. (in press) Seasonal changes of bitterness and pulcherrimine content in gonads of green sea urchin *Hemicentrotus pulcherrimus* at Iwaki in Fukushima Prefecture. *Fisheries Science* 68.

Nicotra, A. & Serafino, A. 1988. Ultrastructural observations on the interstitial cells of the testis of *Paracentrotus lividus*. *International Journal of Invertebrate Reproduction and Development* 13: 239-250.

Noguchi, H., Kawahara, I. Goto, M. & Masaki, K. 1995. Promotion of gonadal maturation by regulating water temperature in the sea urchin *Pseudocentrotus depressus* -II. *Bulletin of Saga Prefectural Sea Farming Center* 4: 101-107. (in Japanese)

Noll, H., Matranga, V., Cervello, M., Humphreys, T., Kuwasaki, B. & Adelson, D. 1985. Characterization of toposomes from sea urchin blastula cells: a cell organelle mediating cell adhesion and expressing positional information. *Proceedings of National Academy of Science USA* 82: 8062-8066.

Olave, S., Bustos, E., Lawrence, J.M. & Carcamo, P. 2001. The effect of size and diet on gonad production by the Chilean sea urchin *Loxechinus albus*. *Journal of World Aquaculture Society* 32: 210-214.

Ozaki, H., Moriya, O. & Harrington, F.E. 1986. A glycoprotein in the accessory cell of the echinoid ovary and its role in vitellogenesis. *Roux's Archives of Developmental Biology* 195: 74-79.

Pearse, J.S., Pearse V.B. & Davis, K.K. 1986. Photoperiodic regulation of gametogenesis and growth in the sea urchin, *Strongylocentrotus purpuratus*. *Journal of Experimental Zoology* 237: 107-118.

Sakairi, K., Yamamoto, M., Ohtsu, K. & Yoshida, M. 1989. Environmental control of gonadal maturation in laboratory-reared sea urchins, *Anthocidaris crassispina* and *Hemicentrotus pulcherrimus*. *Zoological Science* 6: 721-730.

Scott, L.B. & Lennarz, W.J. 1989. Structure of a major yolk glycoprotein and its processing pathway by limited proteolysis are conserved in echinoids. *Developmental Biology* 132: 91-102.

Shyu, A.-B., Raff, R.A. & Blumenthal, T. 1986. Expression of the vitellogenin gene in female and male sea urchin. *Proceedings of National Academy of Science USA* 83: 3865-3869.

Shyu, A.-B., Blumenthal, T. & Raff, R.A. 1987. A single gene encoding vitellogenin in the sea urchin Strongylocentrotus purpuratus: sequence at the 5' end. *Nucleic Acids Research* 15: 10405-10417.

Spirlet, C., Grosjean, P. & Jangoux, M. 1998. Reproductive cycle of the echinoid *Paracentrotus lividus*: analysis by means of the maturity index. *Invertebrate Reproduction and Development* 34: 69-81

Takashima, R. & Takashima, Y. 1965. Studies on the submicroscopical structures of the nurse cells in sea urchin ovary, with special reference to the glycogen particles. *Okajimas Folia Anatomica Japonica* 40: 819-831

Tominaga, A. & Takashima, Y. 1987. Ultracytochemical study on the behavior of the nurse cell giant granules in the sea urchin ovary. *Acta Histochemica et Cytochemica* 20: 569-579

Unuma, T., Konishi, K., Furuita, H., Yamamoto, T. & Akiyama, T. 1996. Seasonal changes in gonads of cultured and wild red sea urchin, *Pseudocentrotus depressus*. *Suisanzoshoku* 44: 169-175.

Unuma, T., Suzuki, T., Kurokawa, T., Yamamoto, T. & Akiyama, T. 1998. A protein identical to the yolk protein is stored in the testis in male red sea urchin, *Pseudocentrotus depressus*. *Biological Bulletin* 194: 92-97.

Unuma, T., Yamamoto, T. & Akiyama, T. 1999. Effect of steroids on gonadal growth and gametogenesis in the juvenile red sea urchin *Pseudocentrotus depressus*. *Biological Bulletin* 196: 199-204.

Unuma, T., Okamoto, H., Konishi, K., Ohta, H. & Mori, K. 2001. Cloning of cDNA encoding vitellogenin and its expression in red sea urchin, *Pseudocentrotus depressus*. *Zoological Science* 18: 559-565.

Walker, C.W. 1982. Nutrition of gametes. In M. Jangoux and J.M. Lawrence (eds), *Echinoderm Nutrition*: 449-468. Rotterdam: Balkema.

Walker, C.W. & Lesser M.P. 1998. Manipulation of food and photoperiod promotes out-of-season gametogenesis in the green sea urchin, *Strongylocentrotus droebachiensis*: implications for aquaculture. *Marine Biology* 132: 663-676

Walker, C.W. McGinn, N.A., Harrington, L.M.& Lesser M.P. 1998. New perspectives on sea urchin gametogenesis and their relevance to aquaculture. *Journal of Shellfish Research* 17: 1507-1514.

Walker, C.W., Unuma, T., McGinn, N.A., Harrington, L.M. & Lesser M.P. 2001. Reproduction of sea urchins. In J.M. Lawrence (ed.), *Edible sea urchins: Biology and ecology*: 5-26. Amsterdam: Elsevier.

Yamamoto, M., Ishine, M. & Yoshida, M. 1988. Gonadal maturation independent of photic conditions in laboratory-reared sea urchins, *Pseudocentrotus depressus* and *Hemicentrotus pulcherrimus*. *Zoological Science* 5: 979-988.

Yokota, Y. & Kato, K.H. 1988. Degradation of yolk proteins in sea urchin eggs and embryos. *Cell Differentiation* 23: 191-200.

Yokota, Y. & Sappington, T.W. (in press) Vitellogen and vitellogenin in echinoderms. In A.S. Raikhel & T.W. Sappington (eds), *Reproductive biology of invertebrates, Progress in vitellogenesis. Part I Yolk protein, structure and biochemical properties*: New York: John Wiley.

The Sea Urchin: From Basic Biology to Aquaculture, Yokota, Matranga & Smolenicka (eds)
© Swets & Zeitlinger, Lisse, ISBN 90 5809 379 4

Fishery and consumption of the sea urchin in Japan

Y. Yokota
Department of Applied Information Technology, Aichi Prefectural University, Aichi, Japan

ABSTRACT: Japan is well known as a country with a large consumption of fish and other sea food. Sources of animal proteins mainly depended on fish, first because of a geographical reason - Japan is surrounded by sea - and secondly owed to religious and historical reasons (eating mammals was forbidden by Buddhism). The thriving of Japanese economy promoted the demand for luxurious food; consequently, consumption of sea urchins in Japan increased . This chapter describes historical background related to sea urchins as well as the contemporary situation of fisheries in Japan and, finally, sea urchin consumption.

1. HISTORY OF THE SEA URCHIN IN JAPAN

It is well known that a large amount of fishes, marine invertebrates and algae are consumed in Japan. The Japanese cuisine, which is known worldwide, has a lot of fish-based recipes, as sashimi, sushi and tenpura. There is a long history of eating fishes and other marine animals.

The sea urchin as a food is generally called sea urchin roe. This, in fact, is composed of ovaries and testes with immature germ cells. However, in this chapter the expression 'sea urchin roe' will be used. Nowadays, the sea urchin is called "uni" in Japanese. This word has been used for sea urchin roe salted for preservation whereas the animal itself has been called "gaze". The sea urchin roe has been eaten since ancient times. Many sea urchin tests were excavated from kitchen middens (shell mounds) in Japan. The oldest description on the sea urchin reported so far is that found in Fujiwarakyo. According to the historical analysis, it was described between 694-710 AD. The oldest recipe of sea urchin roe was also found in an old document of the mid-8th century, in Heijokyo (ancient name of Nara). That recipe was as follows: sea urchin roe and slices of abalone were mixed with some salt, conditioned for a few days and then served. Considering the geographical position of Fujiwarakyo and Heijokyo, sea urchins fishing had been constantly carried out as a ordinary work. Heijokyo and Fujiwarakyo are approximately 100 km distant from the sea. Indeed, the sea urchin roe preserve (uni) was described in Yororitsuryo (ancient law, 757, AD), Ryonoghighe (practical instruction for Yororitsuryo, 834, AD) and Enghishiki (manual for ceremonies and daily life at court, 927, AD). In the Edo era (early modern ages), the sea urchin roe were described in the encyclopedia of foods, herbs and medical plants (named Honcho-Shokkan, 1697) and the sea urchin was cited in Wakan-Sansaizue (encyclopedia of animals and plants, 1712). According to Wakan-Sansaizue, the sea urchin roe from Fukui, Kagoshima, Nagasaki and Fukushima were a fine food. The red-yellowish roe was said to be better than the pale yellowish one. The sea urchin roe from Fukui has been regarded as one of the bests from old times until today. It has been ascertained that the sea urchin roe of *Hemicentrotus pulcherrimus* has been appreciated. The roe of *H. pulcherrimus* produced in Fukui is one of the most expensive uni today.

Since the sea urchin roe was regarded as one of very expensive foods for a long time, its preserve was a gift, or it was consumed on special occasions, mainly in the form of salted preserve. Before the World War II, the fresh sea urchin roe was one of the local gastronomic delicacies, with limited consumption. Eating fresh sea urchin roe has become popular since these 30 or 40 years. Today, many people appreciate it. Sea urchin roe sushi started in the 50s. The social, economical and dietary trends in Japan for these three decades and the demands for luxury food increased sea urchins consumption. Since 1977, the amount of annual production began to decrease. Now the annual production is less than half of that in 1977 (approximately 26,900 metric tons). Since Japan cannot cover the domestic consumption demand, sea urchins have to be imported (the last 4 years import is given in Table 1). This trend in Japan clearly reflects the global fishery business and worldwide resources of sea urchins. Approximately 80% of the global sea urchins production is consumed in Japan.

Table 1. Sea urchin import of Japan

Year	Live sea urchin & fresh roe	Frozen roe	Total
1997	4,366,969 23,773,191,000	1,027,814 2,093,277,000	5,394,783 25,866,468,000
1998	3,843,935 21,031,290,000	1,421,109 3,001298,000	5,265,044 24,032,589,000
1999	3,996,683 20,297,642,000	1,526,262 2,795,138,000	5,522,945 23,092,783,000
2000	3,778,628 18,973,480,000	1,651,940 3,241,808,000	5,430,568 22,215,288,000

The upper and lower lines of each year express the amount in kg and the price in Japanese yen.

According to the annual statistics, live sea urchins and fresh sea urchin roe were imported from S. Korea, N. Korea, China, Hong Kong, Vietnam, Philippine, Norway, Russia, Canada, USA, Peru, Chile, Australia and Palau. The frozen sea urchin roe was imported from S. Korea, N. Korea, Canada, USA, Peru, Chile and Australia. Japan can be thus considered as the largest sea urchin market in the world. In Japan, Tokyo Tukiji is the largest sea urchin market, followed by Osaka and Sapporo, where sea urchins and other fishery products are auctioned.

2. FISHERY AND AQUACULTURE OF SEA URCHIN IN JAPAN

Various species of sea urchins inhabit Japanese territory according to climatic differences, from Hokkaido (north) to Kyushu (south). Average water temperature in August is 17 °C in Hokkaido, 24 °C in the middle area of Honshu and 29 °C in Kyushu. In February, the above-mentioned areas register temperatures of 3 °C, 14°C and 29 °C. Approximately 180 species are supposed to inhabit the areas around Japan. *H. pulcherrimus, Anthocidaris crassispina, Strongylocentrotus intermedius, S. nudus, Pseudocentrotus depressus,* and *Tripneustes gratilla* are Japanese seafood resources (Fig. 1). Sea urchins are essentially collected by diving, by picking-up from a boat and by hand-catching walking in intertidal zones. Diving catch is the most efficient method to collect animals on the seabed over 10 m. Catching from a boat is the most common practice in Hokkaido and northern areas of Honshu Island. This requires skills as well as fair weather conditions and visibility in water. Collecting sea urchins by walking does not require any techniques. This method may damage the seabed, small sea urchins and juveniles. Therefore, some fishermen co-operatives do not allow such way of catching.

Sea urchins live in shallow waters; as their locomotion is slow, it is easy to collect them by hand. Their worst predator is man. Resource management by regulation, artificial propagation,

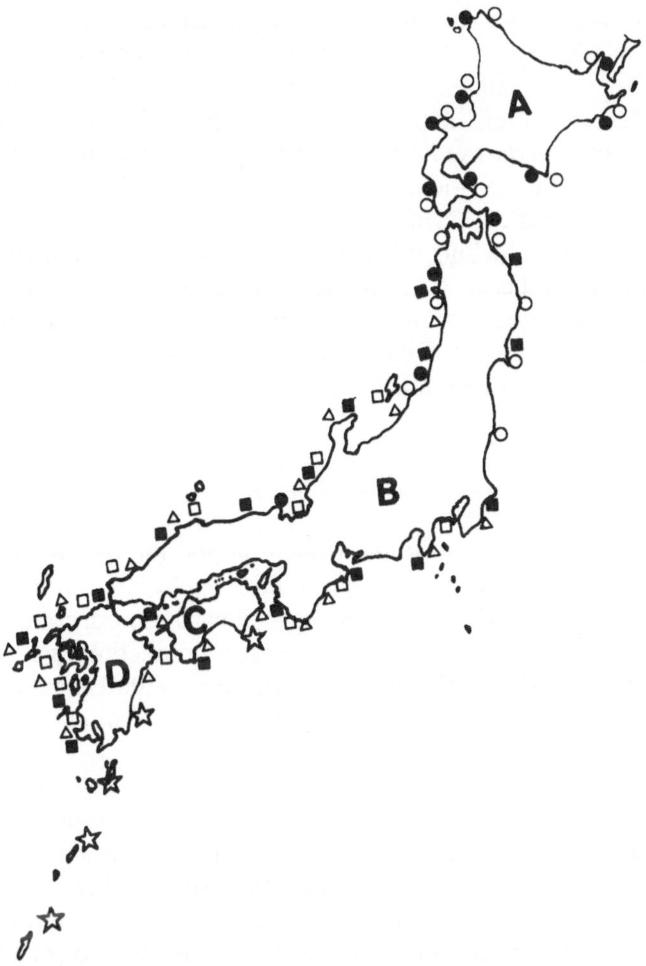

Figure 1. Geographical distribution of sea urchins in Japan. A: Hokkaido, B: Honshu, C: Shikoku, D: Kyushu. (○): *S. intermedius*, (●): *S. nudus*, (■): *H. pulcherrimus*, (Δ): *A. crassispina*, (□): *P. depressus*, (☆): *T. gratilla*.

artificial promotion of growth and development of new fishing areas are supposed to be effective for maintenance and recovery of sea urchin resources. Fishery business identified Japan as a responsible for the protection of such resources.

2.1 *Management by regulation*

Catching stop and catching size are ruled by prefectures, since the breeding season of sea urchins varies according to localities, species, climate and other factors. Co-operative associations' dispositions are often stricter than prefectures' rules (see larger catching size and catching stop in the afternoon).

In Fukui prefecture, sea urchin species for fishery is only *H. pulcherrimus*. *A. crassispina* also inhabits in Fukui, though it is not allowed to be caught. *H. pulcherrimus* can be caught for a month, from 21 of July to 20 of August, by disposition of Fukui prefecture. Sea urchin smaller than 2 cm (diameter) cannot be collected. The breeding season of *H. pulcherrimus* in Fukui goes from the end of December to the beginning of March.

In Hokkaido, *S. intermedius* and *S. nudus* are fished. The biological stop for *S. nudus* goes from mid-July to the end of September in Tokachi, Kushiro and Nemuro districts. The whole month of October is free in Soya district. *S. intermedius* can be caught from July to September in Tokachi, Kushiro, and Nemuro and from September to October in other areas. Since weather conditions are bad in Hokkaido, sea urchin catching is actually impossible in winter (usually from November to middle April). The catching-size limit is defined by a test diameter of 4 cm for *S. intermedius* and 5 cm for *S. nudus*. In the case of *S. intermedius*, approximately 40 % of fishermen's associations apply a larger size-limit than the governmental one. The breeding season of *S. intermedius* in Kushiro district goes from middle-May to middle-August by Akkeshi Marine Biological Station, Hokkaido University. A more effective regulation to sustain sea urchin resources is expected.

In Yamaguchi prefecture, catching of *H. pulcherrimus* and *A. crassispina* is permitted from June to July and from December to April. Some co-operatives catch only on Monday, Tuesday and Wednesday. Each one cares for protection and management of resources restricting the number of catch per day, per year. Other measures take into account fish preserve (fishing prohibited zone), short fishing period, catch restriction according to the test diameter (larger than 40 mm).

2.2 *Aquaculture*

Aquaculture has a long history. During the Roman period the eel was grown in lagoons. In China, a book on the techniques of carp aquaculture was written in ca 500 B. C. In Japan, industrial aquaculture started about 300 years ago. Now the production exceeds 1,300,000 t and costs more than 2,700 billion Japanese yen. Largest productions are as follow: yellow tail (150,000 t), red sea bream (30,000 t), eel (35,000 t), carp (18,000 t), rainbow trout (17,000 t), oyster (250,000 t), scallop (140,000 t), laver (400,000 t), *Undaria* (140,000 t), and *Laminaria* (50,000 t).

Sea urchins aquaculture development started with *P. depressus* by Yamabe (1962). The research of the production of artificial sea urchin seeds was done in *H. pulcherrimus* and *A. crassispina* at the Yamaguchi Prefectural Fishery Station since the end of the 60's. The research at Fukushima Prefectural Fishery Station and Saga Prefectural Fishery Station in *S. intermedius* and *P. depressus*, respectively, also began in the middle 70's. The seed production on a market scale was carried out in 1975. Seed supply of sea urchins - *P. depressus* and *H. pulcherrimus* in Saga Prefecture and *S. nudus* in Fukushima Prefecture – attained the level of a million per year. Over 100,000 of sea urchin seeds are supplied in Nagasaki Prefecture, Fukuoka Prefecture and Okinawa Prefecture. In addition, facilities producing 5 million of juvenile sea urchins (*S. nudus* and *S. intermedius*) of 15 mm diameter for seed release were built in 1983 in Iwate Prefecture.

In Hokkaido, Akkeshi Marine Biological Laboratory, Faculty of Science, Hokkaido University carried out the basic research for artificial seed production of *S. intermedius* (1973). Consequently, fishermen co-operatives led a pilot and experimental collection of sea urchin seeds from the wild. However, the number of collected juveniles for cultivation differed considerably. Therefore, collection from the natural environment was not suitable for mass production. Hokkaido Central Fisheries Station developed this kind of collection and implemented it in the marine area from Shakotan Peninsula coast to the Rishiri Island, Sea of Japan. Research of the artificial sea urchin seed production proceeded simultaneously. Although a demand for sea urchin seed of *S. intermedius* increased, the supply of natural seed for transplantation was tending toward *S. nudus*. Supply of *S. intermedius* was remarkably lower than demand for *S. nudus* and unable to cover its needs. A brochure, "A guide book for the artificial seedling production of *S. intermedius*" (in Japanese) was published by Hokkaido in January 1985. Its revised version was published in March, 1992.

S. intermedius and *S. nudus* are artificially fertilised and grown in the northern area of Japan (Hokkaido, Aomori and Iwate). *T. gratilla* is reared from unfertilised eggs to juveniles in the southern semi-tropical area (Okinawa and Kagoshima). In the south area of central Japan (Wakayama, Shimane, Yamaguchi, Ehime, Fukuoka, Saga, Nagasaki, Kumamoto and Kagoshima), the seeds of *P. depressus* are produced. *A. crassispina* and *H. pulcherrimus* are grown in Iwate and Saga. The largest production of sea urchin juveniles is that of *S. intermedius*, followed by *Pseudocentrotus depressus* and *S. nudus*. Approximately 80% of *S. intermedius* juveniles produced in the facilities are released in the fishery.

Sea urchins aquaculture from the egg to the commercial size in the sea cage is not so frequent in Japan. Sea urchins are periodically collected 2 - 3 years after planting the juveniles (10-15 mm diameter) in the limited fish area. The estimation of appropriate sea urchin density based on the productivity of fish farms and algal inhabitants, and elimination of predatory animals before planting are essential. The major predators of sea urchins in the wild are starfish and clubs. During sea urchin growth, fish farms are regularly inspected and managed. Instruction for releasing sea urchin juveniles (seeds) in fish farms is published by Hokkaido Fishery Experiment Laboratory (in Japanese). The systematisation of periodical release into sorted areas and fishing after a given interval is required in order to promote the sea urchin fishery business.

In Hokkaido, the annual production of *S. intermedius* and *S. nudus* in 1999 was 65,238,000 and 234,000, respectively. Twenty-two facilities are currently working for the production of seeds in Hokkaido. The largest facility is capable of producing 8,000,000 ~ 10,000,000 sea urchin seeds per year. A brochure in Japanese, explaining the methods and techniques to grow sea urchin juveniles from the stage of unfertilised egg, was published by Hokkaido Fishery Experiment Laboratory. An English version is expected. Hokkaido Fishery Experiment Laboratory received trainees from foreign countries by support of Japan International Cooperation Agency (JICA).

In addition to rearing starting from the egg, transplantation is also performed. Sea urchins living in an environment with low photosynthetic activity are generally lean. This is why they are transplanted in fish farms with sufficient algae or cultured in the net-cage with feeding. After gonads are grown, the animals are collected and delivered to the market. These measures to improve sea urchins quality are called "chikuyo" (meaning 'keeping and rearing'). Two antagonistic opinions argue that transplantation promotes the decrease of sea urchin resources and that a population in poor environments does not contribute to the proliferation.

Recently, aquaculturists showed that the small number of parental animals for the collection of gametes causes low genetic diversity in fish farming stocks. Moreover, it was affirmed that the adaptability to environments and resistance to diseases decrease. In *S. intermedius*, artificial seed production from both wild animals and cultured ones is responsible for low genetic diversity and inbreeding, due to the small number of parental animals.

When artificial sea urchin seeds are released, competition between them and natural ones may occur. Artificial seeds should be released in a controllable area and fished at the appropriate catching size. On the other hand, resource management for reproduction must be performed intentionally.

Improvement of seabed and development of new fisheries are also important to raise the production of sea urchins. Since the benthic flora subjected to rapid currents and tides may be poor, productivity could be enhanced by intervening in the sea bottom. Recently it was shown that forest lands can contribute to the productivity of the coastal flora and fauna, as they block sand and mud transport to the sea. Since sea urchins are sensitive to decreases in salinity, heavy flood have been responsible for damaging sea urchin resources. Effluents from forests are supposed to enhance the marine flora and fauna health. Fishermen co-operatives successfully take care for re-forestation activities in order to safeguard littoral productivity.

3. CONSUMPTION OF SEA URCHIN ROE

The sea urchin roe is consumed as fresh roe or processed product. In Japan, fresh roe are eaten both raw and cooked. There are various methods of conservation. The product can be consumed as non-heated (preserve) or heated (steamed, baked). Snack flavoured with sea urchin roe are also available.

3.1 *Fresh roe*

Fresh roe for sushi, sashimi and others are usually prepared as follows:

1. Collected sea urchins are opened with knives or special pliers.
2. Viscera other than gonads are removed.
3. Gonads -both ovaries and testes- are saved and washed in salted water (3% NaCl).
4. Gonads are selected and placed in trays (usually wooden ones) to drain water. (Damaged gonads are removed or used differently.)
5. Gonads are soaked in a potassium aluminium sulphate solution. This process is effective to maintain the texture of roe.
6. The roe is ready for the market.

These procedures have been adopted by other countries exporting sea urchin roe to Japan.

Fresh sea urchin roe are mainly consumed as sushi and sashimi. However, they are also used in recipes of Japanese and European high cuisine. Some examples are given below:

Sea bass and sea urchin roe with cream sauce: Sprinkle a small amount of salt and pepper on the cleaned sea bass fillet (approximately 7 mm thick). Cook the fillet with butter on a low flame. When it is almost ready, add white wine, *fumet de poisson* (fish broth) and cream and poach it. Take out the fish and keep it warm. Poach sea urchin roe in the remained sauce and put it on the fish. Concentrate the sauce, season with salt and pepper and drain it on sea urchin roe and fish.

Terrine of sea bream and sea urchin roe: Put sea bream cleaned and cut in a pan with water, *bouquet garni* and fennel. When boiling, stop the flame and leave it for 5 min. Remove the fish and cut it in small pieces. Put the fish in a blender or a food processor with milk, cream, egg white and a pinch of nutmeg. Homogenize all ingredients and season with salt and pepper. Add sea urchin roe and chopped parsley in the mixture; put it in the buttered casserole for terrine and bake in the oven tray, with hot water, at 160 °C for 30~40 min. Serve cool.

Gratin of egg and sea urchin roe: Put one or two poached egg in a buttered casserole. Put sea urchin roe on the egg and cover them with *béchamel*. Bake them in the hot oven for 5 min. Serve warm.

Spaghetti with sea urchin roe: While spaghetti are boiling, fry cloves of garlic in olive oil on a low flame until they becomes golden. At this stage, remove the garlic and put sea urchin roe in the oil without cooking them. When spaghetti are *al dente*, mix them with oil and sea urchin roe and sprinkle with chopped parsley.

Sea urchin roe in a fish ball (uni-manju): Cut fillets of flounder and scallops into small pieces. Grind thoroughly with a blender or a food processor and mix with eggs, starch and small amount of water. Season the mixture with salt. Wrap sea urchin roe with the mixture and shape them into a ball. Steam the ball and serve in the warm fish broth.

Sushi of uni (sea urchin roe): Cook rice and season with rice vinegar. Make small ovals of rice and roll the upper half of ovals with a small sheet of seaweed (nori) to place sea urchin roe. Put the sea urchin roe on top of the oval.

Abalone with sea urchin roe (awabi-uniae): This is one of the most ancient recipes. Sliced abalone and sea urchin roe are mixed in a bowl or a bottle with some salt. Leave the mixture to rest for several hours in a cool place and serve.

Rice with sea urchin roe (uni-meshi): Cook the rice with Japanese fish broth seasoned with soy sauce. When the rice is almost done, add sea urchin roe on the rice and finish cooking.

Squid or cuttle fish with sea urchin roe (ika-uniae): Clean the squid or cuttle fish and fillet it. Cut the fillet into strings. Put sea urchin roe in a small bowl, ground and mixed with salt and *sake*. Add the cut squid (or cuttle fish) in the mixture and leave them to rest for a while.

Prawn roasted with sea urchin roe (ebi-uniyaki): Mix sea urchin roe (preserve or fresh), egg yolk, *sake* and salt (in the case of fresh one) to homogeneous paste. Clean the prawn, remove dorsal viscera, stick a skewer into prawn to avoid bend. Sprinkle a pinch of salt on the prawn and roast it (not too much). Coat the prawn with the paste and roast again to dry it. Repeat coating and roasting twice or three times. Remove the skewer and serve.

Squid or cuttle fish roasted with sea urchin roe (ika-uniyaki): Clean a squid or a cuttle fish and cut it into small pieces. Stick skewers into the piece to avoid bends. Roast the piece, coat the squid (cuttle fish) with the paste of sea urchin roe and egg yolk (same as above) and finish roasting.

Steamed ground fish meat flavoured with sea urchin roe (uni-kamaboko): Cut fillets with white muscle (sea bream, sea bass or others) and grind them thoroughly with a little salt. When the meat becomes a homogeneous paste, add sea urchin roe and grind again. The smooth paste is shaped in cubes on the wooden board and steamed.

Soup of sea urchin roe and abalone (ichigoni): Cook sea urchin roe and slices of abalone in the fish broth seasoned with salt.

3.2 *Sea urchin roe preserve*

Shimonoseki in Yamaguchi prefecture is famous for the preserved sea urchin roe. The modern method of sea urchin roe preserve was invented by Kyushichi Kido about 140 years ago. The methods to preserve the fish and other marine invertebrates had been established long time ago in Japan. The most famous recipe is Shiokara. This is salted and fermented fish or other marine invertebrates, such as squid, ascidia, octopus and others.

Sea urchin roe preserve is substantially categorised in Shiokara. Mr. Kido suggested the addition of ethanol for a longer conservation. Ethyl alcohol blocks proteins denaturation, microbial growth and autolysis by endogenous hydrolases. It also contributes to decrease the amount of salt added and generates the typical flavour of sea urchin roe preserve. The final concentration of ethanol varies from 5 to 10% depending on the amount of salt. A typical method to preserve roe: cleaned roe are put on a wooden board covered with small amount of salt. Salt is sprinkled further on the sea urchin roe. The roe is gently mixed with sprinkled salt and left to rest for a while. Damaging the roe must be avoided. The final amount of salt should be approximately 10% of the sea urchin roe weight. Pour 95% ethanol in a bottle to 1/6-1/4 volume. Put the roe into the bottle and stir ethanol and the roe gently and thoroughly. This process is repeated three or four times to fill the bottle. Stirring must be done immediately before fixation by ethanol to avoid damaging the roe texture.

Sea urchin roe preserves described below are commercially available in Japan.

Tsubu-uni: cleaned and drained sea urchin gonads are salted and preserved with a proper amount of ethanol to retain the texture of fresh roe.

Neri-uni: cleaned and drained gonads are mixed with seasonings and preserved with an appropriate quantity of ethanol.

Shio-uni: This type of sea urchin preserve does not contain ethanol. Therefore, its duration is shorter.

Unizuke (Mixed with other seafood): The neri-uni is mixed with other seafood, such as finely chopped squid (Ika-uni), jelly fish (Uni-kurage, *Rhopilema esculenta*) and herring roe (Uni-kazunoko). Recently, sea urchin roe preserves are also served in the European dietary style as *canapé*.

Steamed sea urchin roe are a popular seafood (Mushi-uni, usually canned), like baked sea urchin roe (Yaki-uni, usually packed in the shell of abalone or cram) and soup of roe with abalone (Ichigoni, canned). Yaki-uni is made with cleaned sea urchin roe put in the shell and baked in an oven heated with charcoals.

Japanese snacks with sea urchin roe are fava beans (Uni-mame) and rice crackers(Uni-arare) flavoured with sea urchin roe.

3.3 *Nutrition*

Sea urchin roe commercially available contain both ovaries and testes in the process for accumulation of nutrients in nutiritive phagocytes for gametogenesis. In general, the egg contains abundant nutrients and materials for embryonic development prior to feeding, and the spermatozoon contains DNA per cell at the rate higher than somatic cells. Therefore, the sea urchin roe is tasteful and nutritious. Nutritional factors of sea urchin roe are given in the following tables. In Table 3, sea urchin roe contain a relatively high amount of unsaturated fatty acids, equal to 73.9%. The sirloin, from which fat tissues are removed, contains 55.1% of unsaturated fatty acids in the total fatty acids.

Table 2. Nutritional Factors of Sea Urchin Roe (per 100 g)

Factors	Fresh roe	Salted roe preserve
energy (kcal)	120	183
water (g)	73.8	51.8
protein (g)	16.0	17.2
lipid (g)	4.8	5.8
carbohydrate (g)	3.3	15.6
ash (g)	2.1	15.6
Ca (mg)	12	46
P (mg)	300	370
Fe (mg)	0.9	1.1
Na (mg)	220	3300
K (mg)	340	280
Mg (mg)	27	63
Zn (mg)	2.0	1.9
Cu (mg)	0.05	0.1
vitamins		
total A (IU)	1170	350
retinol (mg)	150	15
carotene (mg)	1300	1000
E (mg)	3.6	4.2
B1 (mg)	0.10	0.14
B2 (mg)	0.44	0.65
niacine (mg)	1.1	1.4
B6 (mg)	0.15	0.07
B12 (mg)	1.3	5.4
folic acid (mg)	360	98
panthotenic acid (mg)	0.72	1.32
C (mg)	3	0
cholesterol (mg)	290	280
NaCl (g)	0.6	8.4

from Standard tables of food composition in Japan 5th revised edition, by Science and Technology Agency, Resources Council.

Table 3. Fatty Acid Composition per 100g of total fatty acids

Fatty acid	Amount (g)
myristic acid (14:0)	6.1
myristoleic acid (14:1)	0.2
pentadecanoic acid (15:0)	0.7
palmitic acid (16:0)	14.7
palmitoleic acid (16:1)	10.2
heptadecanoic acid (17:0)	0.5
heptadecenoic acid (17:1)	1.3
stearic acid (18:0)	3.2
oleic acid (18:1)	7.5
linoleic acid (18:1)	2.4
linolenic acid (18:2)	1.5
octadecatetraenic acid (18:4)	7.6
arachidic acid (20:0)	0.8
eicoseic acid (29:1)	8.8
eicosadienic acid (20:2, n-6)	1.9
eicosatrienic acid (20:3, n-6)	0.6
eicosatetraenic acid (20:4, n-6)	2.5
arachidonic acid (20:4, n-6)	7.1
eicosapentaenic acid (20:5, n-3)	16.6
docosenic acid (22:1)	0.7
docosapentaenic acid (22:5, n-3)	0.6
docosahexaenic acid (22:6, n-3)	1.1
tetracosnic acid (24:1)	3.1

Fatty acids were extracted in chloroform-methanol solution and estimated by gas chromatography.
From Standard tables of food composition in Japan 5th revised edition, by Science and Technology Agency, Resources Council.

Table 4. Amino Acid Composition (per 100g of roe)

	Amount
protein	15.8 g
isoleucine	600 mg
leucine	910 mg
lysine	940 mg
methionine	360 mg
cystine	250 mg
pheniyalanine	530 mg
tyrosin	550 mg
threonine	620 mg
tryptophane	190 mg
valine	740 mg
histidine	300 mg
arginine	950 mg
alanine	760 mg
asparagic acid	1100 mg
glutmic acid	1500 mg
glycine	2000 mg
proline	470 mg
serine	570 g

Amount of amino acids were analysed by HPLC after hydrolysis in 6 N HCl.
From Standard tables of food composition in Japan 5th revised edition, by Science and Technology Agency, Resources Council.

REFERENCES

Yamabe A. 1962. On the rearing of *Pseudocentrotus depressus* juveniles. Suisanzoshoku 10: 213-219 (in Japanese).

The Sea Urchin: From Basic Biology to Aquaculture, Yokota, Matranga & Smolenicka (eds)
© Swets & Zeitlinger, Lisse, ISBN 90 5809 379 4

Sustainable land-based mariculture in arid environments

A. Neori, M. Shpigel & O. Zmora
Israel Oceanographic and Limnological Research, National Center for Mariculture, Eilat, Israel

ABSTRACT: The basic natural resources required for an algal biofiltration system are low-cost coastal land and labor, and plenty of sunshine. It costs more to remove pollution by algae than by nitrification-denitrification bacterial biofilters. Unlike the latter, however, algal and invertebrate biofilters generate income (that may offset the increased costs), because they produce highly valuable products and increase diversification and income of the mariculture operation. There are large markets for the biofilter products (seaweed, microalgae, shellfish etc.) of the IOLR-NCM integrated mariculture concept. It is therefore economically viable, indicates that the operation is efficient and utilizes the best know-how on the culture of each individual organism and on the proper integration among the organisms.

1. INTRODUCTION

In the last decade, all growth in world fisheries has been in aquaculture (Chamberlain & Rosenthal 1995; Boyd 1999; see also FAO site at http://www.fao.org). The acceleration of this trend is inevitable. Although capture of food grade fish is predicted to remain stagnant at 60 million tons a year, demand for such fish is expected to surpass 100 million tons by 2010. The difference can be covered only by aquaculture. While freshwater availability for aquaculture diminishes, the role of mariculture becomes prominent.

Seacage mariculture nearshore is widely used, but has limitations of suitable sites, weather, and conflicts with natural coastal ecosystems, shipping, tourism and others. Land-based coastal mariculture will take over as a main fish and seaweed production technology, as is already happening with production of shrimp and freshwater fish. Land-based mariculture can turn sparsely populated coastal deserts into economically thriving zones (Nunes & Parsons 1998), putting to a good and sustainable use the supply of sunlight, land, seawater and readily available workforce. Non arable coastal deserts exist in many areas of the world, including southern and northern regions of Africa and Arabia.

At the Israel Oceanographic and Limnological Research, National Centre for Mariculture (IOLR-NCM), on the coast of the Gulf of Aqaba (Elat), Red Sea, we have developed non-polluting pond culture of marine fish (sea bream and sea bass) and penaeid shrimp to the commercial scale. The ponds are integrated with biofilters of microalgae, macroalgae, and algivores (bivalves, abalone, sea-urchins and brine shrimps). The algae-herbivore biofilters remove most of the pollution from the fishpond effluents and turn it into other desired seafood products (Shpigel et al. 1993, Shpigel &Neori 1996, Neori & Shpigel 1999, Neori et al. 2000). The designs are based on algal sunlight-dependent assimilation of excess nutrients, and their conversion into microalgal or macroalgal biomass (Cohen & Neori 1991, Neori et al. 1991, Neori et al.1996). The algae are then sold as primary commodities or fed on-site to the algivores.

2. SUSTAINABLE INTEGRATED MARICULTURE: SCIENTIFIC SETTING

Algae can be perfect partners to fish and shrimp in mariculture, because they remedy the deleterious effects which animals release in water (Fig. 1). Algae use ammonia and CO_2 excreted to the water by fish and shrimp. Algae replenish the water with oxygen consumed by fish and shrimp.

In modern intensive fishculture, the cost of fishfeed protein constitutes the largest expense (FAO site at http://www.fao.org). Fish metabolize most of their ingested protein-N and excrete it as ammonia and feces (Fig. 2). This wasted protein-N is also the main source of nitrogenous pollution from fish and shrimp culture. It is therefore desirable, economically and environmentally, to maximize the recovery of waste nitrogen by productive biofilters into valuable

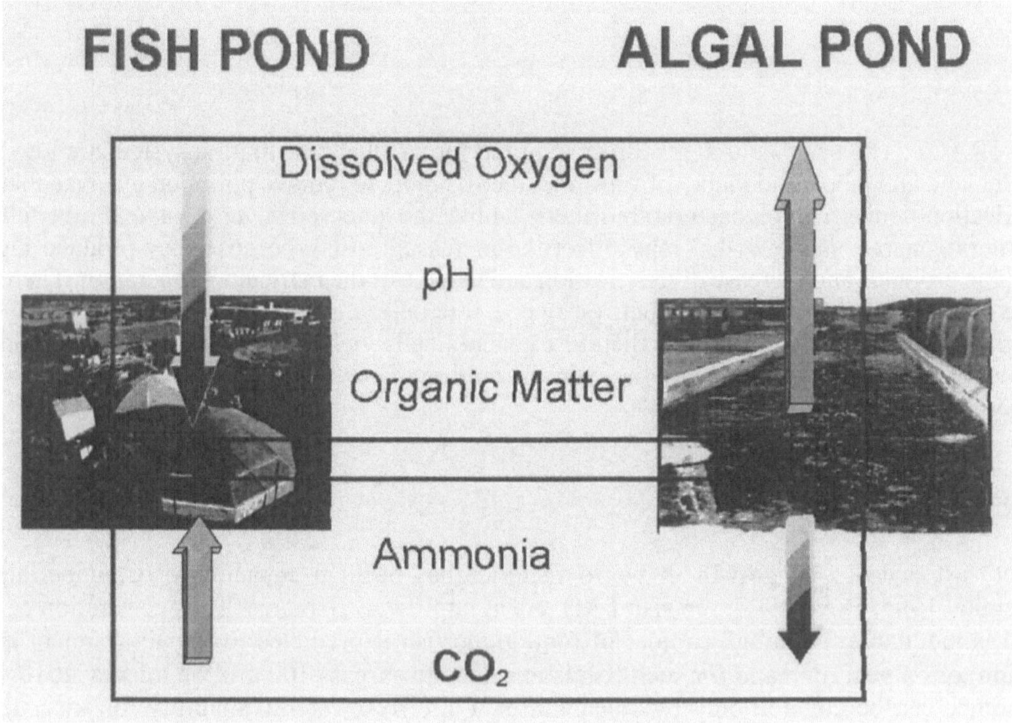

Figure 1. The counterbalancing effects of fish and algae on water quality. Upward and downward arrows indicate rising or lowering effects, respectively, on levels of the indicated water quality components.

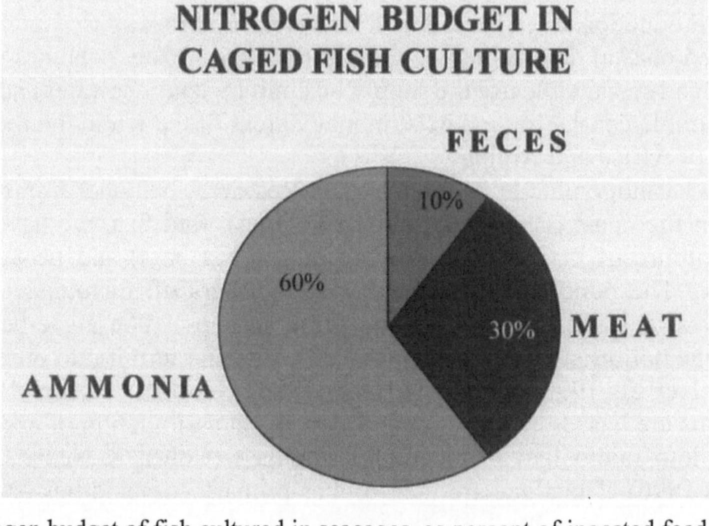

Figure 2. The nitrogen budget of fish cultured in seacages, as percent of ingested feed protein.

products instead of pollution. This is difficult to achieve in seacages. Remediation of pollution, therefore, requires culturing the fish or shrimp in ponds.

Marine fish have been intensively cultured in ponds for years. Sea bream grows in the ponds from 20g to commercial size of 470g in 16 months, with average daily growth rate of 0.6%. Annual mortality averages 10%. The fish are fed on a pelleted diet with average food conversion ratio (consumed dry weight feed per produced live weight fish) of 2, and total annual fish production of >10 kg/ m^3. Intensive pond shrimp farming has also been developed, but is not discussed here.

Two main approaches for sustainable pond mariculture based on biofiltration by algae and invertebrates have been developed at the IOLR-NCM. Both involve recycling of both seawater and nutrient among at least three cultured organisms. Both approaches are based on algal sunlight-dependent assimilation of excess nutrients, and their conversion into microalgal or macroalgal biomass (Fig. 3). The algae can then be consumed, together with other excess organic particulate matter, by algivore invertebrates, such as bivalves, abalone and sea urchins.

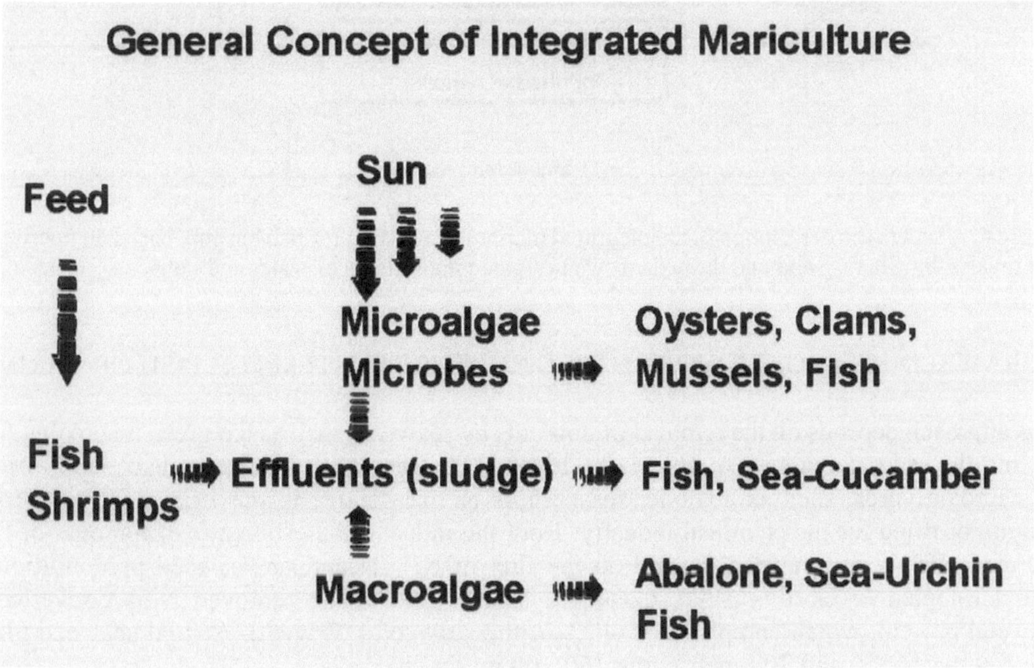

Figure 3. A conceptual diagram of fully integrated algal-based sustainable mariculture.

3. MICROALGAE-BIVALVE/BRINESHRIMP BASED INTEGRATED CUTURE SYSTEMS FOR FISH OR SHRIMPS

This approach is based on the consumption by bivalves of microalgae, which develop in fishponds and in sedimentation ponds that collect their effluents (Fig. 4). The surplus nutrients from the fish culture support dense microalgal populations, and are converted into microalgal biomass. The algae support the growth of bivalves (*Crassostrea gigas* and *Tapes semidecussatus*), that grow on the bottom of the sedimentation ponds or in rearing tanks. Brine shrimp (*Artemia* sp.) can also be produced at high rates on these microalgae. The bivalves, grown in biofilters by our technology, reach commercial size in 18 months, with average daily growth rates of 0.8%. Annual mortality averages 20-30%. Final stocking density is 8-10 kg m^{-2}. Brine-shrimp yields can reach from 1.8 to 3.6 kg m^{-2} y^{-1} in extensive pond culture to 500 kg m^{-3} y^{-1} in intensive tank culture (supported by effluents from intensive land-based fishculture) (unpublished results).

A commercial pilot facility of this design (PGP 1994 Ltd.) has recently begun to operate near the NCM in Eilat, southern Israel.

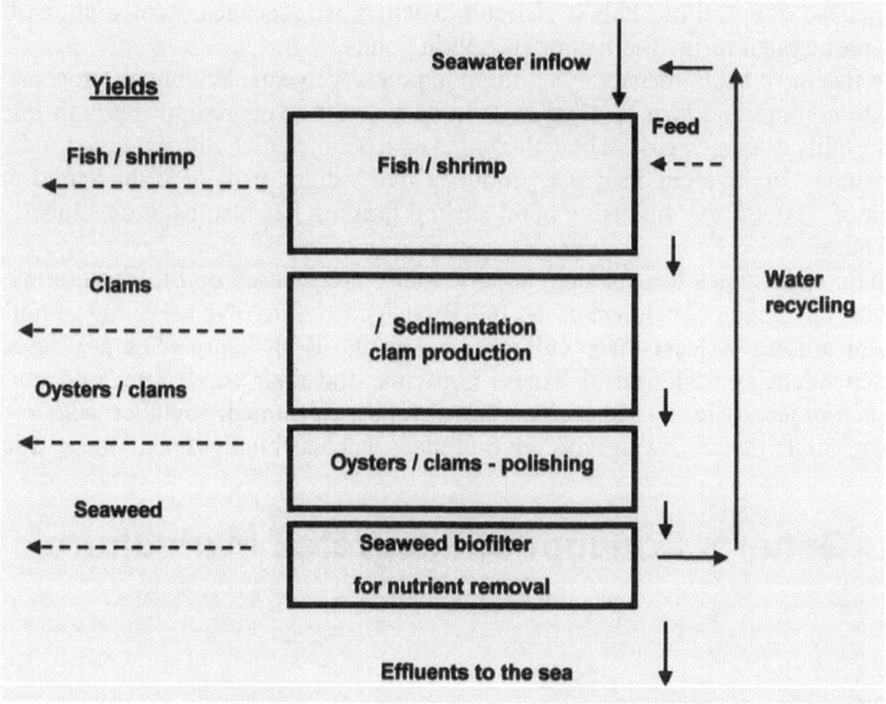

Figure 4. A schematic description of an integrated mariculture system for fish/shrimp, based on microalgae and bivalve biofilters. Solid and dashed arrows designate main flows of water and nutrients, respectively.

4. SEAWEED-MACROALGIVORE BASED CULTURE SYSTEM FOR FISH OR SHRIMPS

This approach depends on the removal of nutrients by seaweed, such as the green macroalga *Ulva* sp. and the red macroalga *Gracilaria* sp. (Fig. 5). The seaweed yield can subsequently be fed to macroalgivores, such as abalone, sea urchins or to humans, in pulverized form to brine-shrimp, or dried for the chemical industry. Both the rate and the efficiency of removal of fish-excreted NH_3-N by seaweed depend on the flux of NH_3-N and on seaweed preconditioning (Vandermeulen & Gordin, 1990, Cohen & Neori, 1991). The removed N is converted to seaweed protein, which comprises 35-45% of the seaweed DW yield. Annual seaweed yields range between 350 and 700-metric tons (50-100 mt dw) ha^{-2}.

Each kg of cultured fish can support by its effluents the yield of 4-8 kg of high-protein seaweed. This crop can sustain the growth of around 0.4-0.8 kg of abalone (Shpigel et al. 1999, Neori & Shpigel, 1999). A fishfarm that annually produces 1000 MT fish can therefore expect to produce 4000-8000 MT of seaweed, which can support the production of 400-800 MT abalone.

Three species of abalone (*Haliotis discus hannai, Haliotis tuberculata and Haliotis fulgens*) have been introduced to commercial size integrated facilities at the NCM. Growth rates attained are similar to or better than those reported for these species by other commercial growers. A commercial facility of this design (Seaor Marine Ltd.) has recently started to operate on the Mediterranean coast of Israel.

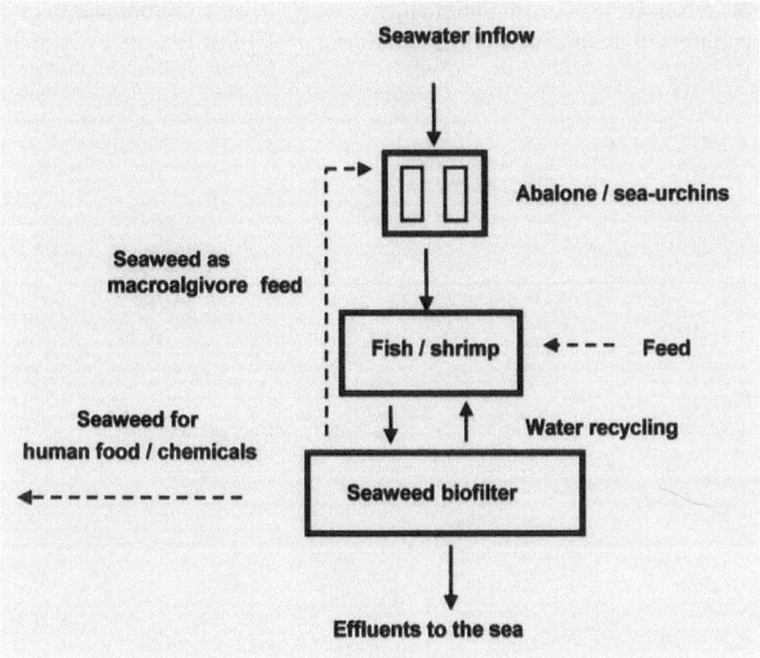

Figure 5. A schematic description of an integrated mariculture system of fish/shrimp and macroalgivores, based on seaweed biofilters. Solid and dashed arrows designate main flows of water and nutrients, respectively.

REFERENCES

Boyd C.E. 1999. Aquaculture sustainability and environmental issues. *World Aquaculture Magazine* June 1999: 10-72.

Chamberlain G. & Rosenthal H. 1995. Aquaculture in the next century: opportunities for growth and challenges of sustainability. *World Aquaculture Magazine* March 1995: 21-25.

Cohen I. & Neori A. 1991. *Ulva lactuca* biofilters for marine fishpond effluents. I. Ammonia uptake kinetics and nitrogen content. *Botanica Marina* 34: 475-482.

Neori A. and Shpigel M. 1999. Using algae to treat effluents and feed invertebrates in sustainable integrated mariculture. *World Aquaculture Magazine* 30: 46-51.

Neori A., Cohen I. & Gordin H. 1991. *Ulva lactuca* biofilters for marine fishpond effluents: II. Growth rate, yield and C:N ratio. *Botanica Marina* 34: 483-489.

Neori A, Krom M.D., Ellner S.P., Boyd C.E., Popper D., Rabinovitch R., Davison P.J., Dvir O., Zuber D., Ucko M., Angel D. & Gordin H. 1996. Seaweed biofilters as regulators of water quality in integrated fish-seaweed culture units. *Aquaculture* 141: 183-199.

Neori A., Ragg N.L.C. & Shpigel M. 1998. The integrated culture of seaweed, abalone, fish and clams in modular intensive land-based systems: II. Performance and nitrogen partitioning within an abalone (*Haliotis tuberculata*) and macroalgae culture system. *Aquacultural Engineering* 17: 215-239.

Neori A., Shpigel M. & Ben-Ezra D. 2000. Sustainable integrated system for culture of fish, seaweed and abalone. *Aquaculture* 186: 279-291.

Nunes A.J.P. & Parsons G.J. 1998. Dynamics of tropical coastal aquaculture systems and the consequences to waste production. *World Aquaculture Magazine* June 1998: 27-37.

Shpigel M. & Neori A. 1996. The integrated culture of seaweed, abalone, fish and clams in modular intensive land-based system. I. Proportions of size and projected revenues. *Aquacultural Engineering*, 15: 313-326.

Shpigel M., Neori A., Popper D.M. & Gordin H. 1993. A proposed model for "environmentally clean" land-based culture of fish, bivalves and seaweed. *Aquaculture* 117: 115-128.

Shpigel M, Ragg N.L., Lupatsch I. & Neori A. 1999. Protein content determines the nutritional value of the seaweed *Ulva lactuca* L for the abalone *Haliotis tuberculata* L. And *H. Discus hannai* Ino. *Journal of Shellfish Research* 18: 227-233.

Vandermeulen H. & Gordin H. 1990. Ammonium uptake using *Ulva* (Chlorophyta) in intensive fishpond systems: mass culture and treatment of effluent. *Journal of Applied Phycology* 2: 363-374.

The Sea Urchin: From Basic Biology to Aquaculture, Yokota, Matranga & Smolenicka (eds)
© Swets & Zeitlinger, Lisse, ISBN 90 5809 379 4

Land-based echinoculture: a novel system to culture adult sea urchins

M.G. Devin
University of Maine, Orono, Maine, USA

ABSTRACT: Green sea urchins (*Strongylocentrotus droebachiensis*) are an important fishery in the Gulf of Maine. Urchin landings have decreased 70% since 1993, while the price per kilogram paid to fishermen has increased from US$1.42 to US$3.03 over the same period. Consequently, considerable interest has arisen to develop sea urchin culture techniques. In this paper a land-based modular sea urchin enhancement system is described. It consists of a series of troughs housed within a PVC frame. The troughs are designed to account for the peculiar morphology and behavior of adult sea urchins, and to minimize cleaning effort, opportunity for disease, and the effects of pre-mature spawning. The system has been tested at two locations. Urchins that are carefully harvested experience low mortality (~2%) in the troughs. Two enhancement experiments demonstrated that the urchin gonad index can be increased 250% in 12 weeks with less than 5% mortality.

1. INTRODUCTION

For the past two decades, sea urchin fisheries have become very valuable and important in the US (McBride 1997) and Canada. Three species of the genus *Strongylocentrotus* are harvested along the West Coast from Alaska to California. In the Northeast, *S. droebachiensis* is harvested from Newfoundland to Massachusetts. Unsexed, whole sea urchins or their processed gonads (roe) are shipped primarily to Japanese markets and also to those of France and Belgium, and sushi markets throughout North America. The processed roe is called uni in Japan, where it ranges from $6/kg to $200/kg based on its quality (Olin 1998). The uni market was worth more *than* $500 millions in 1995. Fresh and processed sea urchin exported from the U.S. exceeded $150 millions that same year (Sono 1996).

With the discovery of the Japanese sushi market, the urchin fishery for the green sea urchin, *S. droebachiensis*, in Maine exploded from a value of slightly over $4000.00 in 1984 to $33 million a decade later. Landing peaked in 1993 at 18,881 metric tons. Since the peak harvest landings have steadily declined. The estimated landings for 2000 are just over 5,600 metric tons (Table 1). Unrestrained fishing in the early 1990s dramatically affected local sea urchin populations, and although, stringent management restrictions have been implemented since 1995, landings continue to drop. Many sea urchin processors believe the overall quality harvested in the Northeast is also declining (Tamaki pers comm).

Sea urchins harvested from the waters off ~~of~~ northeastern North America possess relatively low quality gonads compared to urchins harvested from Japanese waters and the west coast of North America. The inferiority is due to issues that can be classified as either extrinsic or intrinsic factors. Extrinsic factors include processing, presentation of the product and post-processing procedures. These are completely industry controlled and dependent upon intrinsic characteristics. Intrinsic characteristics include roe yield, roe color, taste, texture, and firmness. Coupled together, these factors define roe quality. Urchins harvested from the Gulf of Maine often possess low roe yield (< 10%) and poor color (white or brown), taste bitter, and lack the desired texture and firmness to satisfy the Japanese palette.

Efforts to improve roe quality have come from three arenas: government (fishery regulations), education (fishermen training and experience), and academia (research). Over the past four years, the Maine Department of Marine Resources has reduced urchin landings by cutting the number of fishing days and decreasing the number of licensed urchin fishermen through attrition. Additionally, minimum and maximum harvest size (test diameter) limits have been implemented. There is anecdotal evidence from fishermen and field researchers that these measures have not prevented further decline in the natural urchin populations. Certainly, the value of urchins from northeast North America has not increased relative to species harvested in other locations.

In the infancy of Maine's urchin fishery, urchins were harvested without regard to their roe quality. Tons of poor quality roe were collected, processed and shipped to the Japanese markets. This caused two major problems, which the Maine urchin industry has yet to overcome. First, it established Maine's product as inferior to other regions. For instance, the green urchins harvested from the west coast of North America, although the same species *(Strongylocentrotus droebachiensis*), garners a higher price. Second, the collection of vast quantities of low quality urchins hastened the depletion of the resource. Fishermen have now learned which microhabitats produce the best roe. They are also aware of when roe quality peaks at specific harvest sites. The fishermen's corporate knowledge benefits the fishery two-fold: low quality urchins are no longer needlessly harvested, and the overall quality raw product provided by Maine fishermen has improved.

Academia has provided insight into urchin ecology for the past three decades (e.g. Vadas 1977, Harris 1981, and Martin et *al.* 1988). However with the emergence of a value fishery and the prospect of overfishing, researchers have expanded their focus to include aquaculture. In Newfoundland, Canada, where gonad indices rarely exceed the minimum acceptable level, focus has been on enhancement and ranching (Hooper et *al.* 1997). Other areas of development include feed development (Robinson and Colburne 1997), hatcheries, and economic projections (Pearce 2001), gonad manipulation (Walker and Lesser 1998, Devin and Walker 200, Garrido and Barber 2001,) and characterization of vitellogenesis (Walker et *al.* 1998, Walker et *al.* 2001) and how newly discovered techniques to manipulate urchins gonads can be applied to aquaculture (Walker and Lesser). An experimental sea urchin hatchery has been established at the University of New Hampshire (Harris 2001). The Peacock Canning Company in Lubec, Maine has established a hatchery that produced 50,000 of juvenile urchins from last spawning season. They expect to reach commercial-scale urchin seed quantities in 2003 (Peacock pers comm.).

Even though the quality of wild-harvested roe is often lower than what it was five or ten years ago, the unit value of harvested urchins has steadily increased during the same period (Table 1). These conditions strongly support the development of sea urchin aquaculture in the United States to offset declining landings and lower quality roe. Lawrence (1999) evaluated strategies and life history characteristics of *S. droebachiensis* and determined that it is a likely candidate for aquaculture. Another factor that makes *S. droebachiensis* a logical choice is that of the 21 species of sea urchins harvested in and imported into Japan, *Strongylocentrotus* spp. accounted for between 46% to 72% of total landings of sea urchin species from 1985 to 1992 (Sono 1996). It is a well-studied animal and a ready market exists for the product.

Although researchers and aquaculturists are investigating the culture of all aspects of the green urchin's life cycle, there is particularly strong interest to develop a land-based holding system for adult sea urchins. Such a system could be used to enhance roe quality, thus increasing the value of raw urchin. Additionally, aquaculturists have interest to control gametogenesis, maintain broodstock for extended periods of time, culture sub-adult and adult urchins, and hold live product when market prices dip temporarily. In this paper, a novel sea urchin roe enhancement system, the Acadia Seafood International (ASI) sea urchin enhancement system is introduced.

Table 1. Sea urchin landings, total value, and price for the Maine urchin fishery (Statistics provided by Maine Department of Marine Resources).

Year	Landings metric tons	Total Value US$	Price US$/kg
1987	653	$ 236,391	$ 0.36
1988	2,822	$ 1,758,805	$ 0.62
1989	4,276	$ 3,608,474	$ 0.84
1990	6,000	$ 5,955,975	$ 0.99
1991	9,315	$ 11,158,425	$ 1.20
1992	11,992	$ 15,148,449	$ 1.26
1993	18,881	$ 26,816,313	$ 1.42
1994	17,371	$ 33,083,336	$ 1.90
1995	15,544	$ 35,604,275	$ 2.29
1996	10,001	$ 23,986,181	$ 2.40
1997	8,362	$ 20,178,238	$ 2.41
1998	6,828	$ 17,072,498	$ 2.50
1999	7,001	$ 20,300,142	$ 2.90
2000	5,617	$ 17,018,200	$ 3.03

2. DESIGN CONSIDERATIONS

To design a functional urchin holding system one must first understand some basic sea urchin morphology and behavior. Regular sea urchins, like those harvested commercially, are benthic grazers. They possess a highly developed feeding apparatus known as Aristotle's lantern. It contains five extremely hard, pointed teeth which are adapted to scrape surfaces very efficiently. New tooth material is produced at a rate of about one to 1.5 mm per week (Barnes 1987). This allows urchins to feed on hard surfaces. Several species can bore into rocks, e.g., *Paracentrotus lividus*, and in laboratory settings they have been known to do considerable damage to tank surfaces. Many commercial urchin species have muscles that have the ability to protrude and retract the lantern. This movement enables the urchin to pull and tear in addition to scrape. Urchins with this ability, such as, *S. droebachiensis* are able to consume macroalgae, detritus, and small invertebrates, e.g., mussels. Although algae are the most important food, commercial urchin species are omnivores and consume a wide variety of plant and animal material in their diets (Lawrence 1975). Urchin fishermen know that sea urchins collected from mussel beds in the Gulf of Maine typically contain higher quantities of roe than neighboring urchins not feeding in the mussel bed. This anecdotal evidence correlates with results from Meidel et al. (1999) who found that urchins fed a diet that includes mussel produced much larger gonads than urchins fed plant material exclusively.

Most sea urchins have a planktonic larval phase that lasts two weeks or longer. *S. droebachiensis* will settle and metamorphosis in the laboratory after 21-30 days. Once metamorphosis occurs, juvenile and adult urchins possess no swimming apparatus. Consequently, they are exclusively benthic and never return to the water column. They tend to seek vertical and upside-down surfaces in tanks. For instance, in a standard rectangular tank, more urchins will be found on the sides of tank than on the bottom, with the highest density occurring in the corners. They also tend to congregate in groups and not disperse throughout the surface area available (Fig. 1). Any tank system designed to culture sea urchins should account for their unique feeding, movement, and congregation behavior.

There is a long history of using the sea urchin as a model for embryological studies dating back to the mid-1800s (Gustafon and Wolpert 1963). Consequently, more may be known about the early development of sea urchins than about any other organisms. The reason for this is that proto-development of echinoderms and vertebrates is similar, specimens are easy to obtain from coastal waters, they spawn readily, and yield large numbers of eggs (often > 10^6). The eggs complete meiosis in the ovary. Thus they can be fertilized immediately upon release from the gonad by adding sperm to an egg suspension.

Historically, most urchin studies ceased after embryological development. Raising echinoplutei through metamorphosis was difficult. Ralph Hinegardner (1969) recognized that if larvae could be raised confidently, it would be possible to apply genetic techniques to the study of sea urchin development. He described techniques to raise sea urchins through six phases that start with the fertilized egg and end with the growth of the young urchin to a reproductive adult (Hinegardner 1969). The maintenance of reproductive adults is the phase that interests many potential urchin culturists.

Leahy et al. (1978) were the first to describe a large-scale laboratory maintenance system for gravid sea urchins, which they called the Kerckoff Marine Laboratory (KML) maintenance system. They required a steady source of gametes for studies of animal development that wasn't dependent upon collecting animals from the field for each experiment. Additionally, they hoped to extend the period that females are gravid.

The KML system consists of 15 shallow Plexiglas tanks that measure 102 cm by 163 cm by 25 cm deep. Each tank is insulated with 5 cm thick Styrofoam. The system consists of five levels with three tanks on each level. Each tank holds two perforated Plexiglas housing units (84 cm by 70 cm by 19 cm) that are removable. About 100 adult *Strongylocentrotus purpuratus* can be maintained within each of the 30 housing units. The entire system houses approximately 200 kg of sea urchins. Ambient seawater temperature is maintained by a chiller tank that is cooled by a 10,000 BTU compressor (Leahy et al. 1978).

Some groups of animals were held over two years in the KML maintenance system. Gravid animals were maintained in a fertile state for many months. Leahy et al. (1978) found that the urchins as a group were able to regenerate usable quantities of mature oocytes several times at one to two-month intervals. This system, although expensive to build, inefficient in the biomass that it can hold relative to later designs, and labor intensive to maintain confirmed adult sea urchins can be maintained in a land-based system for long periods and demonstrated the potential for gonad or roe enhancement in a land-based system (Leahy et al. 1978).

Numerous aquaculturists and fishermen in the northeast Atlantic coastal areas have attempted to culture adult sea urchins over the past decade. Efforts focused primarily on improving roe quality. Would-be-urchin-enhancers started with tanks designed to hold other species. It was quickly learned that standard rectangular tanks, including raceways, are very inefficient for culturing sea urchins. The urchins tend to cling to the walls and corners, and even when there are more urchins than surface area, they will congregate and pile on top of one another rather *than* distribute themselves along the bottom of the tank (Fig. 1). They never enter the water column; consequently much of the tank space, i.e., most of the water column, is under-utilized.

Sea urchins are slow moving. For example, *S. franciscanus* off the California coast moves 7.5 to 50 cm per day (Barnes 1987). Their sedentary nature poses a difficult problem for feed

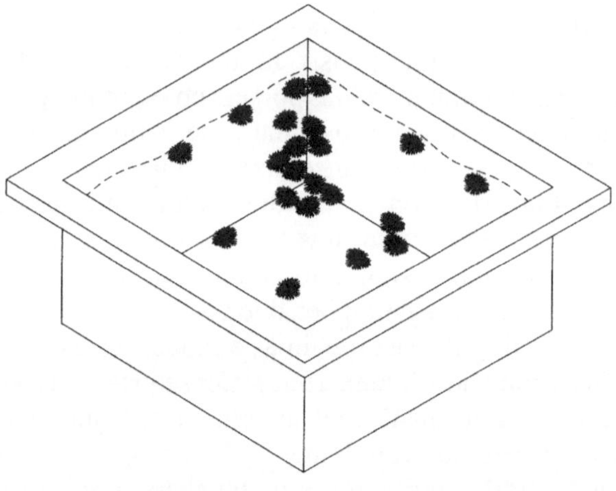

Figure 1. In rectangular tanks, sea urchins tend to congregation on vertical surfaces, in corners and near the air-water interface. Relatively few urchins crawl to the bottom unless they are searching for food.

delivery. They can not quickly collect or catch feed such as finfish or shrimp. Nor are they able to affect water movement like bivalves that filter water. Food must be delivered directly to the animal and in a manner that allows the urchins to grip and manipulate it. Rectangular tanks are not good for this purpose. The vertical walls, where most urchins tend to congregate, provide an opportunity for urchins along the top of the wall to receive feed directly (Fig. 2). All other food falls to the bottom of the tank and the remaining urchins must then make their way from the tank side and onto the tank bottom. This is inefficient and, in many cases, ineffective.

Several researchers have installed baffles or vertical plates to increase available surface area for urchins. This alteration does increase the number of urchins that a tank can accommodate. But feed delivery is difficult because only the top urchins can be fed directly, as previously described (Fig. 2). Tank maintenance becomes more difficult than a tank without baffles, because waste products collect and coagulate between the baffles and is hard to remove when the tank is filled with urchins.

Another approach attempted was to coat the baffles with a food source. Gelatin and agar, among other binders have been used to attach the feed to the baffle surface. Urchins often got their spines stuck in the coated surface, dislodging the feed from the baffle. Much of the feed was lost as the urchins were unable to dislodge it from their spines or it dropped to the tank bottom.

In an effort to augment surface area with a tank, cages have been utilized. This dramatically increases the surface area available to sea urchins. Researchers favor this method because it also increases the number of experimental units in an experiment. However, the cages must be removed to deliver feed, which is labor intensive. Feeding becomes difficult if cages are stacked on top of one another. Detritus, uneaten feed and waste products clog cages causing health issues.

3. LAND-BASED SEA URCHIN ENHANCEMENT SYSTEMS

As early as 1982, LaGall and Bucaille (1989) investigated land-based sea urchin culture because the market value of sea urchins increased as landings decreased. They based the potential for urchin farming on the high price urchins garner in the market, the ability of urchins to

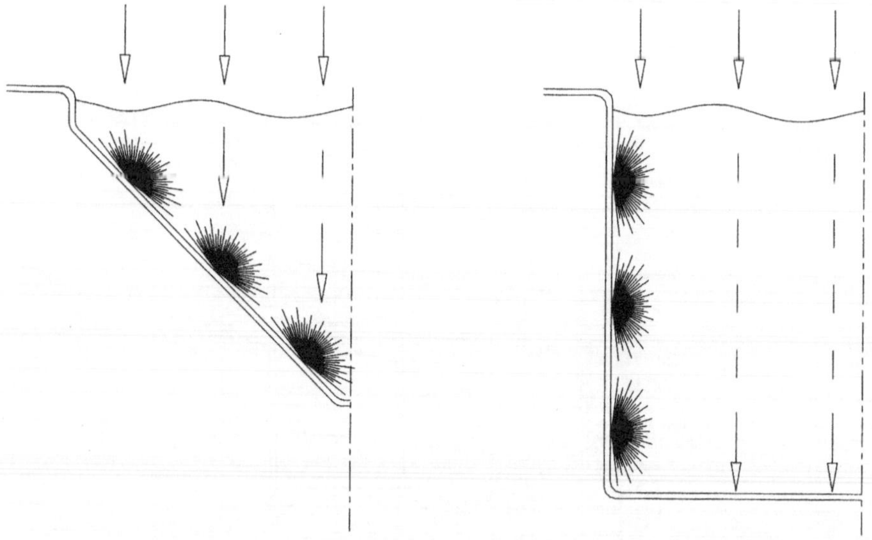

Figure 2. Vertical walls (cartoon on the right), where most urchins tend to congregate, provide an opportunity for urchins along the top of the wall to receive feed directly. Most of the food falls directly to the bottom of the tank. However, when the wall is angled (cartoon on the left), all urchins clinging to the wall receive food directly. This is especially important when using palletized artificial feed.

survive on almost any type of food, and their rapid growth. Over the past 12 years, three unique land-based designs have been introduced specifically for the maintenance and culture of adult sea urchins. They include a tray or toboggan system (La Gall 1990, Grosjean et al. 1998), a cylindrical design (Pennington unpub), and the ASI sea urchin enhancement system, which is introduced in this paper. These designs specifically address culture issues unique to sea urchins including behavior and morphology.

The shallow tray or toboggan design, as introduced by Le Gall (1990) is a series of trays or toboggans stacked on top of one another. Each tray is sloped to ensure that water flows by gravity. The basic design is shown in Figure 3. The top tray is fed from a reservoir tank by a pump. Each successive tray is fed from the tray directly above it until the water returns to the reservoir in a zigzag fashion. Grosjean et al. (1998) built and operated an experimental closed-cycle system on a pilot scale. Each tray or toboggan, as they refer to them, measured 4-m long by 60-cm wide. The depth ranged from 5 to 10 cm with each toboggan holding approximately 500 L of seawater. In their pilot study, urchins were raised from metamorphosis through enhancement in toboggans. Adults were stocked at a density of 250 m^{-1}. If the system holds adults exclusively, it has a capacity of 117 kg (Grosjean 1998). General characteristics are listed in Table 2. A similar system is presently undergoing research and development at the University of Southern Alabama (Watts 1998).

Richard Penfold of Stonington Sea Products, Inc. (Stonington, Maine, USA) has designed a cylindrical tank system in which to enhance sea urchin roe (patent pending). The tank is designed with 2.7 m diameter and 1.6 m depth. Its base is cantered towards a 10 cm central drain well. It is supported 35 cm off the floor with an elbow joint at the waste removal outlet on the underside of the tank. A false bottom provides support for a spiraling wall and 680 kg of additional weight. The spiral wall is 1.55 wide (deep) with a 15.5 cm gap separation between spirals. The spiral wall is held in place by parallel supports countersunk into the wall and bracketed to the wall in such a way that the spiral wall can be lifted from above, to facilitate harvesting, while holding the full weight of the sea urchins.

Water is supplied at the top of the cylinder in the direction of the spiral and cycles around the spiral as it flows towards the drain. The urchins are fed by feeding fronds draped over the spiraling wall. The feeding fronds, which contain artificial food, are designed to mimic

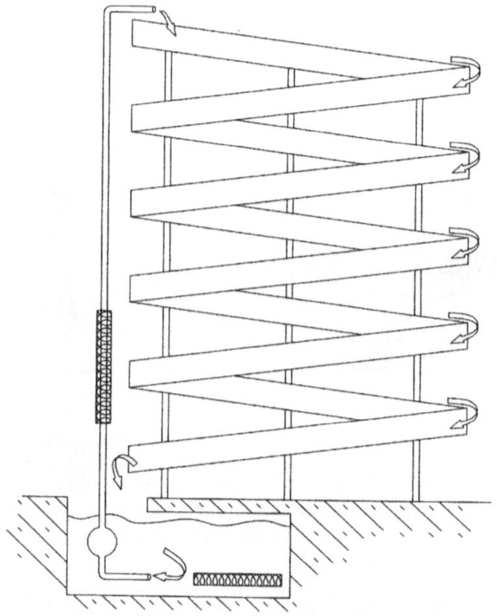

Figure 3. Diagram of a sea urchin culture unit utilizing trays. Water is pumped from the reservoir up to the top tray. It then flows in a zigzag as it returns to the reservoir (Redrawn from LeGall 1990).

kelp. The system holds 8,400 liters. Pumped seawater fed into the system at 70 liters / min replaces 90% of the water every four hours. The system, which has an estimated capacity to enhance 620 kg of sea urchins, costs around $5,500. It has yet to be tested, however a feasibility study demonstrated that the design holds potential. Theoretical design parameters are listed in Table 2.

The trough design introduced in this paper, was initially devised in 1997 by Mr. LeRoy Creswell of Harbor Branch Oceanographic Institute, Florida, USA. He conceived of a sea urchin tank system that included a trough within a tank or raceway. That concept was later abandoned in favor of a system in which the troughs themselves served as the water holding system. Modified rain gutters were initially used to test the concept. Each gutter was approximately 3 meters long and capped at both ends. The bottom of the gutter measured 8 cm wide and the sides were 15 cm. Each side flared out so that it formed a 135° angle with the bottom. At the discharge end of each gutter, a 2.5 cm hole was cut to accommodate a standpipe to control the depth of water in the gutter. Water was supplied by a 4 mm nozzle and flowed unidirectional from the supply to the discharge end of the trough. The system was tested at the Darling Marine Center in 1998.

Table 2. General characteristics and capacities of the three land-based urchin culture systems discussed in the paper. The toboggan and trough systems have been tested on pilot-scale. Both are modular systems.

Design	System Footprint meter2	Water Volume liters	Urchin Capacity kilograms	Water Flow l/min	Cost of System $US	Reference
Toboggan or tray system	3.5	1,500	117	?	?	Grosjean et. al 1998
ASI troughs	7.0	1,240	200	24	$4,000.00	This paper
Circular design	8.5	8,400	390	70	$5,500.00	Unpublished

After several trials using modified rain gutters were conducted at the Darling Marine Center of the University of Maine in Walpole, Maine, U.S.A., a revised trough system was designed and built in 1999 (Fig. 4). The new design includes a drainage channel that runs the length of the trough (Fig. 5). Also, to increase the flushing rate during cleaning a drainage sump was added at the discharge end of the trough (Fig. 6). Broad Bay Tooling, Inc. (BBT, Waldoboro, Maine, USA) engineered a mold and fabricated several test trays.

Each trough is 3.05 m long. The overall length includes a housing channel that is 2.97 m long with a drainage sump attached at one end that adds an additional 8 cm (Fig. 5). The sides of each trough are 21 cm. They are separated by 3.1 cm at the base and fan outward at a 30° angle so that they are 17.8 cm wide at the top (Fig. 7). This angle ensures that all sea urchins are at least partially exposed in the vertical direction which permits simultaneous feeding of all urchins. At the bottom of each trough there is a drainage channel which runs the entire length of the housing channel. The drainage channel is 3.1 cm wide and 3.2 cm deep. The drainage channel collects feces, uneaten food, and detritus. It often becomes anoxic due to the build up of organic matter; however, it is too narrow for an adult sea urchin to enter. The drainage channel is easily cleaned by draining the trough and flushing with water sprayed down the center of the trough from the supply to drainage end.

A drainage sump is located at the discharge end of the trough that measures 18 cm by 8 cm by 21 cm deep (Fig. 6). In the center of the sump is a 38 mm drain with a fitting. A standpipe is inserted into the fitting to control the depth of water in the trough. The large drainage line allows easy and thorough flushing of the drainage channel.

Figure 4. Photograph of an ASI sea urchin enhancement module set-up in the Flowing Seawater Laboratory at Darling Maine Center in Walpole, Maine, USA.

The troughs are rigidly connected in sets of three and called trays (Fig. 8). There is a 2.5 cm lip around the outside of each trough so that when three troughs are affixed together a tray is 69 cm wide. This arrangement allows an average-sized worker to reach across the three troughs. A frame constructed of PVC piping houses the trays (Fig. 4). PCV piping is used because it can support the weight of the troughs when they are filled with water and it does not rust or rot. Maintenance and cleaning of the frame is minimal, and assembly and disassembly is easy and quick, making the system portable if necessary. The ASI PVC frame holds four layers or tiers of trays with two trays on each layer (Fig. 9). The tiers are slightly inclined to promote water flow with the supply end slightly elevated above the drainage end. The frame system accommodates 24 troughs. Each trough supports a volume of 51.6 cubic liters and holds approximately 120 adult green sea urchins (~13 kg biomass), giving the system the ~310 kg of biomass capacity.

The troughs, although physically molded together, each maintain its own water intake, which is supplied through a supply manifold (Fig. 10). The supply manifold is fed by a 5 cm water hose. The manifold in turn feeds each individual trough through a 6 mm nozzle, which can be closed to isolate the trough. Oxygenation occurs through the turbulence caused by the nozzle (Fig. 10). They also drain directly into a drainage manifold, which effectively isolates each individual trough. This design minimizes the effects of disease outbreak and premature spawning. Experience has shown that a flow rate of 1.5 liter*min.$^{-1}$ is adequate to support a trough filled with sea urchins.

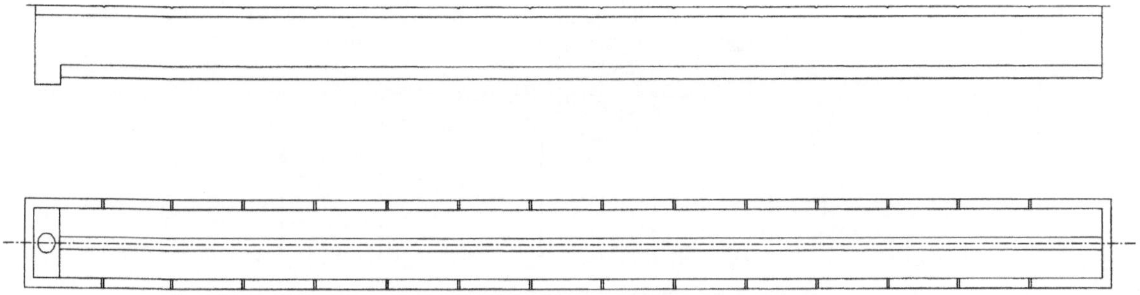

Figure 5. Side view and overhead view of ASI sea urchin troughs. The supply end is to the right and the drainage end to the left. A drainage channel runs the length of the trough.

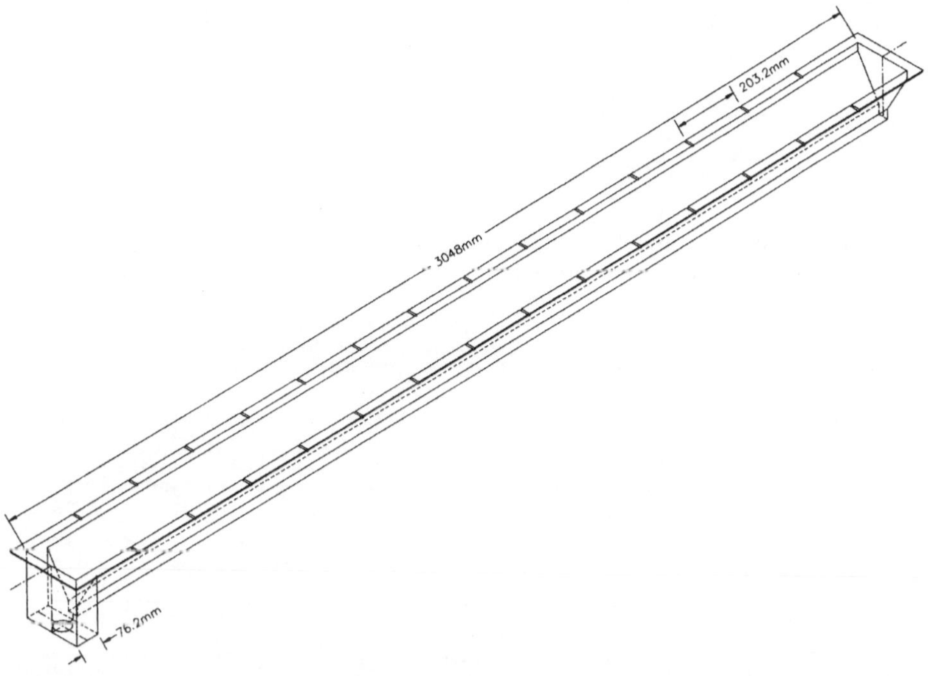

Figure 6. The drainage sump extends 76.2 mm beyond the trough. Uneaten food, feces and debris is flushed from the trough into the drainage sump.

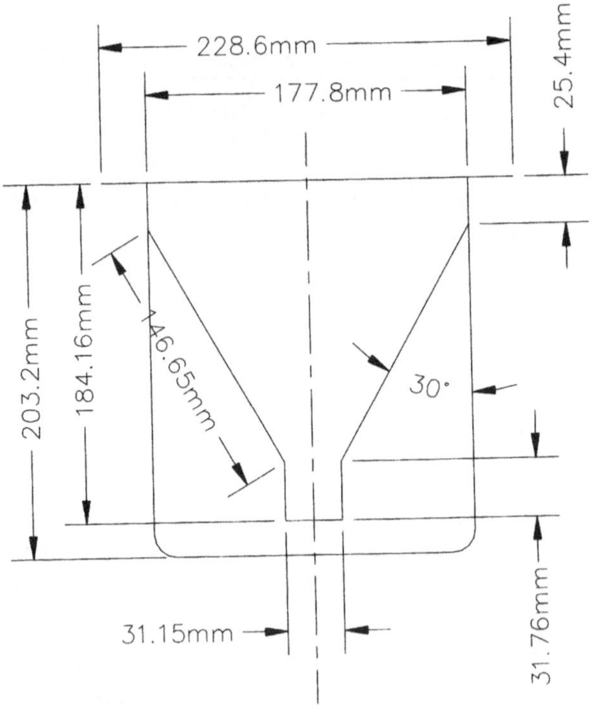

Figure 7. Trough end view. The trough sides are separated by 31.15 mm at the top of the drainage channel and fan at a 30° angle so that they are separated by 117.8 mm at the top.

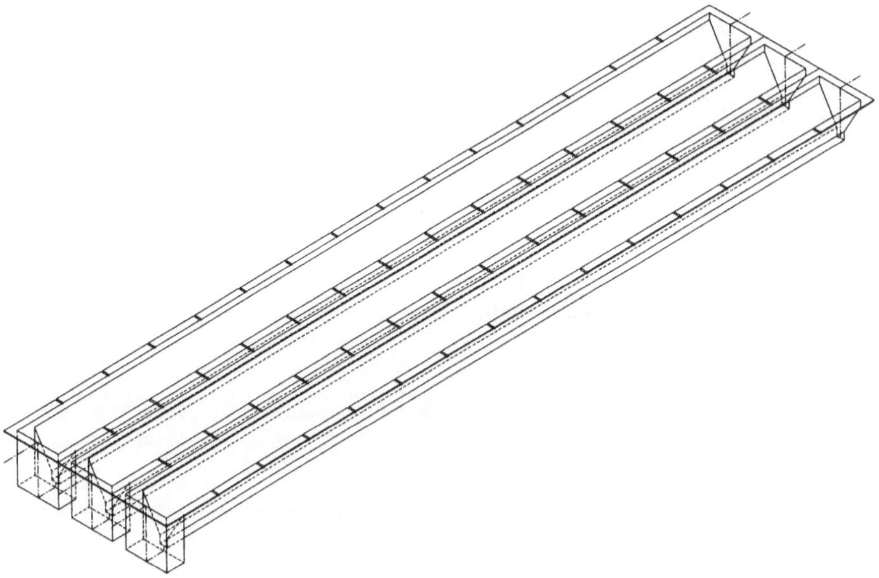

Figure 8. Diagram of three urchin troughs ganged together, which is called a tray. Trays have a width (69 cm) that can be spanned easily by an average sized worker.

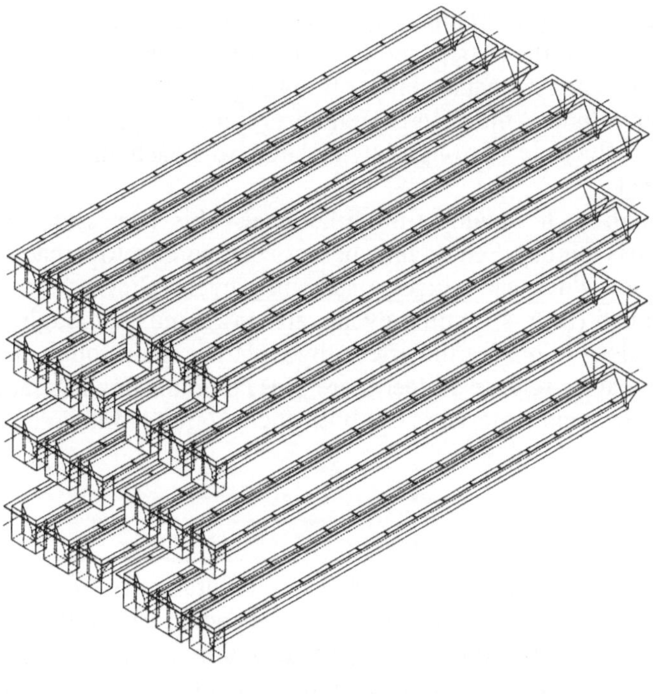

Figure 9. Troughs are ganged together in sets of three and referred to as trays. Modules consist of four levels of trays, with two trays on each level.

Figure 10. Flowing seawater is supplied through a supply manifold fitted with a 6 mm nozzle. Oxygenation occurs through the turbulence caused by the nozzle.

In their natural environment urchins often hide in crevices and under rocks. In tanks they are typically found on the walls as opposed to the bottom of the tank, and will often congregate near the water surface. The ASI enhancement trough optimizes the inclined surface area while minimizing the water volume – the design itself ensures all urchins are on a vertical surface and near the water surface. Because all troughs are isolated from each other, the potential for disease to spread is minimized.

Water flow through each trough is unidirectional. This increases flow rate, which may increase growth (Harris unpub). Sea urchins possess chemoreceptors that sense sperm released into the water column, which can trigger a spawning event. Since the marketable portion of sea urchins is the gonad (roe), spawning events are unwanted. Although this design does not prevent spawning, it does minimize the effects by isolating each trough. The only urchins that are affected by a premature release of sperm are those animals downstream. Cleaning effort is also reduced, since unwanted material is easily flushed from the drainage channel.

4. DISCUSSION

Mortality in ASI sea urchin enhancement modules varies dramatically depending on the method of harvest. Hand collected urchins harvested gently, handled carefully, and maintained below 15° C during transport, usually experience less than 2% mortality. On the other hand, urchins harvested with an urchin drag can experience mortality rates as high as 90%. We conducted a study to determine rates of mortality based on harvest method. An ASI sea urchin enhancement module was set-up at the Peacock Canning Company in Lubec, Maine. On March 26, 2001 green sea urchins were collected off Estes Head in the Cobscook Bay channel (depth 34m) by an urchin drag, brought to the surface, released on deck and shoveled into fish totes. A total of 80 kg of urchins were harvested for this experiment. The surface water temperature was 1° C. The air temperature ranged from 3° to 5° C and seas were calm. The urchins sat on deck for 90 minutes while transported to the cannery. When the urchins arrived at the cannery their overall condition was poor. Many tests were cracked, they suffered significant spine loss, and individuals were not reactive. They were transferred to urchin troughs as carefully and quickly as possible.

Sea urchins were also collected by hand on March 26, 2001 at low tide along the breakwater behind the Peacock Cannery by Mr. Michael Devin and Mr. Hank Stence. Approximately 300 urchins were harvested. The hand-collected animals were carefully removed from submerged rocks and placed in plastic buckets. They were then transferred to the urchin troughs. As a group, they were reactive and looked healthy.

The mortality of urchins collected by the urchin drag and hand-collection was monitored and recorded weekly. The urchins from both sites were similarly sized, although they were not aged. After three weeks, the urchins collected with an urchin drag had sustained 76% mortality, while only 2% of the hand-collected urchins had died (Devin 2001). The hand-collected urchins still looked robust and were active after two months, while the sea urchins that had survived the urchin drag still looked unhealthy. These results are typical of our experience and underscore two important points. First, they demonstrate that the ASI sea urchin enhancement system can adequately maintain urchins. Second, mortality in the troughs is highly dependent upon the collection method. For this reason, we ensure that the urchins used in our system are collected by divers who are properly trained to minimize the stress experienced by sea urchins in the process of collecting and transporting sea urchins into the ASI sea urchin enhancement system.

Numerous studies have shown that using natural and prepared diets can enhance the quality of sea urchin gonads (roe). We tested the ASI sea urchin enhancement module to determine if it could be used as a culture system to improve urchin roe. The combined effects of light and diet on gonad growth and quality were investigated in two experiments that were conducted at the Darling Center in Walpole, Maine in an ASI sea urchin enhancement module (Figure 4). The first occurred in the winter (Nov-Jan) of 1999-2000, when *S. droebachiensis* undergoes the Gametogenesis and Nutritive Phagocyte (NP) Utilization stage (Walker et al., 2000). Animals collected off the coast of the Pemaquid Peninsula in Midcoast, Maine (initial GI = 8.6 +/- 3.5%) were fed either kelp (*Laminaria sp.*) *ad libitum* or a prepared diet (1% BW*day^{-1}). There were two light treatments: Ambient photoperiod and fixed photoperiod (12hr dark/12hr light). Urchins fed the prepared diet and maintained under fixed light (mean GI = 24.6 +/- 4.4%) achieved the largest gonadal index. The second largest gonads resulted from the treatment that received prepared diet and ambient light (GI = 20.8 +/- 3.4%). Fixed photoperiod produced a mean GI that was 3.8% greater than in those animals maintained under ambient photoperiod ($p<.02$). Industry experts were asked to evaluate four criteria of gonad quality: color, firmness, taste and texture. A scale of zero (lowest) to five (highest) was used. Mean scores were computed from panelists grading sheets and then normalized to a scale of zero to 100. The prepared diet yielded gonads of higher quality for each criterion tested according to the evaluators. The fixed photoperiod also produced firmer gonads (Devin and Walker unpub).

The experiment was rerun in the summer of 2000 during the Pre-Gametogenesis and NP Renewal stage, before gametogenesis normally begins in the green sea urchin (Walker et al., 2000). Sea urchins (initial GI = 6.2 +/- 2.1 %) were placed in tanks and maintained under the same four treatments as outlined for the previous experiment. Urchins fed the prepared diet increased their GI to 15.1 +/- 2.6%, while the kelp-fed urchins showed only a small increase. Once again, the prepared diet out-performed the kelp diet in the four criteria analyzed by industry experts. However, in this experiment, there was no significant difference between fixed photoperiod and ambient photoperiod (Students t-test, $t = .67$, d.f. = 22) (Devin and Walker unpub). The mortality observed in both experiments was less than 5%. The results of these two experiments suggest that urchins containing higher quality gonads of both sexes can be produced in the ASI sea urchin modules both during and between normal, annual gametogenesis, with little mortality.

It has been shown that the urchins can be maintained in a healthy condition with low mortality in the ASI sea urchin enhancement modules. With proper diet and physical parameters, such as controlling the photoperiod, roe enhancement in terms of size and other traits is possible.

Any cultivation system designed to specifically raise or enhance adult sea urchins must address critical morphological and behavior issues. They are sedentary animals that have a propensity to congregate on vertical surfaces near the waterline. Any successful system will account for this specific behavior. Tray or toboggan systems are shallow so that the vertical surfaces are very limited. All feed can be applied to all urchins simultaneously. This system has a small footprint compared to the Penfold design and ASI system (Table 2), however the Penfold design is the most space efficient supporting 45.9 kg / m^2, whereas the tray and ASI trough systems only support 33.4 and 28.6 kg / m^2, respectfully. It should be noted that the both the tray and ASI trough systems are stackable and thus could support a higher biomass provided that the overhead space is available. This is particularly true since they require much less water volume than the Penfold design.

A major existing disadvantage of the Penfold design is that the tray and ASI trough systems have been built and tested on a pilot-scale. It has been shown that they are both capable of maintaining and enhancing adult sea urchins, with relatively low mortality. Cleaning and feeding of these two systems is simple and easy. Feeding for the Penfold design includes the feed frond preparation. The feeding fronds are designed to permit the Penfold culture tank to be deeper. The Penfold circular tank has a depth of 160 cm. The tray system and ASI enhancement modules are both less than 15 cm. However, the extra step of frond preparation may prove to be costly, even though the Penfold design has a favorable biomass to system footprint ratio.

As sea urchin fisheries around the World continue to decline, the development of culture methods becomes more important. Aquaculture researchers are presently investigating all phases of the life history of sea urchins. Several systems to culture and enhance adults have been tested. Researchers should now focus their efforts on testing the Penfold design, and improving existing systems, e.g. refine culture techniques and concentrate on increasing revenue generated from them – thus leading to commercialization of land-based echinoculture.

ACKNOWLEDGMENTS

The author thanks Mr. LeRoy Creswell for assistance and general aquaculture advice and Mr. Tom Bernardi of Broad Bay Tooling for discussions on engineering designs, as well as Mr. Tom Bernardi, Jr. and Coorain Devin for help in figure preparation. Dr. Charles Walker is recognized for the histology he conducted for the enhancements studies.

Research and development of the ASI urchin enhancement modules was supported by grants from the U.S. Dept. of Commerce Small Business Innovative Research Program (NOAA Contract No. 50-DKNA-9-90071), the Maine Science and Technology Foundation, and the Maine Aquaculture Innovation Center (MAIC Project 98-12). Travel and publication costs were supported by grants from University of Maine's Association of Graduate Students and Maine Sea Grant (Grant NA96RG0102), a Maine Department of Marine Resources Fellowship to the author, and industry partners. The author also expresses appreciation to Dr. Yukio Yokota for organizing the highly successful Advanced Workshop in Honor of Eizo Nakano (Sea Urchin Workshop).

REFERENCES

Barnes, R.D. 1987. *Invertebrate Zoology*. Philadelphia: Saunders College Publishing.

Devin, M.G. 2001. Mass translocation of low quality deepwater sea urchins to overfished shallow-water sites. In L. Hiemstra (ed.), *Proceedings of The Sea Urchin Culture Workshop, Nanaimo, BC (Can), Sept 24-25, 2001*. Nanaimo: Malaspina University-College.

Devin, M.G. & Walker, C.W. 2001. The effect of light and prepared diet on roe quality in the green sea urchin *Strongylocentrotus droebachiensis*. *Aquaculture 2001*: Book of Abstracts. 143 J.M Parker Coliseum Louisiana State University Baton Rouge LA 70803 USA: World Aquaculture Society.

Garrido, C.L. & Barber, B.J. 2001. Effects of temperature and food ration on gonad growth and oogenesis of the green sea urchin, *Strongylocentrotus droebachiensis*. *Marine Biology*. 138 (3): 447-456.

Grosjean, P., Spirlet C., Gosselin, P., Vaitilingon, D., & Jangoux, M. 1998. Land-based, closed-cycle echinoculture of Paracentrotus lividus (Lamarck) (Echinooidea: Echinodermata): A long-term experiment at a pilot scale. *Journal of Shellfish Research* 17 (5): 1523-1532

Gustafson, R.R., and Wolpert 1963. The cellular basis of morphogenesis and sea urchin development. *International Review of Cytology* 15: 139-214.

Harris, L.G. 1981. Studies of community succession in a sea urchin barrens area following sea urchin removal. *American Zoologist* 21(4): 1019.

Hinegardner, R.T. 1969. Growth and development of the laboratory cultured sea urchin. *Biological Bulletin* 137: 465-475.

Hooper, R.G., Cuthbert, F.M., & Mckeever, T. 1997. Feasibility trials for sea urchin aquaculture using natural feeds. *Bulletin of the Aquaculture Association of Canada. Sea Urchin Culture Workshop*. Sackville, NB (Canada): Tribune Printing.

Lawrence, J.M. & Bazhin A. 1999. Life history strategies and the potential of sea urchins for aquaculture. *Journal of Shellfish Research*. 17(5):1515-1522.

Leahy, R.S., Tutshchulte, T.C., Britten, R.J., & Davidson, E.H. 1978. A large-scale laboratory maintenance system for gravid purple sea urchins (*Strongylocentrotus purpuratus*). *Journal of Experimental Zoology* 204: 369-381.

LeGall, P. 1990. Culture of echinoderms. In: G. Barnebe (ed.). *Aquaculture, Vol 1*. New York: Ellis Harwood.

LeGall, P. & Bucaille, D. 1989. Sea urchin production by inland farming. In: N. De Pauw, E. Jaspers, H. Achefors & N. Wilkins (eds.). *Aquaculture – A Biotechnology in Progress*. Bredene, Belgium: European Aquaculture Society.

Martin, P.D., Truchon, & Harris, L.G. 1988. *Strongylocentrotus droebachiensis* populations and community dynamics at two depth-related zones over an 11-year period. In R.D. Burke, P.V. Mladenov, P. Lambert, & R.L. Parsley (eds.), *Echinoderm Biology*. Rotterdam: Balkema.

McBride, S. 1997. Aquaculture boosts urchin roe production. *California Agriculture* 51 (4) 30-31.

McGinn, N.A. Lesser, M.P. & Walker, C.W. 1999. The influence of estradiol on vitellogenesis in the green sea urchin, *Strongylocentrotus droebachiensis*. *Journal of Shellfish Research* 18 (1): 311.

Meidel, S., Scheibling, R. & Metaxas, A. 1999. Relative importance of parental and larval nutrition on larval development and metamorphosis of the sea urchin *Strongylocentrotus droebachiensis*. *JEMBE* 240:161-178.

Olin, P.G. 1998. Preliminary trials for red sea urchin enhancement using prepared artificial diets. Conf. for shellfish growers: clam and oyster farming, 3-4 Mar 1997, Olympia, WA (USA).

Pearce, C. 2001. Land-based gonad enhancement of the green sea urchins: economic projections. In L. Hiemstra (ed.), *Proceedings of The Sea Urchin Culture Workshop, Nanaimo, BC (Can), Sept 24-25, 2001*. Nanaimo: Malaspina University-College.

Robinson, S.M.C. & Colborne, L. 1997. Enhancing roe of the green sea urchin using an artificial food source. *Bulletin of the Aquaculture Association of Canada: Proceedings Sea Urchin Culture Workshop*.

Sono, C.S. 1996. The Japanese sea urchin market. *U.S. Department of Commerce Report* No. NOAA-TM-NMFS-SWR-030.

Vadas, R.L. 1977. Preferential feeding: an optimization strategy of sea urchins. *Ecological Monographs* 47:337-371.

Vadas R.L., Beal, B., Dowling, T., & Fegley, J.C. 2000. Experimental field tests of natural algal diets on gonad index and quality in the green sea urchin, *Strongylocentrotus droebachiensis*: a case for rapid summer production in post-spawned animals. *Aquaculture* 182: 115-135

Walker C.W., & Lesser, M.P. 1998. Manipulation of food and photoperiod yield an out-of-season gametogenesis in the green sea urchin, *Strongylocentrotus droebachiensis*: Implications for aquaculture. *Marine Biology* 132: 663-676.

Walker CW, McGinn N.A., Harrington L.H., & Lesser MP. 1998. New perspectives on sea urchin gametogenesis and their relevance to aquaculture. *Journal of Shellfish Research* 17: 1507-1514.

Walker, C.W., Harrington, L.H., McGinn, N.A., Lesser, M.P., & Devin, M.G. 2001. Normal and altered gametogenesis in the green sea urchin: Implications for aquaculture. *Aquaculture 2001*: Book of Abstracts. 143 J.M Parker Coliseum Louisiana State University Baton Rouge LA 70803 USA: World Aquaculture Society, 2001. 672.

Watts, S.A., McClintock, J.B., & Lawrence, J.M. 1998. Assessment of Sea Urchins as Fishery and Aquaculture Candidates in the Gulf of Mexico. Sea Grant Project Number R/LR-42, pp 171-186.

The Sea Urchin: From Basic Biology to Aquaculture, Yokota, Matranga & Smolenicka (eds)
© Swets & Zeitlinger, Lisse, ISBN 90 5809 379 4

Sea urchin coelomocytes, the progenitors of vertebrate immune effectors, as bio-indicators of stress and pollution

Valeria Matranga & Rosa Bonaventura
Istituto di Biologia dello Sviluppo del C.N.R.

ABSTRACT: It is well recognized that the marine environment is in danger. If no protection activities are initiated the marine ecosystem will be dramatically affected. The standardization of simple methods which utilize biomolecular techniques is therefore a pressing requirement for monitoring of environmental hazards. The aim of our work has been the introduction of novel cellular biosensors (marine invertebrate cells) and molecular tools (stress proteins) to estimate the health state of ecologically and commercially important marine invertebrates at their natural habitats or in aquaculture. Coelomocytes, the immune cells of the sea urchin, have been shown to respond to environmental and experimental challenges by the activation of a well known stress marker. On the basis of our findings we propose hsp70 as a sensitive marker for monitoring pollutants in seawaters.

1. INTRODUCTION

The long-term goal of decision-making authorities is the monitoring of environmental pollution in seacoast waters. This is of extreme interest for the economy of European countries of the Mediterranean Sea in general and for Italy in particular. One component of the pressing requirements for appropriate management of endangered environments is the analysis of biological responses at all levels. Therefore there is an urgent need for techniques that provide rapid assessment of stress in the earlier stages of environmental pollution, acute as well as chronic responses. However, modern techniques in cellular and molecular biology are only occasionally used in marine stations. Therefore the standardization of simple methods which utilize biomolecular techniques will enable Mediterranean countries to apply modern techniques for monitoring environmental hazards. In this chapter we will briefly describe the general features of the sea urchin immune system. Effector cells are the so-called coelomocytes, free circulating cells found in the coelomic cavity of the sea urchin. Their functions and the molecules they produce will be shortly described. Herein after we will show, by field and laboratory experiments, that sea urchin coelomocytes can be used as a cell laboratory where to test different pollutants. We consider these results of great relevance for other ecologically and/or commercially important marine invertebrates. The application of the techniques we used to monitor pollution are pre-requisite for the appropriate setting of any aquaculture plant.

1.1 *Echinoderm coelomocytes features and functions*

Ceolomocytes are the cells circulating in the coelomic cavity of Echinoderms from which they can be collected by bleeding of the sea urchin after practicing a cut in the peristomal membrane around the teeth (Figure 1). A variety of different functions are displayed by coelomocytes, these include: cell motility, cytotoxicity, phagocitosis, encapsulation, clotting, graft rejection,

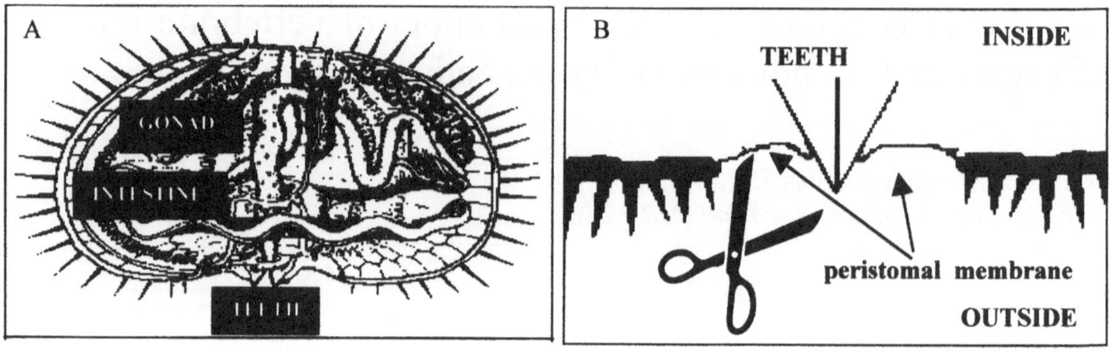

Figure 1. A) Anatomy of the sea urchin: cross-sectional drawing ; B) Coelomocytes collection: bleeding of coelomic fluid which contains coelomocytes is easily achieved through a cut in the peristomal membrane. Modified from http://www-leland.stanford.edu/group/Urchin/contents.html

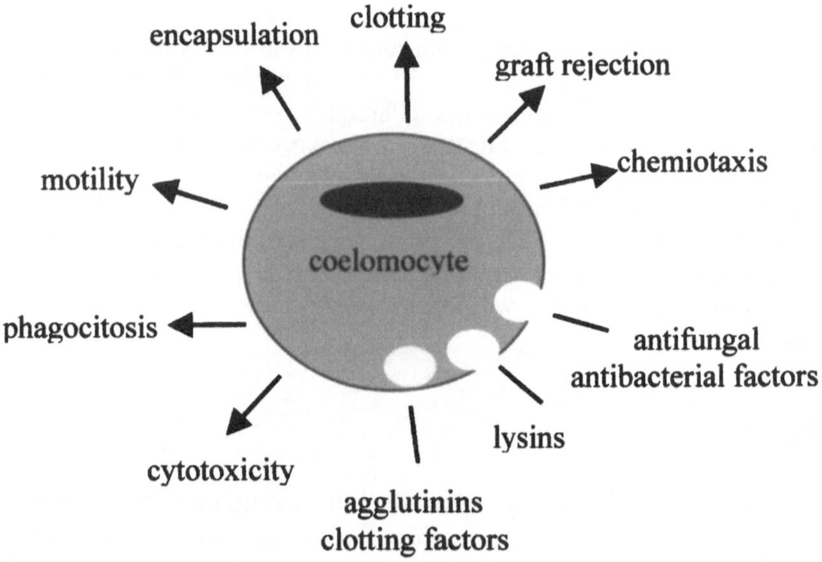

Figure 2. Echinoderm coelomocytes features and functions.

chemiotaxis, as well as the secretion of antifungal or antibacterial factors, lysins, agglutinins and clotting factors (Figure 2, for a review see also Matranga, 1996). It is still under debate if all these function are mediated by a single cell type or not, other authors tend to give to each cell type a specific function. However to little data are present in the literature and need other confirms.

Studies are present in the literature coming from different orders, genera and species of Echinoderms and some times controversial reports are due to the fact that it is difficult to give a correct interpretation to findings coming from different animal sources. In all cases coelomocytes are a mixed cell population which constitutes the first barrier against injuries, attacks of host or pathogens, and cytotoxic agents. It is for these reasons that coelomocytes are regarded as the immune effectors of Echinoderms.

It is possible to purify from a total coelomocyte cell suspension different phenotypes by a simple procedure which is based on different size and mass of cell types and takes advantage of different cell densities. It has then been reported that the proportion of each cell type can vary not only among species, but also individuals of the same species. We have evidence that it is sometimes dependent on the stress conditions to which the sea urchins have been exposed. As we will describe later we found that sea urchin coelomocytes respond to environmental and experimentally induced stresses by an increase in the number of the so-called red spherula cells (Matranga et al., 2000).

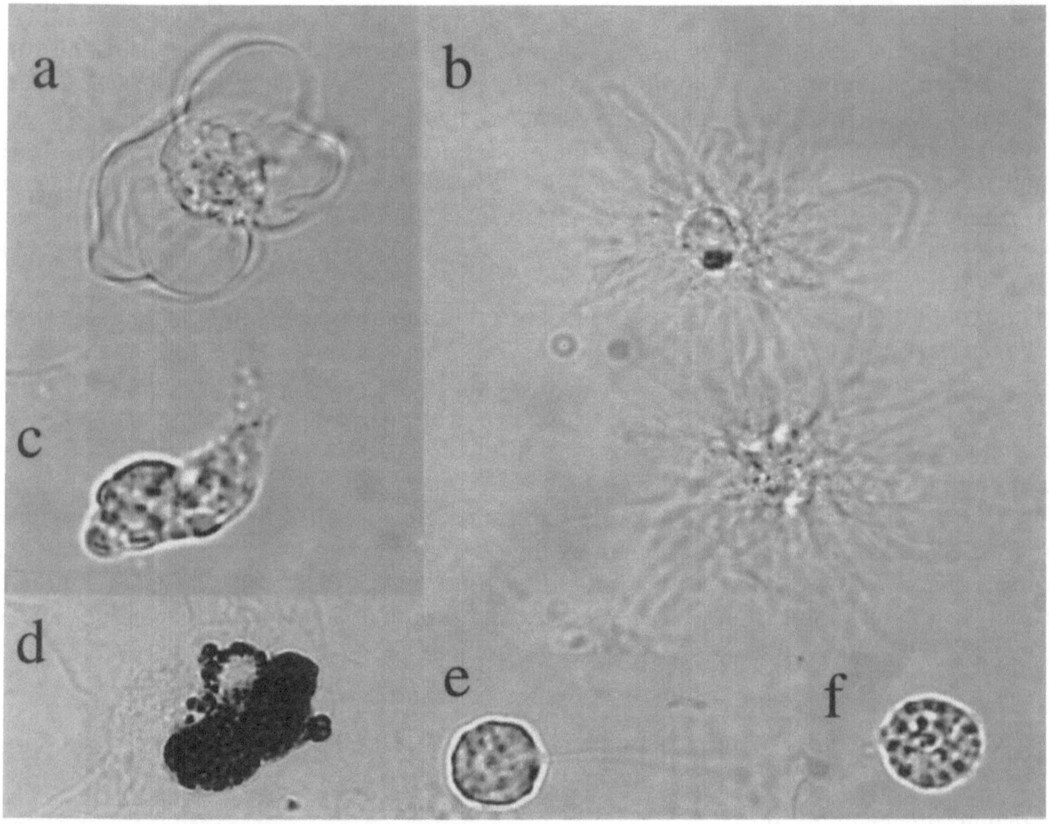

Figure 3. *Paracentrotus lividus* coelomocytes. a) petaloid phagocytes ; b) phillopodial phagocytes ; c) white amoebocytes ; d) red amoebocytes ; e) vibratile cell ; f) spherula cells.

In Figure 3, a panel shows the most frequent cell types found in the species *Paracentrotus lividus*. Petaloid (3A) or phillopodial phagocytes (3B) are the most abundant phenotypes found in the total coelomocyte population and they have been described as different functional states of the same cell type (Edds 1985, Edds 1993, Henson et al., 1999). Round-shaped cells are generally called spherula cells (3C) and their function has been related to cell clumping and encapsulation occurring after host invasion (Johnson 1969). They contain granules of stored proteins and have been shown to be involved in the production of arylsulfatase (Canicattì & Miglietta 1989; Pagliara & Canicatti 1993), a molecule found in the white series of blood cells associated with inflammatory phenomena. Vibratile cells are round-shaped cells with a single long flagellum which, working like a propeller, confers to the cells a rapid and helicoidal motion (3D). Their function is unknown. White (3E) and red (3F) amoebocytes are the fast moving cells which are thought to be involved in recognition, encapsulation and phagocytosis of foreign materials (Smith 1981).

It should be noted however that most of the studies which describe features and functions of different coelomocytes subpopulations are quite old. Sometimes they are contradictory and need more experimental work. Furthermore, some authors have hypothesised that the different cell types observed derive from a putative stem cell in adult tissues of the sea urchin. Subsequently, a maturation process would be initiated by external or internal stimuli. This proposition is very attractive, but still lacks an experimental confirm.

1.2 *Echinoderm coelomocytes molecules*

Ceolomocytes synthesize and release upon stimulation a variety of different molecules such as cytokines, lectins, perforins, lysosomal enzymes, serum proteins, profilins and others. All of them are related to immune functions and have been identified either on the basis of their function, protein isolation or gene expression. The list comprises: released on challenge histamine (Smith & Smith 1985), hemagglutinins (Parrinello et al. 1979), lectins (Canicattì et al. 1992), profilin homologues (Smith et al. 1992), IL-1 (Beck & Habicht 1986), IL-2 (Beck et al. 1989), IL-6 (Beck & Habicht 1996), homologue of complement component C3 (SpC3) (Al-Sharif et al. 1998), homologue of complement system factor B (SpBf) (Smith et al., 1998), immune system specific transcription factors (NFKB) (Pancer et al. 1999), collagen (Smith 1981), globin genes (McDonald et al. 1992), lysozyme (Shimizu et al. 1999). Furthermore, other putative immune response genes have been identified by expressed sequence tags (Smith et al. 1996).

Other not strictly related immune proteins, recently found in sea urchin coelomocytes, are: vitellogenins (Cervello et al. 1994, Unuma et al., 2001); actin (Henson et al., 1999); intermediate filament proteins (Holy 2000); myosin (Sirotkin et al., 2000), Na/K/Cl cotransporter (D'Andrea et al. 2001). The list of proteins and expressed genes identified in sea urchin coelomocytes is growing every day, thanks also to the efforts in the complete genome sequencing of the sea urchin (Cameron et al. 2000).

In conclusion, coelomocytes can be referred as the effector cells of the echinoderm immune system. Because of their immune functions they are considered as progenitor immune cells (Pancer et al. 1999, Gross et al. 1999, Smith et al., 2001). Again, more studies are required for the analysis of gene expression in coelomocytes and for the understanding of specific function of each cell type and the putative cross-talk among them.

2. RESULTS OF FIELD EXPERIMENTS

2.1 *Sea urchin sampling sites*

Sampling areas have to be chosen carefully in order to assess the state of polluted seawater. Therefore, adult specimen are collected from well-known polluted seawaters, and results obtained are compared with those obtained from specimen collected from well-known clean unpolluted seawaters. It is recommended to perform chemical analyses of seawaters first if they

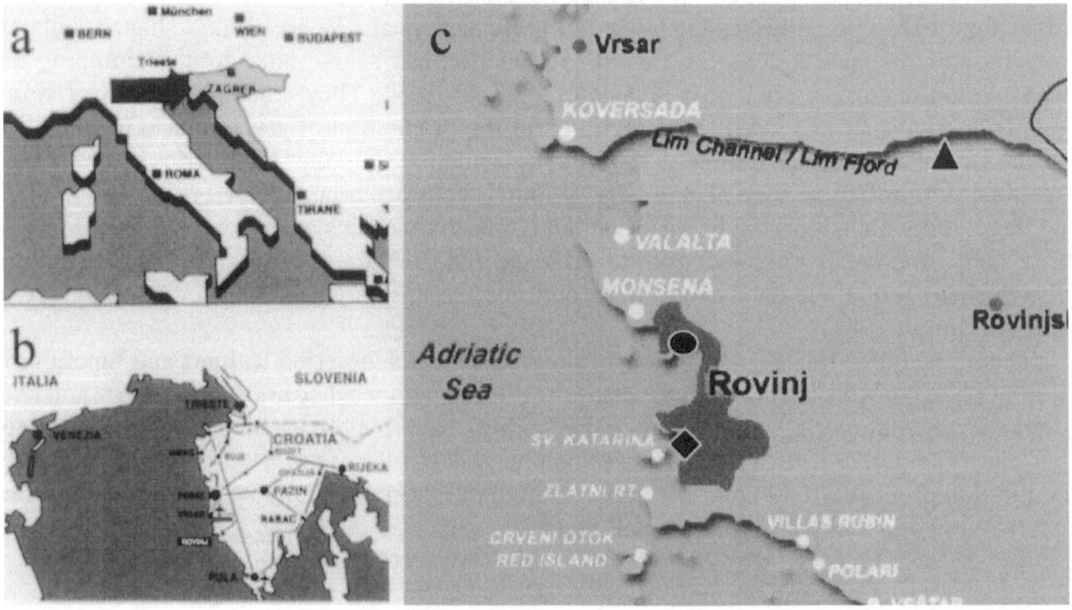

Figure 4. Sampling sites for unpolluted and polluted sea urchins in the Rovinj (Croatia) area of the Northern Adriatic (Mediterranean Sea), b and c are higher magnification of the area pointed by the arrow shown in a. ▲ Unpolluted seawater: Limski Canal; ● Industrial waste seawater: Ruder Boskovic Marine Station; ● Urban runoff: Valdibora.

are possible. In other cases animals are collected from areas suspected to be contaminated or polluted, and results are compared to reference seawaters.

In our studies we have collected sea urchin adults during a Summer campaign in the Northern Adriatic Sea, along the coasts near Rovinj (Istria), and kept them in tanks under constant aeration in circulating seawater, for the minimum time required before the experiment. This is very important because we have noticed that animals kept in tanks for long periods show signs of stress, at all similar to those induced by pollution or experimentally induced stress. Sites known to be uncontaminated, i.e. the Limski Canal (north of Rovinj) were taken as unpolluted, whereas those collected nearby the Ruder Boskovic Marine Station water front were considered as polluted, due to the industrial waste of a cannery (organic material and heavy metals). Sites for the collection of specimen are mapped in Figure 4.

Similarly other sites were chosen along the western coast of Sardinia, in particular near the location called S. Giovanni Sinis, well-known unpolluted area. As sea urchins living in polluted areas, we decided to maintain specimen in seawater collected from polluted areas near the Industrial Zone of the Oristano Gulf.

2.2 *Separation of coelomocytes into cell populations*

As already mentioned coelomocytes can be collected, as a total cell population, by bleeding of individual sea urchins in anticoagulant solution through a cut in the peristomal membrane.

It should be recalled that it is not possible to pool coelomocytes suspensions from different individuals. In fact, sea urchins recognize self from non-self, and mixing cells from 2 individuals will cause clotting. Cells coming from a single individual will eventually clot in any case if the phenomenon is not prevented by the use of an anticoagulant (usually EDTA) or delayed by placing the cell suspension at 4°C.

The method used for the separation of coelomocytes subpopulations is relatively easy and, as we will see in the following, the presence of high amounts of red cells is a clear-cut indicator of pollution or stress.

The volume of the collectable coelomic fluid greatly depends on the size of the sea urchin and on the species used. Using *Paracentrotus lividus*, about 10 ml can be collected per each sea urchin, containing between 1 and 10 x 10^6 cells per ml. Cells are collected in an anticoagulant solution, in order to avoid cell clotting, pelleted and resuspended in the same buffer. The suspension is then loaded and centrifuged on a step gradient as detailed in the following.

2.3 *Experimental procedure*

- bleed sea urchins individually through a cut in the peristomal membrane.
- pour the fluid, usually 10 ml, containing the total coelomocyte population on 30-ml ice-cold 20mM Tris-0.5M NaCl-30mM Ethylenediaminetetraacetic acid (EDTA) pH 7.5 (ISO-EDTA) through a gauze (to trap big clumps or debrids)
- centrifuge cells at 700-800 rpm in ALC for 10 min at 4°C
- resuspend the pellet gently with a Pasteur pipette in 1 ml ISO-EDTA
- load the suspension on a Natrium-Megluminamidotrizoat in ISO-EDTA step gradient, previously prepared as it follows in 10 ml tubes which will not allow cells to stick to their surfaces (siliconized glass or plastic):

	37.5%	25%	18.75%	15%	12.5%
ISO-EDTA (ml)	0.5	0.67	0.75	0.8	0.84
* 76% (ml)	0.5	0.33	0.25	0.2	0.16
Final volume (ml)	1	1	1	1	1

* Natrium-Megluminamidotrizoat

- mix the volume indicated on the table to obtain each step of the gradient
- slowly pour with a Pasteur pipette every step. Take care not to mix the steps
- centrifuge the gradients at 700-800 rpm in ALC for 15 min in a swing-out rotor at 4°C
- collect bands with a Pasteur pipette
- wash cell suspensions in 10 volumes of ISO-EDTA

2.4 *Results of density gradients*

When gradients are observed after centrifugation, cells are found separated on the basis of their density and size, as schematised in Figure 5. The most abundant cell type is constituted by phagocytes which band at the top of the gradient and never cross the 12.5% step. This band is usually referred to as band A (see also Figure 3). It follows a less dense band just across the 15% step, mostly containing rounded-shaped cells and referred to as band B. A thinner band is immersed in the 18.75% step, containing rounded-shaped cells and referred to as band C. It follows a doublet of close bands, referred to as D and E bands, mostly containing smaller rounded-shaped cells. Last band, a very thin one, sometimes reaching the bottom is band F, containing red cells.

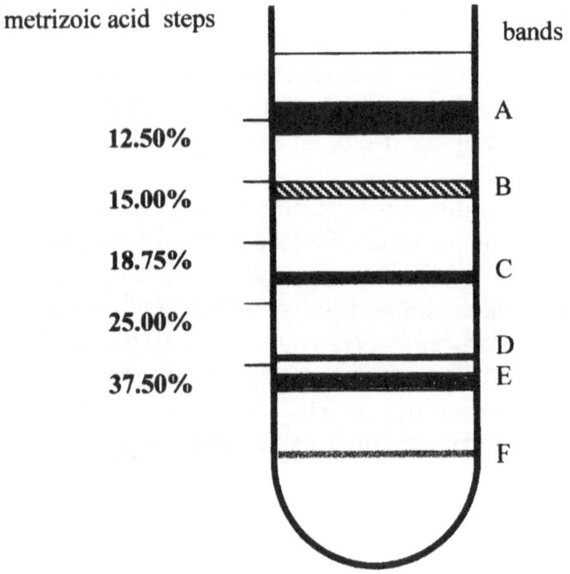

Figure 5. Schematic drawing of *Paracentrotus lividus* total coelomocytes separation into subtypes by a density gradient.

At this point we should clarify that the appearance, in terms of shape and motility, of coelomocytes subpopulations described in our previous paper (Matranga et al., 2000), was greatly affected by the medium we used for their collection and centrifugation into the gradients. We would like to stress here that the real appearance of the cell types is given in Figure 3, a part for their active motion which we have been able to monitor and record in movies and involve mostly white and red amoebocytes, as well as vibratile cells.

2.5 *Polluted seawater increases the proportion of red cells subpopulation*

A clearcut indication of differences found in coelomocyte subpopulations came from comparison between gradients of total coelomocytes cells collected from sea urchins living in polluted and unpolluted sea waters. The major difference was found in band F, previously identified as the red spherula cells (Matranga et al., 2000) which was increased very much in size. This big difference in the quantitative composition of the coelomocyte population was consistent since we found the same result in 9 out of 10 different samplings.

Similarly, when sea urchin coelomocytes separation into subpopulations was performed using sea urchins collected from west coasts of Sardinia, we found consistent results indicating a higher quantity of red cells in those specimen maintained in polluted sea waters than those found in coleomocytes from sea urchins collected in unpolluted sea waters.

2.6 *Sea urchin coelomocytes respond to injuries by an increase in the number of red cells*

In some cases, we found in the wild sea urchins showing morphological abnormalities, such as scars and/or loss of spines. This is probably due to pathologies or accidents. An increased number of red cells was expected to be found compared to "normal" sea urchins coming from the same area. We decided to test this hypothesis and run gradients of coelomocytes collected from "injured" sea urchins found along the coasts of Sicily. We found that, as expected, the number of red cells was highly increased (Matranga et al., 2000).

2.7 *Stress-response proteins in marine invertebrates*

Heat shock or stress response is one of the most highly conserved adaptive responses in nature. In single cell organisms, the stress response is activated to tolerate a variety of stresses including hyperthermia, hyperoxia, hypoxia, and other perturbations altering protein synthesis. This tolerance phenomenon is also extremely important in the multicellular organism, resulting in not only thermal tolerance, but also resistance to stresses of the whole organism. Moreover, recent data indicate that these stress proteins have the ability to modulate the cellular immune response (Mosely, 2000).

Heat shock proteins such as hsp70 are ubiquitous stress-response proteins suggested to be useful biomarkers of exposure for assessment of environmental contamination (Schröder et al., 1999). Genes coding for stress inducible proteins have been cloned from different marine invertebrate species and have been used as probes to monitor stress in different aquatic environments (Müller et al., 1998; Wiens et al., 1998; Schröder et al., 1999)

In the sea urchin embryo, pioneering studies initiated by Giudice's group in the early 80s' showed the synthesis of heat shock proteins in response to shifts in the temperature of seawaters in which embryos were cultured (for reviews see Giudice 1985; Giudice et al., 1999). However, no studies on the presence of constitutive and/or inducible hsp70 protein and their putative overexpression upon physical stress, chemical stress or pollution in adult sea urchin cells were reported.

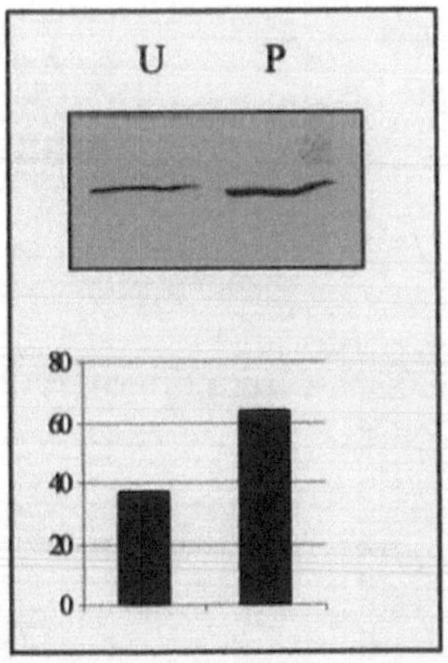

Figure 6. Sea urchin coelomocytes respond to pollution by increasing hsp70 levels.
In the upper panel: immunoblotting analysis of total coelomocyte lysates from unpolluted (U) and polluted (P) sea waters using a monoclonal antibody anti-hsp70. In the lower panel: quantitative analysis of the immunoblotting expressed in arbitrary units. U: sea urchin from Limski Canal. P: sea urchin from Ruder Boskovic water front.

First we tested whether sea urchin coelomocytes were able to respond to pollution by the upregulation of the hsp70 gene and the accumulation of the corresponding protein. Specimens were taken from unpolluted and polluted seawaters and total coelomocytes collected in anticoagulant solution. Cells were lysed, protein concentration was determined by the Lowry method. The same amounts of total proteins were loaded on a 6% SDS-polyacrylamide gel. The expression of hsp70 was determined by immunoblotting using a specific monoclonal antibody which recognizes both constitutive and inducible hsp70. As shown in Figure 6, an increase in the levels of the protein expressed by polluted versus unpolluted sea urchins was found. The quantitative analysis carried out on band intensities showed a 1.7 fold increase of the polluted sample with respect to control unpolluted levels.

Therefore, in agreement with our previous observations (Matranga et al., 2000), coelomocytes collected from sea urchins found in the wild are sensitive to pollution and respond both at the cellular level, increasing the subpopulation of red amoebae, and at the biochemical level, upregulating the expression of the hsp70.

3. INDUCTION OF STRESS IN THE LABORATORY

3.1 *Experiments with whole sea urchins*

Field experiments have the advantage to depict the situation which is going on in the wild, but have the disadvantage that different stress conditions are impossible to be assayed and most parameters are uncontrollable. Our target is to find a molecular tool for the assessment of stress that could be used in pollution studies. Therefore we designed a procedure for shock

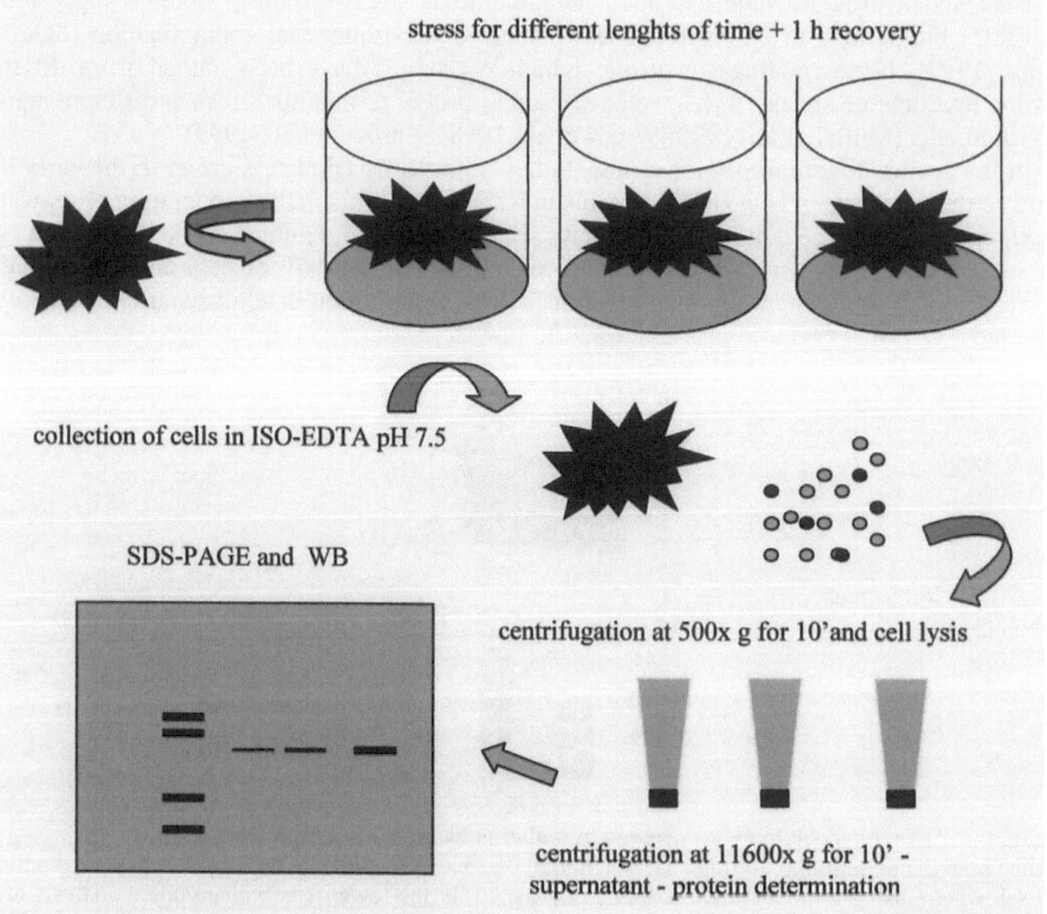

Figure 7. *In vivo* experiments on sea urchin coelomocytes.

experiments in the laboratory and found that the hsp70 could be a good stress marker for sea urchins. The experimental protocol is summarized in Figure 7.

3.2 *Types of stress:*

A certain number of laboratory stress types could be used on sea urchins, here are described:

- *temperature stress:* for these experiments the sea water in which sea urchins will be immersed should be maintained at the temperature to be tested long time in advance.
- *chemical stress, chronic effect:* in this case different concentrations of the substances whose biological effects have to be tested are added to the seawater. However in this case long incubation period are needed and the effect measured would be a chronic effect.
- *chemical stress, acute effect:* the other possibility is to inject the substance to be tested into the sea urchin in this case an acute effect could be measured.
- *environmental seawaters (chronic or acute effects)*: sea urchins are immersed in or injected with seawaters collected from the area to be monitored, to establish the presence of toxic substances. Their concentrations must be sufficient to induce chronic or acute biological effects.

3.3 *Induction of temperature stress on sea urchins*

Individuals of *Paracentrotus lividus* can be temperature-stressed in the laboratory and the expression of hsp70 can be determined by immunoblotting. Here is a scheme of the procedure to be followed. Sea urchins are stressed for variable periods: 30', 1h and 2h. Remember to keep at least one sea urchin at the permissive temperature (usually 16-18°C), to be used as negative control. It is also very important to give at least 30-60 minutes of recovery to the animals by placing them in seawater kept at the permissive temperature.

Start **2 hrs**(1) Cold Stress	Start **1 hrs**(1) Cold Stress	Start **30 min**(1) Cold Stress	Stop cold stress Start **1 hr recovery**(2)	Collect coelomocytes(3)	Lyse for 20 min(4) Centrifuge for 10 min(5) End exper.(6)

1. place sea urchins in cold water at 4 °C
2. place sea urchins, after the stress, in sea water at 16 °C for recovery, at least 1 hour
3. collect the coelomic fluids from stressed and control sea urchins, after practising a cut in the peristomal membrane, by immediately pouring on 30 ml ice-cold ISO-EDTA. Centrifuge cells at 700-800 rpm in ALC refrigerated centrifuge for 5' and discard the supernatant. Usually a medium size *P. lividus* sea urchin contains 10 ml of coelomic fluid. In case the volume of fluid collected is different, dilute in ISO-EDTA accordingly
4. lyse the pellets in 500ml of a ISO-EDTA 1:10 dilution, supplemented with a protease inhibitors cocktail: 1ml MIX (Antipain 1mg/ml, Leupeptin 1mg/ml, Benzamidine 500mM), 1ml Aprotinin (1mg/ml), Pepstatin 1ml (1mg/ml), 0.25ml PMSF (200mM) with occasional pipetting at 4°C for 20 minutes
5. centrifuge the lysates at 12000 rpm in ALC refrigerated centrifuge for 10' and recover the supernatant at 4°C
6. keep aliquots at -20°C for Lowry determination
7. freeze-dry aliquots of the volume corresponding to 20 mg for immunoblotting
8. use the same procedure for obtaining and processing coelomocytes from sea urchins immersed in or injected with polluted sea waters

3.4 *Detection of stress proteins (hsp70) in "temperature stressed" coelomocytes*

For the analysis of the hsp70 stress marker on samples subjected to temperature stress, coelomocytes were collected and lysed. Equal amounts of proteins, determined by the Lowry

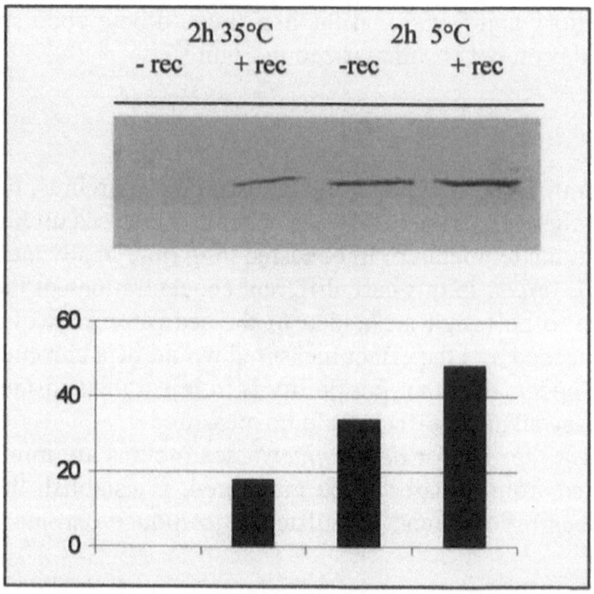

Figure 8. Sea urchin coelomocytes respond to temperature stress by increasing the levels of hsp70 expression. In the upper panel: immunoblotting analysis, using a monoclonal antibody anti-hsp70, of total coelomocyte lysates from sea urchins maintained for 2 hours at 35°C or 4°C, without (-rec) or with (+rec) a recovery period of 1 hour at the permissive temperature of 18°C. In the lower panel: quantitative analysis of the immunoblotting expressed in arbitrary units.

method (Lowry 1951), were loaded on SDS-PAGE and analyzed for the expression of hsp70 by immunoblotting. Result of a typical experiment is shown in Figure 8. It indicates the detection of hsp70 in 2-hours heat (35°C) and cold (4°C) stressed sea urchins, with or without 1 hour recovery period at 16°C. The quantitative analysis of the levels of expression was also determined by scannings of the immunoblotting.

3.5 *Coelomocytes in vitro cultures*

We found very difficult to compare results of stress induced in the laboratory (temperature, pH, others) using sea urchin coelomocytes coming from different individuals. This is due to several reasons:

1. sea urchins may have been taken from different sites
2. different basal levels of hsp70 expressed by different individuals
3. different pathologies among different individuals

All these different cases lead to different basal levels of expression of hsp70, which therefore render sometimes the comparison of results very difficult. It is for this reason that we designed another "*in vitro*" approach, schematized in Figure 9.

Using this protocol, we verified the ability of coelomocytes to respond to temperature stress by overexpression of the hsp70 protein. As shown in Figure 10, we found an increase in the levels of the protein, as assessed by immunoblotting using monoclonal antibody against hsp70, which reached a maximun as early as 0.5 hours after the stress, with 6-fold increase in the levels compared to coelomocytes cultured in *vitro* at the permissive temperature of 18°C. One and two hours of stress gave a 5-fold and 4.2-fold increases in the hsp70 levels respectively. After 4 hours of culture at 4°C the response was minimal, probably due to the suffering conditions of cells *in vitro*.

Similarly we found that coelomocytes respond to acid pH and heavy metal ($CdCl_2$) treatment when cultured *in vitro* (Matranga et al., 2002), leading to the study of environmental stress.

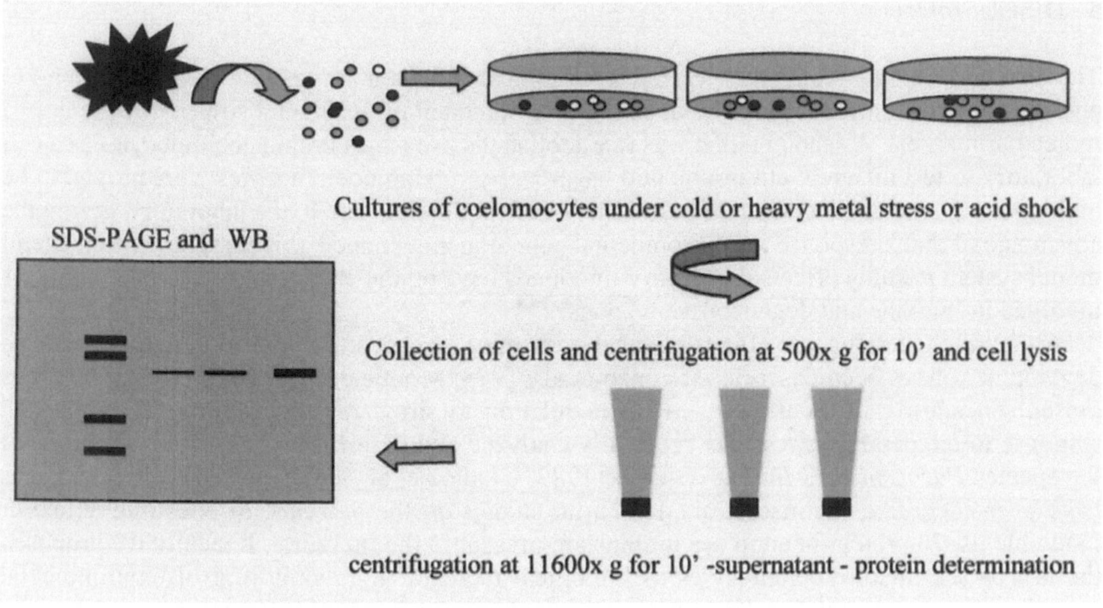

Figure 9. *In vitro* coelomocytes short term cultures.

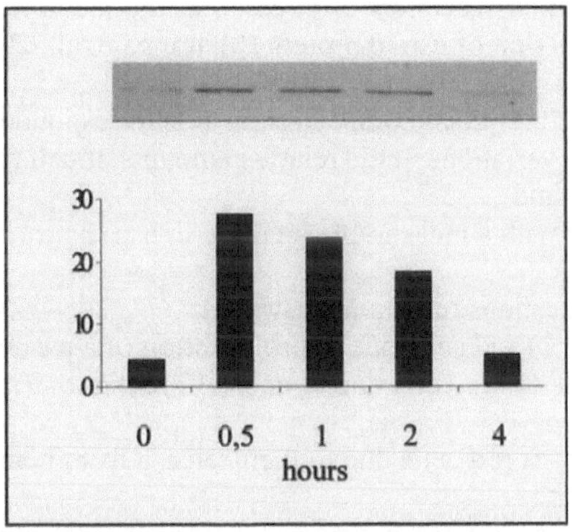

Figure 10. Coelomocytes respond to *in vitro* cold stress by hsp 70 overexpression.
In the upper panel: immunoblotting analysis, using a monoclonal antibody anti-hsp70, of lysates from total coelomocytes cultured at 4°C for the indicated period of time. All cultures were given a 1 hour recovery period at 18°C. Lower panel: quantitative analysis of the immunoblotting shown expressed in arbitrary units.

4. COELOMOCYTES CULTURED *IN VITRO*

A further extension of our studies was the attempt to maintain coelomocytes for long-term cultures *in vitro* and to assay the expression of stress markers after induction with heavy metals and other stressors. Coelomocytes were shown to be good markers of stress up to 6 hours in culture. They respond by the over-expression of hsp70 (Matranga et al., 2002). However in order to maintain cells in culture for longer periods, the medium adopted had to have a different composition from that utilized in previous studies. One of our present goals is to define a new formula of a medium for long term experiments which enables to test the chronic effects of drugs and/or pollutants. Another long-term goal is the isolation and characterization of the first sea urchin adult cell line.

5. DISCUSSION

The first aim of this study was to find a biological indicator to be used as a stress marker and to characterise the response induced by environmental or accidental stresses by the use of molecular markers. A second target was to establish the use of sea urchin coelomocytes as a cell laboratory to test different environmental hazards. Sea urchin coelomocytes were proven to be highly sensitive to environmental stress factors both in the field and in the laboratory, giving the advantage of quick response to environmental contaminants. Indeed, this presents also an useful model system to study effects that toxic compounds have on the elevation of molecular markers involved in immune and degenerative responses.

Although the teratogenic effects of various chemical agents and drugs on sea urchin embryo development have been described (Sconzo et al., 1995; Morale et al., 1998), no literature was present on adult cells of the sea urchin in relation to stress response. Similarly, while the response to temperature stress has been fully analyzed and described in sea urchin embryos of the species *Paracentrotus lividus* (Giudice 1989, Giudice et al., 1999) and the involved genes have been identified (Sconzo et al., 1992), no studies on the presence of constitutive and/or inducible hsp70 protein in adult sea urchins are present in the literature. Recently we proposed the use of sea urchin coelomocytes as biological indicator for monitoring of environmental pollution. It was shown for the first time in this system that coelomocytes respond both to environmental (polluted seawaters) or experimentally induced (temperature) stress by the over-expression of stress proteins (hsp70). Variations in coelomocytes subpopulations were observed in sea urchins living in polluted seawaters or stressed in the laboratory. These variations are correlated to increased expression of hsp70 protein (Matranga et al., 2000).

Here we have shown that sea urchin coelomocytes are able to respond to stress:

i) at the cellular level: with variations in the relative amounts of cell types, such as an increase in the red spherula cells, and
ii) at the biochemical level: with a measurable increase in hsp70 protein levels.

At the cellular level some questions remain unanswered:

i) is the increase in number of red cells due to proliferation of a particular subset of cells?
ii) is the increase in number of red cells due to cell differentiation (maturation) of stem cells pre-existing in the coelom?
iii) is the increase in number of red cells due to their release from hematopoietic areas present in the sea urchin tissue?
iv) are there other specific markers responsive to environmental and experimental induced stress?

At the biochemical level, studies are still necessary to understand the nature of the pigment contained in the red cells and its possible implication in the stress response.

Furthermore, coelomocytes were proved to be responsive to environmental stress, such as urban runoff, industrial waste, injuries and pathogens (this paper and Matranga et al., 2000). Coelomocytes respond to experimentally-induced stress, such as variations in temperature (this paper and Matranga et al., 2000), acid medium and heavy metals (Matranga et al., 2002). Therefore we propose that the accumulation of heat shock proteins can be used as a marker to evaluate the biological damage due to chronic exposure of an organism to physical (non permissive) conditions or sublethal levels of pollutants.

Although we have learnt that sea urchin coelomocytes can be used as bio-indicators, much care has to be taken when comparing results of "in vitro" and "in vivo" experiments, since each type of experiment has advantages and disadvantages. Herein there is a scheme to recapitulate some of them.

The finding that biochemical mechanism underlying multi-xenobiotic resistance of marine invertebrates, mussels (Kurelec, 1992) and sponges (Müller et al., 1996), is similar to the mechanism of multidrug resistance (MDR) in vertebrate tumor cells, opens the way to the use of sea urchin coelomocytes as cellular laboratories where to investigate on diseases that have as target vertebrate blood cells.

Table 1. Comparison between *in vivo* and *in vitro* experiments

in vivo	*in vitro*
different sea urchin personal history	same individual used
differences among individuals	identical cell type composition
different number of cells	same number of cells
physiological conditions	culture conditions stress
unknown conc. of injected drugs	morphology monitoring

Finally, it should be considered that other environmental damages can be measured by the use of sea urchin coelomocytes as bio-indicators. This is the case of the UV radiation. It has been calculated that 50% of red light and 20% of the ultraviolet light is absorbed at 1 meter depth. By 10 meters, all red light is absorbed, while 20% of the UV radiation remains. It is well known that UV radiations cause damage to DNA and/or protein in a variety of organisms (Epel, 1999). Many invertebrates, including sea urchins, spawn and/or develop in the 1 meter zone and thus are susceptible to genetic damage (Epel, 1999).

Experiments are in progress in our laboratory to determine the effects of heavy metals and UV radiation on the expression of hsp70 and other stress markers.

All our results confirm that coelomocytes are the progenitors of vertebrate immune effectors (Smith & Davidson 1992).They also reaffirm the importance of studies on Echinoderms, a phylum in the direct evolutionary line leading to the vertebrates (Fig 11). In fact we should remember that sea urchins, hemichordates and other echinoderms occupy a key phylogenetic position as the only non-chordate deuterostomes. Finally, the understanding of stress and immune phenomena in the sea urchin will explain how adaptive immunity worked in early evolution of vertebrates.

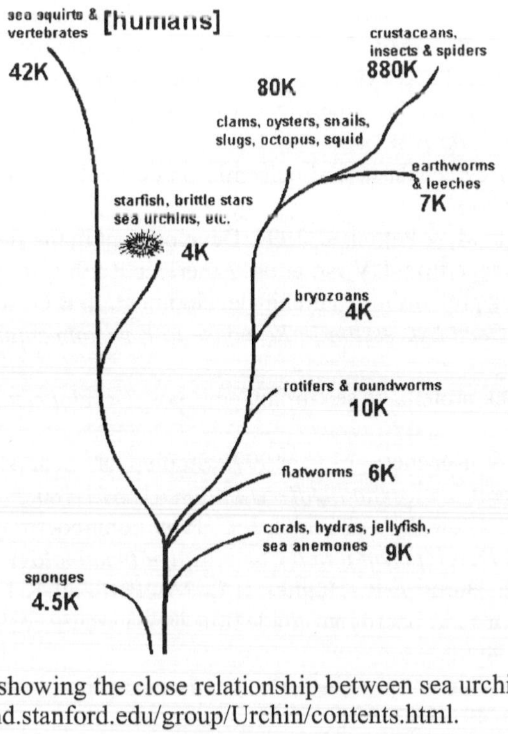

Figure 11. Phylogenetic tree showing the close relationship between sea urchins and humans. Taken from http://www-leland.stanford.edu/group/Urchin/contents.html.

ACKNOWLEDGMENTS

The authors wish to thank Prof. H.C. Schröder and Prof. W.E.G. Müller for collaborative research in the frame of the EU RTD Project: "Development and field validation of biosensor methods for the assessment of the effects of pollution and solar UV radiation on commercially and ecologically important marine invertebrates". Many thanks go to the staff of the Ruder Boskovic Marine Laboratory of Rovinj, Croatia, for their hospitality during part of the work.

The authors are grateful to other members of the group for many insightful discussion and helpful criticisms. This work has been fully supported by an EU program, Contract N° EVK3-CT-1999-00005. R.B is payed by a CNR Research Contract in the frame of the EU grant.

REFERENCES

Al-Sharif, W.Z., Suyer, J.O., Lambris, J.D. & Smith, L.C. 1998. Sea urchin coelomocytes specifically express a homologue of the complement component C3. Journal of Immunology, 160: 2983-2997.

Beck. G, & Habicht, G. 1986. Isolation and characterization of a primitive interleukin-1 like protein from *Asteria forbesi. Proceeding of National Academy of Science* USA, 83: 7429-7433.

Beck, G. & Habicht, G. 1996. Characterization of an IL-6-like molecule from an echinoderm (*Asteria forbesi*). *Cytokine*, 8: 507-512.

Beck, G., O'Brian, R. & Habich, G. 1989. Invertebrate cytokines: the phylogenetic emergence of interleukin-1. *Bioessay,* 11: 62-67.

Cameron, R.A., Mahairas, G., Rast, J.P., Martinez, P., Biondi, T.R., Swartzell, S., Wallace, J.C., Poustka, A.J., Livingston, B.T., Wray, G.A., Ettensohn, C.A., Lehrach, H., Britten, R.J., Davidson, E.H. & Hood, L. 2000. A sea urchin genome project: sequence scan, virtual map, and additional resources. *Proceedings of National Academy of Science USA*. 97: 9514-9518.

Canicattì, C., Pagliara, P. & Stabili, L. 1992. Sea urchin coelomic fluid agglutinin mediates coelomocytes adhesion. *European Journal of Cell Biology* 58: 291-295.

Canicatti, C. & Miglietta, A, 1989. Arylsulphatase in echinoderm immunocompetent cells. *Histochemical Journal* 1989 21: 419-24.

Cervello, M., Arizza, V., Lattuca, G., Parrinello, N, & Matranga, V. 1994. Detection of vitellogenin in a subpopulation of sea urchin coelomocytes. *European Journal of Cell Biology* 64: 314-319.

D'Andrea-Winslow, L., Strohmeier, G.R., Rossi, B., & Hofman, P. 2001. Identification of a sea urchin Na(+)/K(+)/2Cl(-) cotransporter (NKCC): microfilament-dependent surface expression is mediated by hypotonic shock and cyclic AMP. *Journal of Experimental Biology* 204(Pt 1): 147-56.

Edds, K.T. 1985. Morphological and cytoskeletal transformation in sea urchin coelomocytes. In W.D. Cohen (ed) *Blood cells of marine invertebrates: experimental systems in cell biology and comparative physiology*: 53-74. New York: A R Liss.

Edds, K.T. 1993. Effects of cytochalasin and colcemid on cortical flow in coelomocytes. *Cell Motility and Cytoskeleton* 26: 262-73.

Epel, D., Hemela, K., Schick, M. & Patton, C. 1999. Development in the floating world: defenses of eggs and embryos against damage from UV radiation. American Zoologist 39: 271-278.

Giudice, G. 1985. Heat shock proteins in sea urchin development. In B.G. Atkinson & D.B. Walden (eds), *Changes in Eukaryotic Gene Expression in Response to Environmental Stress*: 115-133. New York: Academic Press.

Giudice, G. 1989. Heat shock proteins in sea urchin embryos. *Development Growth and Differentiation*, 31: 103-106.

Giudice, G., Sconzo, G. & Roccheri, M.C. 1999. Studies on heat shock proteins in sea urchin development. *Development Growth and Differentiation* 41: 375-380.

Gross, P.S., Al-Sharif, W.Z., Clow, L.A. & Smith, L.C. 1999 Echinoderm immunity and the evolution of the complement system. *Developmental and Comparative Immunology* 23(4-5): 429-442.

Henson, J.H., Svitkina, T.M., Burns, A.R., Hughes, H.E., MacPartland, K.J., Nazarian, R. & Borisy, G.G. 1999. Two components of actin-based retrograde flow in sea urchin coelomocytes. *Molecular Biology of the Cell* 10(12): 4075-4090.

Holy, J. 2000. Intermediate filament proteins in echinoderm coelomocytes. *Comparative Biochemistry and Physiology* B127(4): 491-504.

Johson, P.T. 1969. The coelomic elements of sea urchins (*Strongylocentrotus*) III. In vitro reaction to bacteria. *J. Invertebr. Pathol.* 13:42-62.

Kurelec, B. 1992. The multixenobiotic resistance mechanism in aquatic organisms. *Critical Reviews in Toxicology* 22: 23-43.

Lowry, O.R., Rosebrough, W.J., Farr, A.L. & Randall, R.S. 1951. Protein determination measurement with the folin phenol reagent. *Journal of Biological Chemistry* 193: 265-275

Matranga, V. 1996. Molecular aspects of immune reactions in Echinodermata. In W. E. G. Müller & B. Rinkevich (eds), Invertebrate *Immunology, PMSB Series Vol. 15*: 235-247. Heidelberg: Spinger.

Matranga, V., Toia, G., Bonaventura, R. & Müller, W.E.G. 2000. Cellular and biochemical responses to environmental and experimentally induced stress in sea urchin coelomocytes. *Cell Stress & Chaperones* 5: 113-120.

Matranga, V., Bonaventura, R. & Di Bella, G. 2002. Hsp70 as a stress marker of sea urchin coelomocytes in short term cultures. *CMB* 48 (3) in Press

McDonald, G.D., Davidson, L. & Kitto, G.B. 1992. Amino acid sequence of the coelomic C globin from the sea cucumber. *Journal of Protein Chemistry* 11: 29-37.

Morale, A., Coniglio, L., Angelini, C., Cimoli, G., Bolla, A., Alleteo, D., Russo, P. & Falugi, C. 1998. Biological effects of a neurotoxic pesticide at low concentrations on sea urchin early development. A teratogenic assay. *Chemosphere* 37: 3001-3010.

Moseley, P. 2000. Stress proteins and immune response. *Immunopharmacology* 25: 299-302.

Müller, W.E.G., Steffen, R., Rinkevich, B., Matranga, V., Kurelec, B. 1996. The multixenobiotic resistence mechanism in the marine sponge *Suberites domuncula*: its potencial applicability for the evaluation of environmental pollution by toxic compound. *Marine Biology* 125: 165-170

Müller, W.E.G., Batel, R., Lacorn, M., Steinhart, H., Simat, T., Lauenroth, S., Hassanein, H., & Schröder, H.C. 1998. Accumulation of cadmium and zinc in the marine sponge *Suberites domuncula* and its potential consequences on single-strand breaks and on expression of heat-shock protein: A natural field study. *Marine Ecology Progress Series* 167: 127-135.

Pagliara, P. & Canicattì, C. 1993. Isolation of cytolytic granules from sea urchin amoebocytes. European Journal of Cell Biology 60: 179-84.

Pancer, Z., Rast, J.P. & Davidson, E.H. 1999. Origins of immunity: transcription factors and homologues of effector genes of the vertebrate immune system expressed in sea urchin coelomocytes. *Immunogenetics* 49: 773-786.

Parrinello, N., Rindone, D. & Canicatti, C. 1979. Naturally occurring hemolysins in the coelomic fluid of *Holuthuria polii. Developmental and Comparative Immunology* 3: 45-54.

Sconzo, G.,Scardina, G., Ferraro, M.G. 1992. Characterization of a new member of the sea urchin *Paracentrotus lividus* hsp70 gene family and its expression. *Gene* 16: 353-358.

Sconzo, G., Romancino, D., Fasulo, G., Cascino, D. & Giudice, G. 1995. Effect of doxorubicin and phenytoin on sea urchin development. *Pharmazie*, 50: 616-619.

Schröder, H.C., Batel, R., Lauenroth, S., Hassanein, H.M.A., Lacorn, M., Simat, T., Steinhart, H., & Müller, W.E.G. 1999. Induction of DNA damage and expression of heat shock protein HSP70 by polychlorinated biphenyls in the marine sponge *Suberites domuncula* Olivi. *Journal of Experimental Marine Biology and Ecology* 233: 285-300.

Shimizu, M., Kohno, S., Kagawa, H, & Ichise, N. 1999. Lytic activity and biochemical properties of lysozyme in the coelomic fluid of the sea urchin *Strongylocentrotus intermedius. Journal of Invertebrate Pathology* 73(2): 214-222.

Sirotkin, V., Seipel, S., Krendel, M. & Bonder, E.M. 2000. Characterization of sea urchin unconventional myosins and analysis of their patterns of expression during early embryogenesis. *Molecular Reproduction and Development* 57(2): 111-26.

Smith, V.L. 1981. The Echinoderms. In N. A. Ratcliffe, & A. T. Rowley (eds) *Invertebrate blood cells Vol 2*: London: Academic Press,.

Smith, S.L. & Smith, A.C. 1985. Sensitization and histamine release by cells of the sand dollar *Mellita quinquesperforata. Development Comparative Immunology* 9: 597-603.

Smith, L.C. & Davidson, E.H. 1992. The echinoid immune system and the phylogenetic occurrence of immune mechanisms in deuterostomes. *Immunology Today*, 13: 356-362.

Smith. L.C., Britten, R.J. & Davidson, E.H. 1992. SpCoell: a sea urchin profilin gene expressed specifically in coelomocytes in response to injury. *Molecular Biology of the Cell* 3: 403-414.

Smith, L.C., Chang, L., Britten, R.J. & Davidson, E.H. 1996. Sea urchin genes expressed in activated coelomocytes are identified by expressed sequence tags. Complement homologues and other putative immune response genes suggest immune system homology within the deuterostomes. *Journal of Immunology* 156(2): 593-602.

Smith, L.C., Shih, C.S. & Dachenhausen, S.G. 1998. Coelomocytes express SpBf, a homologue of factor B, the second component in the sea urchin complement system. *Journal of Immunology* 16: 6784-6793.

Smith, L.C., Clow, L.A. & Terwilliger, D.P. 2001. The ancestral complement system in sea urchins. *Immunological Review* 180: 16-34

Unuma, T., Okamoto, H., Konishi, K., Ohta, H. & Mori, K. 2001. Cloning of cDNA encoding vitellogenin and its expression in red sea urchin, *Pseudocentrotus depressus*. *Zoological Science* 18: 559-565

Wiens, M., Koziol, C., Hassanein, H.M.A., Batel, R., Schröder, H.C. & Müller, W.E.G. 1998. Induction of gene expression of the chaperones 14-3-3 and HSP70 by PCB118 (2,3',4,4',5-pentachlorobiphenyl) in the marine sponge *Geodia cydonium*: novel biomarkers for polychlorinated biphenyls. *Marine Ecology Progress Series* 165: 247-257.

The Sea Urchin: From Basic Biology to Aquaculture, Yokota, Matranga & Smolenicka (eds)
© Swets & Zeitlinger, Lisse, ISBN 90 5809 379 4

Management of an aquaculture company: evaluation of economic and technical aspects

V. Balsamo
Ittica Mediterranea S.r.l, Marsala, Italy

ABSTRACT: Marine aquaculture in the Mediterranean Sea is a sector which has grown very rapidly in the past few decades. Mariculture fish in the Mediterranean Sea is characterised by the dominion of two species: Sea bass (*Dicentrarchus labrax)* and Sea bream (*Sparus auratus*). They represent 40% and 58% of the total production, while the remainder 2% is constituted by the so-called new species (*Dentex dentex, Pagrus pagrus, Diplodus sargus).* The target of lowest production costs has pushed more and more farmers to improve rearing technologies. Within this context, managers' expertise in human resources as well as cost managing should represent an aid in achieving this aim.

1. INTRODUCTION

Aquaculture can be defined as an activity to enhance the economic potential of the marine environment. In general, aquaculture can be practised wherever water resources allow it. However, when such activities are established in a given zone, climatic, geomorphologic, environmental and social conditions of the area must be assessed. Moreover, the ecological context has to be taken into account in selecting the most appropriate rearing method and identifying the most suitable structures (Ravagnan G., 1992).

The climate of the area is a primary factor, together with water temperature, annual rainfall and wind conditions.

The following elements below are particularly important:

Environmental:
- quality and amount of the water resources
- physical characteristics of the suitable areas
- possibility to establish aquaculture plants in given area

Social:
- production facilities
- employment
- dietary habits
- existence of a scientific and technical institution

The evaluation of all these factors will allow to identify the species and the most proper rearing method, besides to arrange the relative staff for scientific research, business management and labour skills. They principally concern the species of aquatic organisms which are to be reared and the rearing method to apply.

A plant might be organised following two possibilities: either choosing a certain species and then looking for suitable conditions and places, or starting from an area and choosing the species accordingly.

What is attractive to investors, indeed, is that rearing is remunerative and technically reliable. As an alternative, they may consider the opportunity to exploit water and territory resources in view of a profitable activity. Once species are selected, the rearing method to be adopted, i.e. intensive, half-intensive or extensive, has to be chosen. Each farm adopts different working proceedings according to the species, as well as to the site and water availability. Even if the basic rearing operations (feeding, selection, etc.) are always the same, different methods could be applied even to the same organism based on the amount and quality of water. This means that each method should be tailor-made according to the specific areas.

2. DETERMINING FACTORS FOR THE ESTABLISHMENT OF A PLANT

2.1 *Criteria*

The criteria to evaluate, in selecting a location for an aquaculture farm, can be divided in three categories:

1. Physical/biological/chemical parameters:
   water temperature, salinity, oxygen, pH, turbidity, pollution, suspended solids, trophic levels, pathogenic organisms
2. Marine environmental parameters:
   sea depth, wave motion, current, sea bottom, atmospheric conditions (air temperature, rainfall, irradiation and wind)
3. Social and economic parameters:
   law, security, access, markets

2.2 *Water quality*

All the factors listed above concern the water quality, which is crucial in establishing the farm and in obtaining good results. Industrial areas nearby are to be avoided, as waters in such areas can be contaminated with toxic substances i.e. ammonia, nitrates, heavy metals and phenol compounds. Besides, waters must have the right pH, temperature, oxygen and salinity for the species to be reared.

Under normal conditions for most teleosts' life, pH of seawater should be from 7.5 to 8.5. Dissolved oxygen in water is a function of temperature and salinity. These values can vary considerably due to the influence of the biological factors in coastal areas, in presence of chlorophyllase synthesis. This balance can be changed at intensive rearing farms, in zones where water availability is limited. The resulting loss of oxygen may affect life quality of the organisms.

Most aquatic organisms do not have a thermoregulatory capacity. Then, metabolism is closely bound to the thermal level of their environment. A curve might be drawn to indicate the correlation between growth and temperature in order to identify the optimum parameters.

Changes in salinity may occur in estuarine areas, or in brackish waters. Despite the lower salinity values, they do not represent any problem to euryhaline species, which are able to adapt to such changes.

2.3 *Marine environmental factors*

Marine weather conditions are fundamental in choosing the type of the farm to be established and have to be taken into serious consideration, especially when offshore cages are going to be placed. Waves can damage the framework of cages; they act as an important stress factor for the reared fish and can determine high mortality due to excessive motion. Assessment of exposure to wave action will be necessary in choosing the installation site, such as bays, headlands or islands. Another important factor in planning a farm is represented by sea currents. Indeed, current intensity and direction are crucial to establish the position of pipelines for clean seawater and discharge of wastewater.

2.4 *Social and economic factors*

Social and economic factors are also important in order to identify elements to design a farm model for productivity and profitability purposes. Other items are related to basic infrastructures, such as the reliability of power supply, fresh water and communication systems. Transport and easy access to the market are also to be considered.

Equipment, technical assistance and labour skills must be available, too. Social and juridical aspects are also relevant to specific areas.

The environmental impact may affect areas with low hydrodynamics levels. The lack of legislation may also interact with environmental and biological elements in developing a rational model of a farm. This should guarantee operating safety, the best qualitative and quantitative production and the maximum profitability.

3. AQUACULTURE METHODS

There are two criteria to classify the various aquaculture methods. The first is based on the density of the species and the second on the alimentary diet of the reared animals. According to the first criterion, an intensive fish farming implies a high density of individuals per volumetric unit; an extensive farming has a low density, whereas a semi-intensive one is an intermediate type.

Evidence and experience indicate that the rearing density is closely related to the amount of food and other factors. When the number of fish individuals reaches the limits posed by the natural feeding resources in a given environment, an appropriate amount of food should be added. The rearing system can also be classified based on food supply.

In particular, it can be:

- extensive, when food supply is totally drawn from the environment.
- intensive, when food supply is totally provided by the farmer.
- semi-intensive, when food supply is provided both by the environment and by the farmer.

3.1 *Extensive rearing*

Managing an extensive rearing farm requires a simple methodology.

Some of the elements involved are:

- the ratio between the number of fish and the area
- choice of the fittest species
- optimum sizes for harvesting
- water management, i.e. water supply
- sufficient knowledge of the species to be reared, their behaviour, their alimentary habits, the optimum environmental conditions, their reactions to the adverse conditions.
- the frequency of tides
- the frequency, intensity and direction of winds.

3.2 *Half-intensive rearing*

In a half-intensive rearing there is a minimal external supply of feeding.

3.3 *Intensive rearing*

The most common intensive rearing systems are the land-based farm (usually in a concrete tank) and the offshore farm (usually in cages).

Tanks can be realised in a many forms and materials. The most common tanks are circular (plastic) or rectangular (concrete). They are often sunk in the earth and sometimes have reinforced concrete walls. In this kind of farm, water is supplied by means of a pipe placed on the sea bottom. Pumps can be submersed or water-scooping. Water can be used in catabolite and excrement removal, gas / heat transport and maintenance of an optimum environment for the reared animals.

Animal growth is directly affected by water temperature. The solubility of oxygen in water is determined by temperature, hydrostatic pressure and salinity. The oxygen needs of an organism also depend on its swimming activity and its health state. Even though many fish can survive in low oxygen conditions for short periods, oxygen levels under a certain threshold have negative effects on fish growth and their general state. The amount of oxygen can be increased by fresh water, enriched by pure oxygen, or by mechanical agitation of the surface.

The management of a land-based aquaculture farm requires to keep the basins bottom clean, in order to prevent the reproductive cycle of many parasites. Property inside the basin is maintained by using brushing machines and a suction pump.

Concerning the feeding, various methods exist, such as manual, by automatic and by tractor feeders. The first technique is applicable in a small farm where the amount of daily feed to supply is small. When substantial amounts of feed are to be administered, the other two methods are preferable. The tractor feeder is the best, as it requires less maintenance and because an operator can give feed up to 7,000 kg.

Sea farming in cages has recorded an explosive development in the last few years in many countries. This trend is confirmed by salmon aquaculture in Finland, Sweden, Ireland, Norway and Scotland (Harache, Y & Paquote, M., 1996). In the Mediterranean Sea, Greece explored sea farming thanks to the European Community financial support. Recently, an investment program has been developed and now Greece is the major producer of sea bass (*Dicentrarchus labrax*) and sea bream (*Sparus auratus*).

What makes this rearing system interesting is the minimum investment required, compared to land-base systems, which need continuos water supplies.

Further aspects of open-sea cages systems are:

- land development
- uncontrolled coastal urbanism
- high value of coastal areas
- a growing attention towards aquaculture, with particular regard to intensive farming.

In Italy the development of mariculture is focusing on the creation of open-sea farms. However, this development is facing some difficulties. Although several plants are present today in Italy, most of them are not highly productive, due to the lack of integrated technologies. In open-sea fish farming, cages are designed for extreme sea conditions. In this kind of farms, automated technologies are needed. The cages consist of synthetic nets, defining the water volume of the rearing place.

Generally they have:

- a framework supporting the net, defining the rearing volume and providing the necessary buoyancy.
- a system of chains or rigid elements connecting the net vertically and at the bottom of the rearing volume.
- an anchorage system.

Developing operations in farm management requires equipment and support systems to simplify and reduce manpower needs, e.g. by automatic feeding, monitoring and inspection systems or computer-controlled management systems involving the operator's intervention only in particular situations. Instruments are located on a boat to reach the cages or on a fixed or floating platform near the cages.

4. FEEDING

The aim of aquaculture is to increase the productivity, decreasing sale prices and operating margins without raising production costs. One of the highest costs concerns the feed. A feed should guarantee an optimum growth and, as a result, a high-quality aquaculture product. In the past, the feed was supplied in pressed pellets, obtained by finely ground raw materials. This process was not the best for feeds with a high energy content, since they dissolved quickly in water, had a poor digestibility and a high environmental impact. Now, almost every feed firm is producing by extrusion technique. The pellets are made by passing them through the extruder for a few seconds and the high temperature of the machine itself compacts the ingredients.

The characteristics of extruded feed compared to pressed products are a better digestibility, an improved stability in water, a smaller quantity of dust and less pollution. In addition, the extrusion technique made it possible to formulate feeds with a greater fat content and thus a higher energy content. The feeding table shows the daily ratio in percentage for the reared biomass. It also indicates how many kilos per day must be given. This table is based on seawater temperature and fish size. Therefore, information on the reared biomass in a cage or a basin is very important in farm management. When the oxygen level in water decreases below 4.5 mg/l, fish become increasingly stressed and an food assimilation decreases, too. In these cases, a feeder should rearrange the feed quantity to normalise rearing conditions.

In the last few years, many intensive aquaculture farms in basins have been providing monitoring systems to regulate the oxygen supply. These monitoring can help to solve one of major problems regarding land-based farms.

When high-energy feeds are used, the feed distribution method is crucial to the fish health. The feeding table is always theoretical and must be adjusted to the rearing conditions.

When a technician prepares a feeding table, he must take several factors into account, such as water temperature, weather conditions, oxygen concentration, property of the basin, rearing density, fish size, pathologies in course. The feeding tables are worked out in order to obtain a low food conversion rate (FCR) and an optimum growth. A correlation exists among daily ratio, conversion and growth. The conversion index is the relationship between the increased biomass of the animal and the amount of feed given. In the near future, feed producers' target should formulate feed compounds capable to achieve results already obtained in reared salmonids. Over-feeding must be avoided, in order to allow the fish to completely metabolise all the swallowed feed. Over-feeding may cause an increase in the food conversion rate (Saroglia M & Ingle E., 1992).

Another very important factor regarding feeding is the number and frequency of daily meals. In fact, while juveniles must be continuously fed to repletion, bigger fish need a reduced number of meals proportionally to their size. This is made in order to feed all organisms in a basin aiming at equal growth effects.

5. PHYTOPLANKTON PRODUCTION

The cultivation of micro-algae represents the first rearing phase for every marine species. Micro-algae indeed represent the first ring of the alimentary chain. They are used in sea bass and sea beam rearing, namely *Isochrysis sp.*, *Nannochloropsis sp.* and *Nannochloropsis oculata* and /or *Chlorella sp.*

The growth of the algae populations undergoes three principal phases: logarithmic, stationary and death phase (Saroglia M & Ingle E., 1992). In the logarithmic phase, the increasing rate of the algae population is constant and it depends on the availability of nutrients in water, light and temperature. In the stationary phase, the growth rate of the population is slackened, because of an intense metabolic activity. Algae reach such a density as they screen the light. The agitation of the cultivation allows exposure of algae to light. In the death phase, algae cells progressively die, because of their high concentration and the nutrient consumption.

The massive cultivation techniques must be designed considering these three phases besides the following elements: carbon dioxide, pH, turbulence and medium.

Temperature: since temperature between 18 and 20°C is a very important factor for algae growth, phytoplankton environments are kept at a constant temperature using heat pumps.

Light: the rearing place should be kept constantly lightened by neon lamps close to the alga containers (flasks, balloons, envelopes, tanks).

Carbon dioxide: this is an indispensable source of carbon for algae. A mixture of air and carbon dioxide in the cultivation is recommended. CO_2 does not become a limiting factor.

pH: the pH value of algae cultivation tends to rise, because of CO_2 consumption. The growth rate of the algae is generally higher at neutral pH, a condition to support a greater availability of inorganic carbon for the algae.

Turbulence: a constant turbulence of the cultivation aids the light exposure of all the cultivated algae cells.

Cultivation medium: it is generally used in liquid form. Although seawater contains sufficient nourishment concentrations for algal growth, these must be raised by adding a suitable cultivation medium in the massive algae productions.

The algae must be kept in aseptic environment in order to avoid contamination. All the material used for the manipulation must be sterilized by autoclaves. Algae are usually kept in a temperature-controlled container. Another common procedure is to inoculate them on agar media. The algae are purchased from Research Institutes or suitable International Banks, which guarantee their absolute purity. New algae are inoculated in 100 ml vessels. In a month, they grow and can be inoculated in a 250 ml flask. Maturity of an algal culture means that a maximum concentration of cells has been reached. The time required for a subsequent maturation will be shorter than the first time as algae are more concentrated. Algae in 100 ml and 250 ml containers require a month to grow. 15 days are needed for the second line of flasks and 7 days are necessary for all the subsequent volumes.

The concentration of algal cells must be periodically counted to follow the cultivation growth. This measurement can be done using a blood cells counter (hemocytometer). The ratio of lighted surface to volume is definitely larger in smaller volumes. The *Nannochloropsis sp.* usually reaches concentrations of 20-40 million cells per ml, while the *Isochrysis sp.*15-25 million cells per ml. These numbers are obtained in bigger volumes. Concentration is equal to 80 million for the green algae and 50 million for the brown algae.

Staff exposed to contact with the phyto-zooplankton must observe the following rules:

Operators must not have previously worked or passed through other environments.
All safety proceedings (such as pedal bath and hands sterilisation) must be thoroughly observed.
All material has to be sterilised prior to use.
Flasks, balloons or envelopes needed for inoculation must not be scattered.
The mouth of test-tubes and flasks must also be fire-sterilised.

All the criteria as above reduce the possibility of contamination to very low levels in a sterile environment.

6. PATHOLOGY

Diseases resulting from intensive rearing represent one of the greatest problems in aquaculture activity. Research on the fundamental defence mechanisms of fish and other marine organism against the pathogenic host was carried out (Le Breton, A., 1998). The study of fish disorders requires a wide knowledge about environmental conditions, aquatic medium, physiology and immune system of fish.

Aquaculture technicians should familiarise with basic fish pathologies to be able to give a detailed report of a preliminary examination to a veterinarian.

Alimentary and feed industries already know cleaning and sterilisation proceedings. In aquaculture however these practices have not been introduced sufficiently because of the recent intensification of production, leading to rear too many individuals in a same tank. This is a very stressful condition, as it triggers various pathologies (Le Breton, A., 1998).

Sterilisation modes can be divided into two categories:

- usual: normal sanitary precaution, such as sterilisation of trucks for fish transport.
- repressive: direct interventions against a given polluting agent or a disease.

Sterilisation removes morbid agents responsible for infections and contagious illness but it also eliminates all biological agents (Roberts R.J., 1990). It is used, for example, for the plankton containers. To be effective, the environment or material must be carefully cleaned beforehand.

The various stages of the cleaning process and subsequent sterilisation are listed:

1. Mechanical removal of all organic material and dust. This procedure is usually carried out with a high-pressure water hose or sponges with a rough surface.
2. Cleaning the surfaces with surface-active agents to remove any oily films.
3. Clean water rinsing to remove the detergent and any residuals.
4. Sterilisation of the environment and/or material.
5. Sterile water rinsing to remove all chemical residues used for the sterilisation itself.
6. Only healthy fish and materials having a suitable certification are allowed in fish farming. Unauthorised persons and materials are not allowed. Clothes (including boots, gloves and shoes) and equipment must be previously sterilised. The so-called sanitary barrier in every feed industry consists in antiseptic pedal bath at the entrance of each department to sterilise boots. The most appropriate disinfectant must be an active substance inert to organic material and resistant to decomposition. Commonly used chemicals are iodous substances, quaternary ammonium salts, sodium hypo-chloride and formaldehyde. These solutions must be periodically changed. It is recommendable to rotate the various products.
7. Cleaning operations to remove excrements on the basins bottom must be performed at least once a week.
8. Dead fish must be daily removed from the basins.
9. The filters must be cleaned and disinfected, as the cages nets.
10. Equipment must be disinfected before passing from one basin to another, or at least in the presence of pathologies.
11. The health state of hatcheries must be assessed before purchasing fish stocks.
12. Fish-eating birds should be scared off by means of acoustic devices and/or physical barriers (Schiavo, A., 1997).

Pathogens must be removed from water by using proper agents, chemical and/or physical. It is also necessary to remember that residues from such treatments can be toxic for larvae even after neutralisation.

We can distinguish water treatments into two methods: chemical and physical. Chemical disinfectants commonly used in farms rearing sea basses and sea breams are chlorine and ozone.

Physical methods have an immediate effect. UV rays have a germicidal action at 250 nm of wavelength. Their low toxicity makes them quite simple to use. Another method is microfiltering. Filtering at 10-30 mm is a widely-used process since it blocks parasites. A lower filtering (0.2 mm) also acts as a barrier against bacteria. Filtering systems are based on sand, on a cartridge, or drums.

6.1 *Prophylaxis*

The pathogenic agents of fish diseases are viruses, bacteria, fungi, dinoflagellate algae and protozoa.

Dangers caused by a prolonged use of chemical treatments in fish diseases (Wardle R., 1998) led pharmaceutical firms to study specific vaccines against the most frequent pathologies like those deriving from bacteria, such as Vibriosis and Pasteurellosis. The available vaccines have been prepared based on inactive bacteria. Recently, polyvalent vaccines, protecting fishes from several diseases are available on the market (Wardle R., 1998). Administrating a vaccine to a fish is an extremely delicate practiceand has to be carried out by qualified technicians by immersion, feeding or intramuscular injection. The most common method is immersion of the fish in a vaccine-containing solution. Prolonged antibiotic treatments or massive mortality rates resulting from some pathologies may be more costly to a fish farm than the high vaccination costs. In addition, it has been proven that vaccination against a few specific bacteria can improve the general state of health of some species due to increasing resistance. Vaccination is not a final solution to diseases occurring in fish aquaculture farms. The acquired immunity is not permanent but gradually declines over a certain period.

7. ECONOMIC FACTORS

Aquaculture is a sector which has grown very rapidly in the past few decades in Europe. The initial development of this industry by the northern European countries was soon transferred to the Mediterranean regions, with extraordinary results. The technological improvement and the favourable environmental conditions provided the basis for this industry. The EU sponsored this productive sector (Harache, Y & Paquote, M., 1996).

Sea bass and sea bream are the main species reared in the Mediterranean; their production increased exponentially in the last decade, from 8,000 tons to nearly 100,000 tons in 1999. During the same period, the total production of Sea bass has increased from 3,400 tons to 42,000 tons.

The sea bream production sharply increased, passing from 4,100 tons in 1990 to more than 55,000 tons in 1999. Greece remarkably contributed to this exponential growth (as shown in Figure 4). In 1990 production levels are similar in European countries. At the end of 1999, Greek production of Sea bass and Sea bream exceeds the total amount of all the other countries together.

Greece success is owed to environmental conditions and financial support by EU. But this exponential growth has caused a drop in prices.

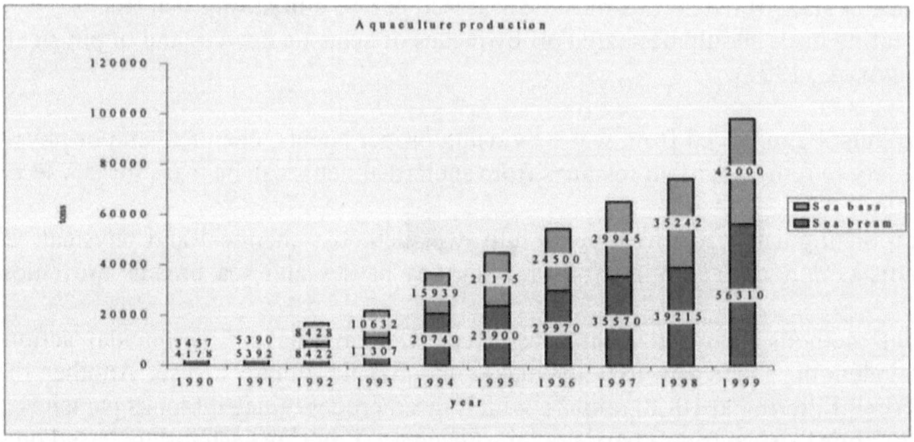

Figure 1. Increase of Sea bass and Sea bream production (by Associazione Piscicoltori Italiani).

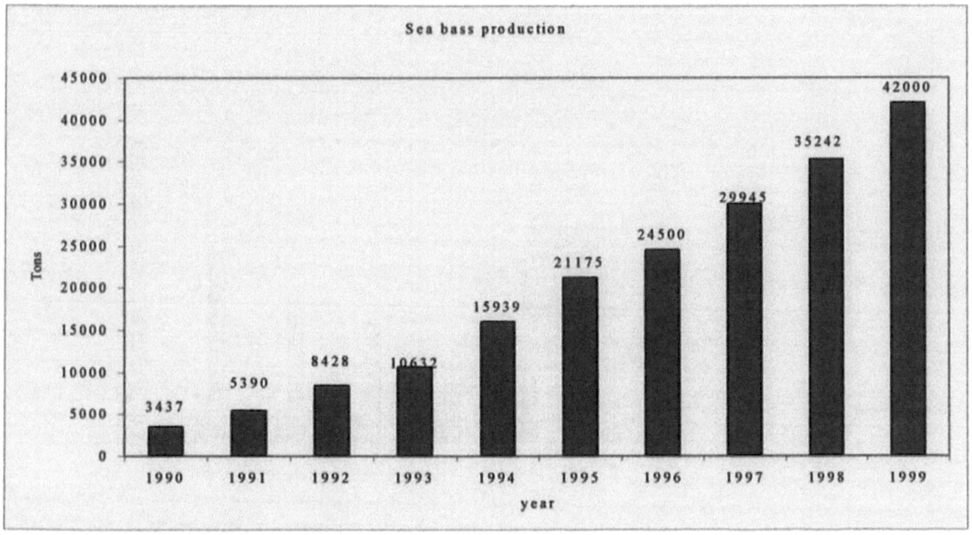

Figure 2. Increase of Sea bass production.

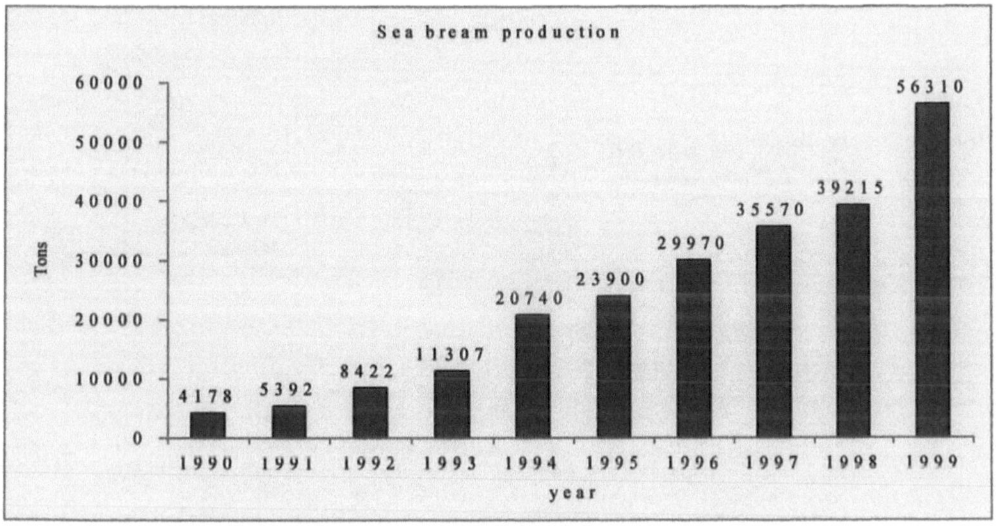

Figure 3. Increase of Sea bream production (by Associazione Italiana Piscicoltori).

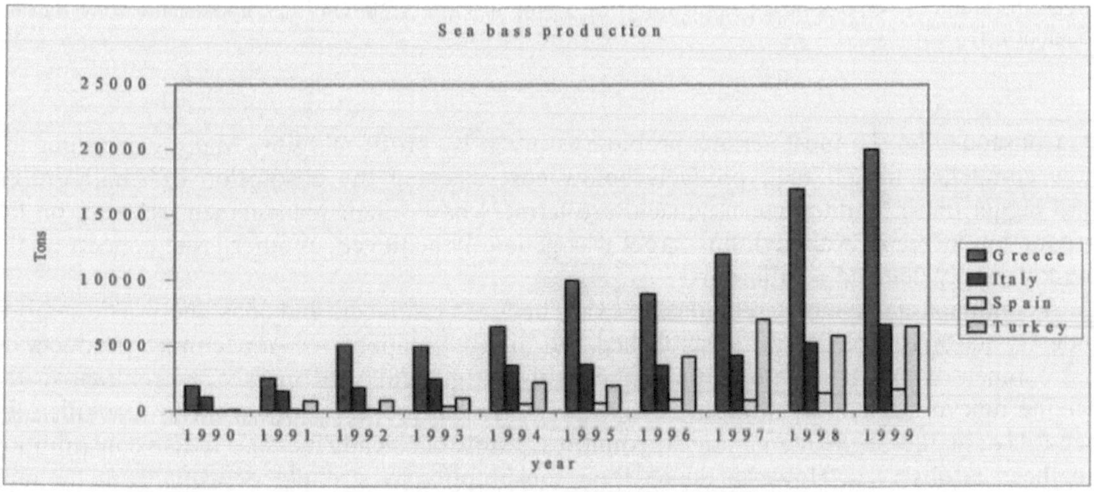

Figure 4. Mediterranean production of Sea bass. (by Associazione Italiana Piscicoltori).

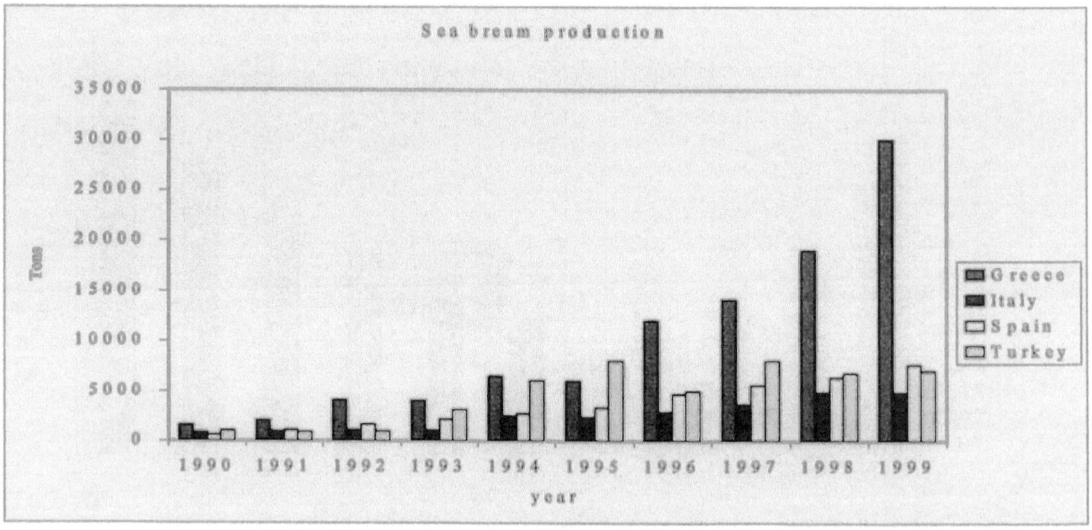

Figure 5. Mediterranean production of Sea bream. (by Associazione Italiana Piscicoltori).

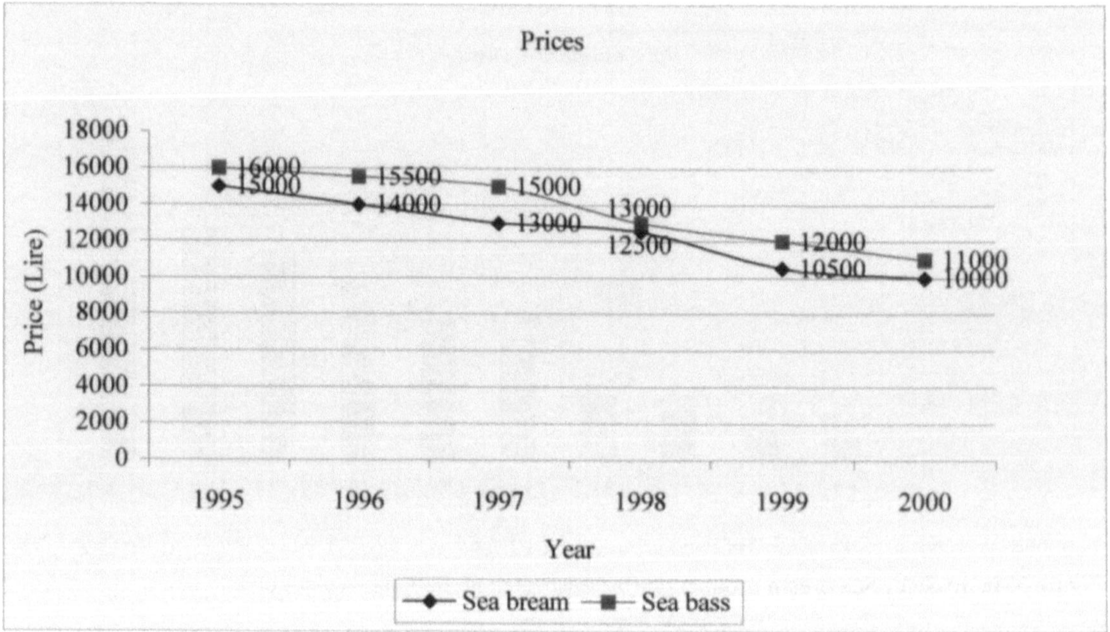

Figure 6. Trend of Sea bass and Sea bream price during last years. Italian lira (by Associazione Italiana Piscicoltori).

Consequently, the most serious problem is given by profit margins. Many companies are now compelled to sell their products below cost, causing the absorption of small groups into bigger ones. Mediterranean aquaculture farmers based their management strategy on the production increase. Although this target was generally achieved, another issue concerned the market supply/demand equilibrium.

A continuos and dynamic relationship exists between producers and consumers. The seafood industry today is globally well-established and highly competitive. Aquaculture products do not compete with other seafood, but rather with the general food market. In contrast to the Mediterranean mariculturers, northern European aquaculture operators seem to recognise the benefits of market segmentation. In EU countries, a slow but steady increase in fish consumption per head is observed. However, since food consumption is strongly dependent on national cultures:

- level of consumption per head
- local favourite species
- geographical origin of imports
- distribution of products in supermarkets.

Fish is important for a healthy diet. A general increase in fish products (fresh, frozen or other) consumption is observed. Aquaculture products advantages are:
- continuous supply
- controlled product
- regularity, homogeneity and quality of the product.

Thanks these advantages, the catering sector (hotels, restaurants, etc. is increasingly purchasing Mediterranean aquaculture products. Despite this advantage, this positive attitude towards fish consumption did not eliminate many traditional negative perceptions from consumers. Quality characteristics (freshness, etc.) of the available products represent the major consumers' concern.

Another issue with which Mediterranean aquaculture has to faced is the low pre-packed fish availability in stores. In order to match the current market and consumer trends, some producers are trying to develop more acceptable products in order to follow the market and fish food consumption trend.

7.1 *Marketing strategy*

An appropriate marketing strategy should be taken into consideration to meet this demand. Points to follow in order to promote the sector are:
- improvement of fish farm image
- quality certification
- association of specific products to a given company
- wider range of cultured species.

Producers must develop promotional actions to inform consumers on seafood products. It is important for producers to realise that the target is the consumer (Papageorgiou P., 1999). Advertising campaigns should emphasise the advantages of aquacultured products as:
- genuine products
- fresh and constant supplies
- high technology
- controlled quality
- competitive prices
- fish-food advantages
- healthy Mediterranean diet

Another important factor is to promote the image of the company associated to the product. This can be done by tagging fish to ensures the provenience. Information should be provided about company name, place of origin, fishing date, name and size of fishes. Such information can also be labelled on fish pre-package. Demand on pre-packed products (filleted) is expected to rise because of modern lifestyle. Mediterranean cultured fish is becoming part of the alimentary habits in countries as Germany, the Netherlands, North European Countries, etc. (Papageorgiou P., 1999). During the last couple of years, both consumers and retailers are emphasising the importance of quality certification. Despite of a variety of initiatives at national and international level for quality-certified aquacultured products, these are almost exclusively national certificates recognised to particular countries (Monfort, M.C., 1998). These certificates are mostly concerned with salmon and trout. With regard to the Mediterranean species, quality branding is almost non-existent, thus creating a serious competitive disadvantage, particularly in non-Mediterranean countries.

Recently, eco-labelling is gaining some importance. More and more consumers are asking for environmentally-friendly practices. Therefore, a company adopting eco-labelling on its productive cycle will be advantaged.

Table 1. New species planned for aquaculture in the Mediterranean Sea

Non-Sparids	Sparids
Solea solea	*Dentex dentex*
Umbrina cirrosa	*Diplodus puntazzo*
Sciaena umbra	*Diplodus sargus,*
Solea vulgarís,	*Diplodus vulgaris*
Thunnus tynnus	*Pagellus erythrinus*
Coryphaena hippurus	*Pagrus major*
	Pagrus pagrus
	Seriola dumerilii

Mediterranean producers have also increased the range of cultured species. Generally, "new" species can be divided into Sparid and non-Sparid. Table 1 lists the main species in which efforts are aimed at their commercial cultivation.

At present, some of these "new species" are going to be cultured on a commercial scale (i.e. *Diplodus puntazzo, Pagrus pagrus and Dentex dentex*) but the other ones are still at the experimental or pilot stage. All of them posed several problems during larval rearing (e.g. cannibalism) and also during growth (lack of an appropriate feed, pathologies). The market price is the same for Sea bass and Sea bream: this suggest that rearing "new species" is not a very convenient strategy of marketing diversification.

8. CONCLUSION

Reproducing and rearing stages of Sea bass and Sea bream are well known, even if new and unknown pathologies are coming out. Exponential growth of reared fish lowered fish prices, with compression of profit margins. Introduction of new products should be planned in order to increase the number of consumers. Diversification of reared species can only be a temporary solution in view of market expansion.

Sea urchin could replace sea bass and sea bream in the market demand. In this way, aquaculture would really introduce a "new species", provided that sea urchin researchers support farmers with their know-how.

REFERENCES

Harache, Y & Paquote, M. 1996.The development of marine fish farming in Europe: a parallel with salmon aquaculture. *Sea bass and Sea bream culture: problems and prospects; Internnational Workshop, Verona, 16-18 October 1996*, Verona: Italy.

Le Breton, A. 1998. An overview of the main infectious problems in cultured Sea Bass. *Fish* health management in Mediterranean mariculture. International Workshop, Athens, 12-15 *November 1998,* Athens: Greece.

Monfort, M.C. 1998. Sea bass and Sea Bream production explodes in Europe. *Fish Farming International. Dec. 98*. pp. 35-39.

Papageorgiou, P. 1999. Marketing development for new Mediterranean aquaculture species: enterprise strategies. *Mediterranean marine aquaculture finfish species diversification. International Workshop, Saragoza, 24-27 May 1999*, Saragoza, Spain.

Roberts, R.J.1990. *Fish Pathology*. In Cassell Ltd. London: England.
Schiavo, A. 1997. *Malattie e prevenzione nell'acquacoltura comunitaria*. Giuseppe La terza edizioni. Bari: Italy.
Sbroglia, M & Ingle, E. 1992. *Tecniche di acquicoltura*. Ed agricole, Bologna: Italy.
Wardle, R. 1998. Fish vaccination. *Fish health management in Mediterranean mariculture. International Workshop, Athens, 12-15 November 1998*, Athens: Greece.
Ravagnan, G. 1992. *Vallicoltura integrata* Ed agricole, Bologna: Italy.

The Sea Urchin: From Basic Biology to Aquaculture, Yokota, Matranga & Smolenicka (eds)
© Swets & Zeitlinger, Lisse, ISBN 90 5809 379 4

The role of the public in sustainable aquaculture

H.O. Halvorson & F. Quezada
Policy Center for Marine Biosciences and Technology, University of Massachusetts Boston, and Biotechnology Center of Excellence Corporation, Waltham, USA

ABSTRACT: Seventy percent of the world's conventional commercial fish species are now fully exploited, overexploited, depleted or recovering from depletion. This dramatic crash in the capture world fisheries production has led to problems in food distribution, balance of payments, employment, and ecological depletion. The food sector and related industries have looked to aquaculture to help offset in part the reduced availability of product from natural fisheries. Aquaculture itself brings with it a host of issues and challenges of various kinds in the areas of fish reproduction and nutrition, fish disease, environmental considerations and others. Science and technology approaches to these problems can contribute much to increasing the sustainability of aquaculture. Public support of aquaculture will thus also require public support and understanding of the science and technology platform upon which its movement toward sustainability is based.

1. INTRODUCTION

The breeding programs relating to terrestrial farm animals and plants in agriculture have revolutionized this industry over the past few hundred years. Similarly, in the "fish-farming" area, the new technologies that utilize modern biology to increase sustainable aquaculture and preserve biodiversity also provide a promise to address the problems referred to above. However aquaculture is somewhat less understood by the general public and itself has not been sufficiently subject to public discussion and approval. For public involvement in the expansion of aquaculture to provide added value to the decision-making process, it must go beyond having public acceptance or acquiescence as an end-goal in itself.

Currently, there is public concern over maintaining the existing pool of genetic information. If science based aquaculture is to respond to the various fisheries crises with innovation, the knowledge gap between public understanding and the orientation of the scientific community must be bridged. Strategies must be developed for achieving this. Release of transgenic species to the environment and handling exotic species, are useful case studies. Illustrations will be given of communication bridges to the public and ways to involve the public in making policy decisions. The main point of this paper is that informed involvement of the public contributes to the sustainability of modern aquaculture.

2. THE CURRENT CHALLENGE

With the continuing depletion of natural fisheries and the growing U.S. consumer demand for fish, there is an increasing percentage of importation of aquaculture-raised fish. Despite recent growth, aquaculture supplies only 10-15% of US seafood needs; the value of US aquaculture production ranked only tenth in the world in 1990. Similar trends are seen in other developed countries. Yet fish farming, the fastest growing sector of the world food economy, will surpass cattle ranching as a food source by the end of this decade (Brown, 2000). Aquacultural output, growing at 11 % a year, has climbed from 13 million tons of fish produced in 1990 to 31 million tons in 1998, according to an analysis written by Lester Brown, chairman of the institute, which tracks environmental trends around the world. On the other hand, global beef production four-fifths of it from rangelands, one-fifth from feedlots hit a high of 53 million tons in 1990 and has leveled off since then. While meat production is concentrated in industrialized countries, about 85 percent of fish farming are in developing nations. The world's leader is China, where aquaculture began 3,000 years ago. China accounts for more than two-thirds of world aquacultural output.

As the aquaculture activity increases, the obstacles also increase. Some of these obstacles are physical and environmental in nature and relate to the nature of the aquaculture activity itself. Other obstacles are legal and political. The public plays a key role in both the manifestation and the creative solution of these latter obstacles. There are several types of examples of deterrents have prevented the growth of marine aquaculture. These include regulatory restrictions, often promulgated for reasons other than aquaculture, public concern over environmental impacts, use conflicts over valuable shoreline property, and lack of local, state or federal support.

There are many specific examples that could be cited from documented public discussion. A brief review of the world press provides many headlines (Table 1). The following three examples will be discussed here:

Table 1. Aquaculture Headlines in the World Press.

Norwegian-bred salmon may have been genetically modified and exposed to cancer-causing vaccines.
Suspects in the mystery of Scotland's vanishing salmon: Fish farms, seals and global warming are all blamed for what some see as a crisis.
The Scottish Parliament is to launch an enquiry into the effects sea cage fish farming has on the environment.
A deadly virus that decimated salmon farms in New Brunswick has been detected in two more Canadian fish pens.
Maine Salmon Face Upstream Battle as Species; Returns Dwindle as Aquaculture Booms
Chris Charlton of Langley has been dealing with disturbing images of huge, grotesque and misshapen genetically engineered salmon swimming in BC waters.
Halting sacrilege in the Highlands: Courtroom success for environmentalists in Canada is a reminder of the task faced by those combating Scottish salmon farming.
U.S. Agrees to E.U. Labeling Requirements for Genetically Modified Foods

"Do escaped farm salmon jeopardize native populations?" A major row has broken out in Canada over a proposal to import Norwegian salmon to stock fish farms (MacKenzie, 2000). The government's Office of Aquaculture and the farming industry have dismissed fears of escaped salmon wreaking havoc, saying they will enrich the genetic diversity of wild stocks. In Montreal the government's fisheries scientists warned that escapees could jeopardize the survival of Canada's wild salmon. Canadian salmon farmers want to import breeding stock from Norway's highly successful farms. "They want the same opportunity to improve their stock as other livestock producers," says Dave Conley, a spokesman for the Office of Aquaculture. The office wants the government to lift its ban on importing farmed salmon.

"Scottish salmon farming revolution that has left the seas awash with toxic chemicals." Carrell (2000) writes that Environmentalists have accused the salmon farming industry of poisoning shellfish stocks, thus creating toxic algae blooms around the coast which threaten the survival of wild salmon stocks. Fish farms have also been accused of using illegal toxic chemicals, leading to criminal inquiries by environmental regulators. Mass escapes of farmed fish have also led to claims that these will irrevocably damage the country's wild stocks. The Scottish Environment Protection Agency (Sepa) is preparing to tighten its already strict rules on the location of salmon farms, their use of restricted chemicals, and their discharge of waste into the seas. Under a stricter system of regulations aiming to combat an increase in pollution incidents, the agency has threatened to revoke licenses and relocate fish farms.

"Greens wary of new fish farm rules." Environmental groups say new aquaculture regulations unveiled October 5th in Victoria, British Columbia will have little clout and may lead to the lifting of a provincial fish farm moratorium that has been in place since 1995 (Gofish.com News, 2000). Lynn Hunter, an aquaculture specialist with the David Suzuki Foundation, told the Canadian Press that the government "is holding hands with the industry" "They are supposed to be acting for the benefit of all British Colombians, not the fish farming industry," she said. The provincial government implemented the new regulations in response to two highly publicized escapes of more than 35,000 salmon into waters off Vancouver Island. The rules require every BC fish farmer to submit a written fish escape and response plan to the government by Oct. 31. The reports must also include predator avoidance plans, inspection and maintenance schedules for containment structures, and descriptions of existing and proposed containment structures. Fisheries Minister Corky Evans says the government is not lifting the moratorium and has no interest in going back to the days fish farming was driven by the stock exchange.

Concerns have also been expressed from governmental institutions, which add to public concern. For example, the decline in pelagic stocks in the Northeast Atlantic is the "single biggest environmental issue" facing the aquaculture industry in Europe. This is the view of Michael O Cinneide, the manager in charge of environmental monitoring at Ireland's Marine Institute (Cinneide). About 50,000 tons each of mackerel blue whiting and horse mackerel are landed into Ireland every year, making a total of 150,000 tons of pelagic species. Much of this is landed into Killybegs and transported to Mayo, where it made into fishmeal for the Irish aquaculture industry. Although stocks of mackerel are relatively healthy at present (about 2.5 million tons in the NE Atlantic) there are concerns about declining stocks of blue whiting and horse mackerel. In the future, consumers in Europe will become aware of this, and environmental groups will become aware on it, and they will want to see the industry - whether Norway or Ireland - using an input that is a sustainable and managed input. In its capacity, the Marine Institute has an important role to play in that we are the Department of the Marine on the management of those stocks," he said.

3. PUBLIC ATTITUDES

Three areas typically associated with public concerns about aquaculture include: (a) organic waste pollution generated by intensive aquaculture activities, (b) genetic "pollution" through the possible mixture of exotic species (or transgenic species) with native species, and (c) land use conflicts, particularly as perceived by abutting property owners. Each of these areas bring forth legitimate issues which have been addressed with varying degrees of success depending on the nature and extent of the specific aquaculture activity in question, and, more importantly, the nature, extent and quality of informed public involvement.

Although the documented articulation of public opposition to aquaculture only occasionally reaches the general press, the concerns of local resident groups are more often expressed at open hearings held as part of the permitting process or other regulatory procedures. Regulations themselves have come under review in states, which are making a deliberate effort to foster commercial aquaculture. Increasingly public interests groups have been involved. For example, recently an environmental group filed Clean Water Act suit against salmon farms as reported by Emery of the Associated Press (2000). A national environmental group plans to file a lawsuit accusing the operators of three of the state's salmon farms of releasing pollutants in violation of the Clean Water Act. The U.S. Public Interest Research Group (PIRG) said Atlantic Salmon of Maine, Connors Aquaculture and the Stolt Sea Farm discharged pollutants such as fish waste and chemicals every day at the mouths of Maine rivers without applying for the required federal discharge permits. But a state aquaculture advocate said the farms are treated uniquely under the Clean Water Act and are not required to get federal discharge permits. The National Environmental Law Center, on behalf of U.S. PIRG and four of its members in Maine, sent a formal notice letter to the companies. The group accused the farms of discharging fish waste and food and a chemical product used to control sea lice infestations. The group also said the farms allow captive-bred salmon to escape and disrupt the habitat and health of wild salmon.

To be sure, aquaculture development is stifled by regulations that are unnecessarily cumbersome and sometimes conflicting, inconsistent and difficult to coordinate among state, local and federal levels. However, it is also argued here that modern, science-based regulations can themselves be important instruments of public involvement. It is when regulatory formulation and policy-making are devoid of science-based content that public misperceptions may arise and stalemates may occur.

Public perception and public involvement in aquaculture are key inputs and necessary components of its viability and thus invite a new definition of sustainable aquaculture. The avoidance of conflict itself is not a goal nor does conflict avoidance necessarily contribute to aquaculture development. It is not a matter of simply getting acquiescence from the public. Building an aquaculture industry with little or no informed public involvement in policy making for aquaculture is like building a house on sand. Indeed, creative outreach measures are needed to enlist the interest and energies that will be required to ensure that everyone who stands to benefit from commercial aquaculture also be part of its development.

At the same time it has been argued that fishermen should also be more proactive. Pendleton, President of the Northwest Atlantic Marine Alliance (NAMA) and Simonitsch, Chairman of the NAMA Board of Trustees, reported (1999) on recommendations on the "The Sustainable Fisheries Act". They wrote, "The fishing industry should claim the conservationists' agenda as its own. Fishermen should become the leading advocates for marine resources. The industry should be proactive instead of defensive and reactive and take charge of the issues. We should be bridging the chasms that separate us from government and conservationists".

4. PUBLIC TRUST DOCTORINE

Several years ago, several of the authors' colleagues, Archer and Bowen, published a book titled The Public Trust Doctrine and the Management of America's Coasts (Archer et al. 1994) Some of the points covered in this work are applicable here.

"The traditional principles of the public trust doctrine are: (1) all tidelands and lands under navigable waters were owned by the original thirteen states at the time of the American Revolution, as successors in sovereignty to the English Crown, and each subsequent state was endowed with similar ownership rights at the time of its admission into the Union; (2) the states own these lands subject to a 'public trust' for the benefit of all their citizens with respect to certain rights of usage, particularly uses related to maritime commerce, navigation, and fishing; and (3) all lawful grants of such lands by a state to private owners have been made subject to that trust and to the state's obligation to protect the public interest from any use that would substantially impair the trust. Moreover, any such conveyed lands must be used by their private owners so as to promote the public interest and so as not to interfere unduly with the public's several rights under the public trust doctrine".

Regarding the influence of English Common Law they write: "Under seventeenth- and eighteenth-century English common law, coastal lands were subject to two different ownership interests. The Crown held title to the coastal waters and lands beneath them by virtue of its sovereignty. This interest was called the jus privatum, which was a transferable property interest held by the Crown or by private persons claiming through it. The 'jus privatum, was, however, subject to the public trust rights of the English people, who held the "jus publicum"- the common right to use these public trust lands and their resources for certain traditional purposes necessary to individual survival and livelihood, including navigation, commerce, and fishing".

Regarding Aquaculture: "Aquaculture is defined as the propagation, cultivation, and harvesting of aquatic plants and animals. Under an expanding concept of the public trust doctrine, these activities may readily be included. The doctrine has traditionally covered the public's right to harvest fish and shellfish. It is only a small step to broaden the doctrine to encourage activities, such as the cultivation of fish that promotes continued abundance. This is particularly appropriate because courts have almost universally held that title to the fish within state waters is held by the state in trust for the people of the state. In administrating this trust, the legislature may pass laws regulating the taking and use of fish for the purpose of both protection and harves".

"The idea that a state may authorize the use of public trust lands for the propagation and cultivation of fish is not new. In the early 1900s the Supreme Court of Florida held that, while public trust lands may not be sold outright to individuals, the state may grant to an individual limited rights to use public trust lands when such use will encourage industries and the development of natural and artificial resources. The court considered 'planting and propagating oysters or shell-fis' an activity that enhanced and the rights and interests of the whole people Therefore, lands held within the public trust could be made available to a private person for undertaking these activities".

5. SCIENCE AS A TOOL

If science is to serve as a tool in working to solve some of the environmental challenges for aquaculture, then we must also be prepared to address certain obstacles relating to public attitudes toward science itself. A lack in public understanding of science is itself a barrier and thus can limit the use of science as a tool in sustainable aquaculture development. Scientific approaches to sustainable aquaculture development afford new and changing insights relating to the public concerns listed above.

Over the past two centuries, genetic improvements have been achieved largely through breeding programs involving hybridization, selection and mutagenesis. In agriculture, breeding programs have been developed for terrestrial farm animals in most countries and terrestrial plant production is generally based on highly improved varieties. Today it is not possible to base terrestrial animal production on wild, unimproved stocks, and yet a large part of the aquaculture industry continues to do that. It has been argued that the current knowledge about phenotype and genetic parameters for economically important traits in aquaculture species is sufficient to start breeding programs for Atlantic and Pacific salmon, rainbow trout, tilapia, catfish and several carp species. Genetic engineering is the most effective method for genetic improvement of specific traits that are controlled by single genes (Halvorson & Quezada, 1999). The success of genetic manipulation techniques with seaweed and marine animals will depend upon the economic gains to the aquafarmers, who thus far have depended entirely on the empirical aspects of naturally occurring genotypes.

Sustainability depends on successful reproduction of the life cycle in captivity. Numerous new techniques in biotechnology have been introduced to induce spawning, control the sex, regulate morophogenesis and accelerate growth, among others. Induction of ovulation and spermination of domestic broodstock are achieved by injection or polymer implant of pituitary hormones, as gonadotropin, synthetic peptides and gonadotropin releasing hormones. As frequently one sex grows faster, sex control is possible by endocrine manipulation during early development. Smolting can be accelerated in salmonids through use of a short photoperiod, by prolactin and somatatropin in salmonids, and in larval abalone metamorphosis is acceleration by treatment with gamma-aminobutyric acid like molecules. Thyroid and steroid injections stimulate growth in a number of fish.

A promising mechanism to control released organisms is to utilize fish that are sterile and therefore cannot establish wild populations or interbreed with wild fish. The common method is to manipulate their chromosomes by thermal, pressure or chemical shocks to newly fertilized eggs so that they produce three sets of chromosomes instead of the usual two. This technique has been used extensively with bivalves leading to superior growth and suppression of gamete formation. Triploid rainbow trout continue to grow long after their diploid counterparts began to divert their energies to reproductive activities. Similar results have been reported for the tench and Atlantic salmon. However triploidy has not always resulted in superior performance. While triploidy is the current method of choice other techniques are being developed to produce sterility in marine animals. A promising approach involves cloning of aquacultured fish by tetraploidizing precursor cells, combined with gynogenic induction of spawned eggs and sex reversal. Cloning wild genotypes has benefits for stock enhancement, conservation, restoration and hatchery production. The ongoing breeding of tetraploids with diploids leads to sterile or partly sterile triploids.

Public concern has also been raised that the use of fishmeal as a diet for aquaculture only further contributes to the declining wild catch yields. The response has been a search for plant substitutes. A recent example is Chinese fish decision to switch to soybean meal (Fossbakk, 2000). For the past eight years, according to the American Soybean Association (ASA), the Association has been teaching Chinese fish farmers how to increase their production by switching from the grass and manure diet to diets based on the use of soybean meal as the primary source of protein. . Fishmeal has traditionally been the principal protein used in fish diets. Chinese officials have said on Nov. 13 of this year that the production of soybean based aquaculture feeds increased from zero in 1990 to more than 110 million bushels by 1998. Due to the rapid growth of the aquaculture industry it is expected that by 2005 the demand will pass 257 million bushels. Chinese farmers have more than 3000 years experience with aquaculture and producing fish on a diet of grass and manure from hogs and chickens. They also began the practice of polyculture, often placing as many as seven species of fish in a single pond. The ASA have been working with many Chinese farmers to encourage them to switch from polyculture to monoculture production.

5.1 *An approach to achieving public involvement*

Although there is clearly no single approach to achieving public involvement, we can outline selected principles of such an approach. They are as follows:

(a) Trust the public. Invite their involvement. Assume they have an interest.
(b) Broaden the definition of "stakeholders". Stakeholders are not just the "squeaky wheels" but rather should include more than the shore residents and abutters. Ultimately the concept of stakeholders must include the consumers, direct and indirect employees and others.
(c) Regularize communications with the public. An ongoing and regular relationship must be built on shared interests in the industry, the environment and the societal good.
(d) Lower the threshold for entry into public involvement. Sometimes the public fails to be engaged if it is too much trouble. We need to make it less onerous for the public to become involved.
(e) Use feedback as a reinforcing mechanism. Keep people informed.
(f) Share successes with the public as well as obstacles and challenges.

How do we increase awareness for the need for the development of marine aquaculture?

Policy decisions are no better than the scientific, economic and other technical advice on which they are based. In her Kennedy School of Government study of advice to state legislators, Jones et al. (1997) reported that most states do not handle science technology issues very well. Analysis of the information from eleven states showed that the most important issue mentioned by legislators and their staff was trust in the validity of the information. Next was the issue of process. Information should be accessible, timely, expressed in non-technical language, and in a convenient format.

There are several approaches one could consider to address the lack of awareness regarding aquaculture. First, one could increase the efforts to get he spectrum of information up to a threshold level to get it into the political currency. Secondly, one could take a more practical approach by building coalitions around an issue that could effect the outcome. In both cases, the arguments include: environmental considerations (sustainable aquaculture, protecting the seas), job creation and job retention (fish processing and marketing industry), and international experiences and competition. The audiences to be addressed are varied and each should be met on their own ground and each involves its own language.

In developing a strategy for increased awareness, one must consider that varying levels of technical information are available at the different levels of decision-making in government (local, state, and federal). Different coalitions may be needed at different levels of government. A promising approach would be to form coalitions between professional organizations (who have the expertise) and organizations representing the public (i.e., potential consumers and environmental organizations), each of which has national and grass roots infrastructures. The credibility afforded by such coalitions has a greater chance of increasing public awareness to impact public policy. This process can best apply to the current public concern over "Genetic Enhancement".

We need to balance the risks with the benefits by a process in which the public is involved and has confidence. Since this process can frequently take about 10 years or more, it is important to start early.

We need commonly accepted guidelines that include containment (both physical and biological) procedures to protect the public and that will also permit testing ofnew technologies to achieve genetic enhancement. These guidelines should be open to periodic review and subject to change as evidence may indicate.

Consensus Conferences should be organized in which relevant Professional Scientific Societies, Environmentalist Organizations, who represent the public, and appropriate government organizations and other stakeholders can meet and seek common goals.

If these are successful the meetings can review and agree on the relevant facts about this new technology and the risks of the containment procedures available.

Such a group could recommend risk assessment experiments for given procedures and point out areas where our scientific information inadequate.

The care and husbandry of many marine animals places a premium on scientific and public access to knowledge of life cycles, distribution and health of marine animals.

The worldwide distribution of marine laboratories is valuable resources for such databases. NAML and MARS offer the public's only opportunity to see marine animals and already collaborating in sharing such data.

We readily admit that no one has the answer to how we go about efficiently involving the public. We do have some useful examples in which release of genetically modified organisms to the environment and in which aquaculture activities have been the basis of decision making by consensus building including broad representatives of the public. These case studies can serve as examples of how we can achieve public participation through the participation of selected publics and decision makers in addressing how science can help with the crises of the collapse of the capture fisheries.

There are several examples I would site. The first On May 23, 1984 Representative Albert Gore, Jr., who was Chairman of the House Science and Technology Subcommittee on Investigations and Oversight, requested that the American Society for Microbiology convene "a national symposium to examine scientific issues involved in releasing genetically engineered organisms into the environment." He suggested that the purpose of the symposium be "to help identify the scientific issues of relevance and to stimulate debate in the appropriate scientific disciplines." He further recommended that the symposium result in the publication of a collection of papers by and for scientists, a summary for a lay audience, and an overview of methods for producing genetically engineered organisms. The meeting was a dramatic success in which molecular biologists and environmentalists, who represent the public, struggled for days before a consensus emerged (Day, 1985). Day remarked they (regulators) must also weigh the estimated risks in the context of a real world in which far more drastic things are done every day on a global scale in the name of agriculture. In summarizing the workshop, Adelberg (1985) concluded "It is up to the scientists to provide the best estimates of benefits and risks and up to the regulators to weigh these estimates wisely when deciding how much regulation is needed". The final result has been a decade of selectively introducing this technology into present day agriculture.

The second example involves the Maryland Roundtable Oyster Mediation (McCoy, 1993). The Secretary of Natural Resources, Torrey C. Brown, brought together various parties: aquaculture industry, Waterman Association, state legislators, regulators such as the Department of Natural Resources, Environmental Protection, various environmental groups, and other concerned individuals and institutions. With the assistance of an arbitration mediator, over a period of more than a year, the parties were able to communicate over complex issues and develop an action plan that sets aside three zones beginning at river mouths and going upstream. This led to a renewal of the industry in Maryland and an improvement in water quality of the Chesapeake Bay watershed. To transfer this strategy, it was thought necessary to do three things. First, identify aquaculture opportunities unique to this area, assess available technologies, address regulatory problems, and prepare educational materials about aquaculture for regulators, practitioners and the general public. Second, recognize and address major deterrents to marine aquaculture, e.g. water quality and competing uses for coastal waters. Third, involve all the relevant constituencies - public and private - in the process of developing recommendations that will accommodate all relevant views and concerns.

The third case is the Sea scallop Working Group in Massachusetts (Johnson & Halvorson, 1996). A Sea Scallop Working Group (SSWG) representing diverse organizations have been meeting since December 1994 to explore critical issues, define possible options for action, and develop an industry-driven, bottom-up approach to sea scallop aquaculture, and to take advantage of the indigenous and abundant giant scallop in Massachusetts waters. A workshop was held which involved many individuals representing all aspects and disciplines that would be associated with the development of a sea scallop aquaculture industry in Massachusetts and used external experts to develop an overall consensus action plan. The document resulting from this Workshop puts forward recommendations from the perspective of potential sea scallop farmers tempered by the advice and guidance of professional scientists, government managers, regulators, lawyers, environmentalists, and economic development specialists.

In January, 2001, we will host a workshop in the Boston area: "Marine Aquaculture and the Environment: A meeting for stakeholders in the Northeast" (Web site). We have use conflicts between existing public and private activities and native aquatic species, which inhabit areas that may be impacted by the aquaculture infrastructure deployment. Coastal user groups range from property owners to maritime recreation and conventional fisheries activities. Furthermore, we have nuisance species and populations based on non-point source pollution issues, which are problematic for both sides. We have policy needs in developing use patterns and identifying potential opportunities. We need policy guidelines on nutrient and biological loading thresholds from finfish operations as they relate to the carrying capacity of the host ecosystem so that we can develop ecologically sustainable aquaculture. We will also need policy structure for the lease of public lands under the ocean, which will support future aquaculture development. To accomplish these objectives and change the future of marine aquaculture in pursuit of Sustainable Seas we must work together. There is a strong need for stakeholders from the aquaculture industry, environmental groups, government officials and the seafood industry to begin a collaborative initiative to develop environmentally sustainable aquacultural practices in the Northeast.

The questions to be addressed are: What are the issues of concern? What is known? Where are the gaps in our knowledge? How do we prioritize and deal with these gaps? How should decision-making about aquaculture development proceed - - given what is known and not known?

To address the questions, key stakeholders (industry through environmentalists) will be brought to "the table" to participate in public discussion where the relevant scientific and technical information, including gaps in our knowledge, is available to all. It is the intent of this Meeting of Stakeholders to have the relevant stakeholders address these decisions collectively and develop a common map and common language relating to the critical issues that need to be addressed.

A two and a half day workshop is planned in which four working papers will be prepared and distributed in advance of the workshop organized around relevant topics which include: (a) Impacts on habitats: determining what is acceptable, (b). Interactions of cultured and wild species, (c). Effects on other wild life: Marine animals, Birds, Invertebrates, and (d). Mechanisms for stakeholders consensus building. Assembled groups will address these issues and make recommendations.

6. FOCUSING ON REAL ISSUES

Congressman George E. Brown of California, former chair of the House Committee on Science and Technology, and a staunch supporter for a public voice in public policy, firmly believed that we should focus our attention and activities on the real issues (Brown, 1988). In his opinion they are:

Protecting our environment.
The need for sustainable resources.
The leveling of the food supply in the face of continued population growth.
The depletion of our fossil reserves.
The problem of reemerging diseases.

Note that sustainable aquaculture applies to three of these issues.

These illustrations of communication bridges to the public and ways to involve the public in making policy decisions may serve as useful mechanisms to establish sustainable aquaculture.

REFERENCES

Adelberg, E. 1985. Scientific Issues. in *Engineered Organisms in the Environment*: 233-235. Washington, ASM.

Archer, J., Connors, D.L., Laurence, K., Columbia, S.C. & Bowen, R. 1994. *The Public Trust Doctrine and the Management of America's Coasts*. Amherst: University of Massachusetts Press.

Brown, G.E. 1988. in Golden W.T. (ed), *Science and Technology Advice to the President, Congress and Judiciary*. 65-70. New York.; Pergamon Press.

Brown, L.R. 2000. Fish farming may soon overtake cattle ranching as a food source. Worldwatch Issue Alert. 3.

Carrell, S. 2000. Scottish salmon farming revolution that has left the seas awash with toxic chemicals. The Independent (London), October 2.

Cinneide, M.O. www.intrafish.com.

Day P. 1985. a Perspective on the Problem. in *Engineered Organisms in the Environment*: 2-10. Washington, ASM.

Emery, M. 2000. Environmental group to file Clean Water Act suit against salmon farms. Associated Press. April 27. www.intrafish.com.

Fossbakk, T-E. 2000. Chinese fish farmers switch to soybean meal. SeaWeb & The Ocean Project, November 13. www.fis.com.

Halvorson, H.O. & Quezada, F. 1999. Increasing public involvement in enriching our fish stocks through genetic enhancement. Genetic Analysis and Biomedical Engineering. 15: 75-84.

Jones M., Guston, D.H. & Branscomb, L.M. 1997. *Informed Legislature: Coping with Science in a Democracy*. Landham: University Press of America, Inc.

MacKenzie, D. 2000. Do escaped farm salmon jeopardize native populations? New Scientist, February 6.

McCoy, H. 1993. Maryland Oyster Roundtable Action Plan, December, Arbitration, Mediation, and Aquaculture. Aquaculture 21: 41-47

Pendleton, C. & Simonitsch, M. 1999. Commercial Fisheries News. February.

Rollin, J. & Halvorson, H.O. 1996. Policy Center for Marine Biosciences and Technology. Journal of Marine Biotechnology : 4, 1-4.

"Greens wary of new fish farm rules B.C. regulations require fish escape and response plans" Gofish.com News October 05, 2000. www.GOFISH.COM.

Web site http://www.es.umb.edu/mae01.

© Swets & Zeitlinger, Lisse, ISBN 90 5809 379 4

Developmental table of *Hemicentrotus pulcherrimus*

(from Ph. D. dissertation of Dr. S. Noguchi)

Hemicentrotus pulcherrimus is one of the most common sea urchins found in the shallow seas around the Japanese coast. It has been a favorite subject for the studies of embryology and cytology, and it has also been one of the marine resources of food in Japan. In the past, the late development of this species was investigated by Mortensen (1937), Kume (1927), and Onoda (1936), but they could not follow the development of larva after the mid 8-armed stage on. Using of the rearing apparatus (Hinegardner 1969), larvae of this species are possible to metamorphose in large numbers in laboratory. In this chapter, development from the unfertilized egg to young sea urchin reared at 17 °C are shown. Bars are 100 mm.

1, unfertilized egg. en, egg nucleus. 2, fertilized egg (3 min after fertilization). Fertilization envelope (fe) has formed. n, zygotic nucleus. 3, 2-cell stage (90 min). 4, 4-cell stage (3 hr). The second cleavage occurs parallel to the egg-axis, animal-vegetal pole axis (A-V axis) which is morphologically discriminated in the embryo at the 16-cell stgae at first. 5, 8-cell stage (4 hr). 6, 16-cell stage (5 hr). Unequal division occurs in the vegetal half and the embryo has three different sized blastomeres, the mesomeres (meso), the macromeres (mac) and the micromeres (mic). A, animal pole. V, vegetal pole. 7, 16-cell stage embryo viewed from the vegetal side. 8, 28-cell stage (6 hr). Divisions of the micromeres are delay. 9, 28-cell stage (vegetal view). 10, 32-cell stage (7 hr, vegetal view). The micromeres unequally divide to give rise to the larger and smaller micromeres (l-mic, s-mic). 11, early morula stage (9 hr, surface view). 12, The early morula focussed in center of the embryo. 13, late morula stage (20 hr). Embryo begins to rotate by cilia (cil) within the fertilization envelope. 14, mesenchyme blastula stage (25 hr). The embryo dissolves the fertilization envelope and swims out. From the vegetal pole, the primary mesencgyme cells (pmc) ingress. 15, mesenchyme blastula viewed from the vegetal pole. 16, early gastrula stage (30 hr). The cell layer at the vegetal pole begins to invaginate. The scale bar in 1 is available for 1-4, and in 5 for 5-16, respectively.

17, mid gastrula stage (40 hr). The invagination (gastrulation) progresses to form the archenteron (arch). From the tip of the archenteron the secondary mesenchyme cells (smc) are liberated. The scale bar is available for 17-22. 18, mid gastrula (vegetal view). 19, late gastrula stage (50 hr). The archenteron elongates to reach the animal pole of the embryo to complete gastrulation. Two triradiate spicules occur in center of the cell clusters of the primary mesenchyme cells (pmc). 20, late gastrula (side view). Tip of the archenteron decline toward anterior side of the embryo, and then contact with the stomodium invagination to open the mouth (the presumptive mouth region is indicated by an arrow). 21, prism stage (65 hr), ventral view. The archenteron tip forms bilobed spaces, the left and the right coelomic vesicles (l.cv, r.cv). 22, side view of the prism larva. The mouth opens. The archenteron constricts into three parts to give rise to the esophagus (es), the stomach (st) and the intestine (i). The blastopore (bp) becomes anus. The scale bar in 17 is avallable for 17-22.

23, early pluteus stage (two-arm pluteus; 78hr), dorsal view. From each triradiate spicule, body rod (br), antero-lateral rod (alr) and vetral transvers rod (vtr) elongate, respectively. Postoral rod (por) branches from one arm of the triradiate spicule. 24, two-arm pluteus (left side view). 25, early 4-arm pluteus stage (4 days, ventral view). Antero-lateral rods (alr) elongate supporting the antero-lateral arms (ala). 26, early 4-arm pluteus stage (4 days, right side view). The digestive tract begins to function, the mouth open wide, esophagus (es) becomes muscular, the stomach develops into a large sac having long cilia, and the slender intestine curves along the stomach to the larval anus on the ventral side. The scale bar in 25 is available for 23-26.

27, late 4-armed pluteus stage (7-days old, ventral view). The preoral ciliated band (pr.cil.b) is formed in front of the mouth. Pigment cells (pig) are distributed uniformly on the body surfaces. The right and left coelomic vesicles (r.cv, l.cv) are adhering closely to the stomach (st) and the esophagus (es), making slender sacs. The body of this larva is supported by the body rods (br), the right and left post oral rods (r.por, l.por), the right and left antero-lateral rods (r.alr, l.alr) and the ventral transverse rods (vtr). The body rods (br) are straight and their posterior margins are broad with some side processes. On the ventral side, the ventral transverse rods (vtr) elongate toward the center of the body having slightly pointed ends without a connection each other. Some mesenchyme cells (me) gathered at the base of the left and right post oral arms (l.poa, r.poa) on the dorsal side of embryo. In the center of the gathered mesenchyme cells (me), the granular crystal of calcite is formed.

28, early 6-armed pluteus stage (9-days old, dorsal view). The epithelium at the bases of the of post oral arms (poa) on the dorsal side, begin to protrude antero-diagonally. In this protruded parts, the granular crystals formed in the center of gathered mesenchym cells (me) grows up to triradiate spicules, which are the rudiments of the right and left postero-dorsal rods (ru. r.pdr, ru.l.pdr). Another cluster of the mesencyme cells appears on the dorsal side near the surface of esophagus and a triradiate crystal, a rudiment of the dorsal arch, is formed in the center of the gathering cells (not shown).

29, mid 6-armed pluteus stage (15-days old, dorsal view). The left coelomic vesicle separates from the wall of esophagus, and constricts to form the anterior and posterior coelomic vesicle. The left anterior coelomic vesicle further divides into a hydrocoel (hy) and an axocoel (ax). The right coelomic vesicle (r.cv.) constricts a little later than the left. A pore canal, which has elongated from the vicinity of a joint of the hydrocoel and axocoel, opens on the dorsal surface as a madreporic pore (m.p). However, since the opening is small and larval body is opaque, it is extremely difficult to recognize the pore without paying special attention by light microscope.

30, late 6-armed pluteus stage (20-says old, dorsal view). The epithelium lying between the left post oral and the left postero-dorsal arms (l. poa, l. pda) begins to invaginate as an amniotic invagination (am.in). The invagination of the amniotic epithelium becomes deeper and comes into contact with the hydrocoel (hy) which has grown remarkably. Meanwhile, the right coelomic vesicle divided into anterior and posteior parts (r.acv, r.pcv), but the boundary between them was not clear. The anterior ciliated bands (epaulet type; a.cil.b) begin to appear on the ventral side of the bases of the post oral arms and on the dorsal side of the bases of the postero-dorsal arms (pda). As the body rods (br) begin to be absorbed from their posterior tips, the posterior area of the body becomes round.

31, early 8-armed pluteus stage (25-days old, ventral view). The right and left preoral arms (r.proa, l.proa) are made by elongation of the dorsal arch (da), increasing the total number of larval arms to eight. The posterior ciliated bands (po.cil.b) begin to form on the right and left sides at the posterior edges of body (r.po.cil.b, l.po.cil.b) and develop encircling in the posterior area of the larva. The hydrocoel (hy) is completely covered with the floor of the amniotic invagination, and give rise to the so-called "Echinus rudiment".

32, mid 8-armed pluteus stage (about one month old, dorsal view). After the absorption of the body rods, the posterior area of the larva becomes a flat hill. The floor of amniotic invagination bulge out forming five radial bulges. Each bulge is a rudiment of the primary tube foot (ru.ptf). With the growth of the rudiment, the left post-oral, left postero-dorsal and left antero-lateral arms (l.poa, l.pda, l.ala) gradually become shorter beginning from the anterior tips. About the 32nd day after fertilization, the larva reached 0.9mm in body length and 0.5mm in body width, and it can be recognized as a white speck with naked eyes.

33, mid 8-armed pluteus stage (about onc month old, left side view). The rudiments of five primary tube feet (ru.ptf) seem to be pentagonal. A pair of ciliated epaulets on the dorsal and ventral sides extends toward the medial line of the body and joined with each other. Consequently, on the dorsal and ventral sides, two transverse ciliated bands (a.cil.b (v), a.cil.b (d)) are formed. With further development, the primary tube feet (ptf) with round tips start to move. Both on the posterior and on the right side of the larva, the knob-liked protrusions of the epithelium appeared In these protrusions, the astral spicules were recognizable by using the polarizing plates (not shown). These were the rudiments of quadrate dorsal spines (d.sp-1 in Fig. 35).

34, late 8-srmed pluteus stage (about 40-days old, dorsal view). The larva is just before metamorphosis. The arms on the left side of larva become shorter. The larval spines (lar.sp) growing on the floor of the invagination pointed their tips towards the left side of the body and move slowly in the invagination. On the other hand, the primary tube feet those tips transformed from round to flat, begin to move actively. Because of the increase in the body weight, it is difficult for the larva of this stage to keep floating by their ciliary movement alone, and they tend to sink to the bottom of the vessel when the agitation is ceased. Since the echinus rudiment grows so much in size that the stomach, which is greenish in color, is pushed towards the right side of the larval body. Some skeletal meshworks, rudiments of genital plates (ru.gpl) are observed in the inner side of main body. When the larva is squashed between a glass slide and a cover glass, the small skeletal meshworks, astral spicules and rudiments of spines are observable in the echinus rudiment.

35, larva immediately after metamorphosis (45-days old, ventral view). On the 45th day, larvae sink to the bottom of the vessel to metamorphosis and begin to extrude the primary tube feet (ptf) and the spines (v.sp) to exterior of the body. While the left side of the larva is expanding because of the emergences of tube feet and spines, the antero-postero axis of the larva gradually bend to the right side of the larva. At the same time, the anterior area containing the antero-lateral and the post oral arms are twisting anti-clockwise when viewed from the ventral and rear sides.

36, young sea urchin, immediately after metamorphosis (45-days old). The primary tube feet (ptf) and spines (v.sp) thoroughly emerge to the exterior, and the metamorphosis is completed. It takes about one hour for the completion of emergence of the tube feet and the spines. The subsequent stage after the metamorphosis is termed "Young sea urchin." The young sea urchin creeps around on the substratum by means of its tube feet, spreading the quadrate dorsal spines (d.sp-2) and the ventral spines (v.sp) towards the left side of the larval body. During these changes, all of the larval arms were being absorbed together with the skeletons and the epithelia.

37, young sea urchin immediately after metamorphosis (45-days old, dorsal view). Young sea urchin turns the left side of its body downward after a little while. As the result, the right post oral and right postero-dorsal arms (r. poa, r. pda) are situated on the upper side of the young sea urchin. Since the future anus opens on the upper side and the future mouth opens on the lower side, it is apparent that the dorso-ventro axis of the young sea urchin shifts 90 degrees from the larval axis through the metamorphosis. As development proceeds, since the above-mentioned bending of antero-postero axis to the right side of the larval stage increases, a part of the dorsal area of the young sea urchin now consists of the posterior area of the larval stage.

38, young sea urchin (47-days old, dorsal view). Both the larval arms and the ciliated bands are completely absorbed. The young sea urchin measured 1mm in diameter including the spines. On the dorsal side, 6 or 9 quadrate dorsal spines (d.sp-1) appear, all of which developed on the body surface other than the echinus rudiment. In the ambulacral zone (amb.z), there were two quadrate dorsalspines (d.sp-2) on the aboral side and a primary tube foot (ptf) on the lateral side. In the interambulacral zone (i-amb.z), a total of 4 ventral spines (v.sp) are formed ; one at the upper row, two at the middle row and one at the lower row.

39, young sea urchin (48-days old, ventral view). The major part of the ventral side is occupied by mouth organ (teeth, to; buccal plates, b.pl). In each ambulacral zone of the ventral side, a pair of the secondary tube feet (stf) develops between the primary tube feet (ptf), and the number of the tube feet becomes to 15. The young sea urchin in this stage still lacks the mouth and the anus.

REFERENCES

Heinegardner, R.T. 1969. Growth and development of the laboratory cultured sea urchin. *Biological Bulletin* 137:465-475.

Kume, M. 1927. On the development of the sea urchin found at Misaki. (In Japanese) *Dobutugaku Zassi* 41:100-105.

Mortensen, Th. 1937. Contributions to the study of the development and larval forms of Echinoderms. III. *Memoir of Royal Academy of Science Letter Denmark* 7: 1-65.

Onoda, K. 1936. Notes on the development of some Japanese echinoids, with special reference of the structure of the larval body, Report III. *Japanese Journal of Zoology* 6: 637-654.

LIST OF ABBREVIATION

a.cil.b: anterior ciliated band
a.cil.b(d): anterior ciliated band of the dorsal side
a.cil.b(v): anterior ciliated band of the ventral side
ala: antero-lateral arm
alr: antero-lateral rod
amb. z : ambulacral zone
am. in : amniotic invagination
anus: anus
arch: archenteron
at: apical tuft
ax: axocoel
bc: blastocoel
bl: blastomere
bp: blastopore
b. pl : buccal plate
br : body rod
cil: cilia
da : dorsal arch
d. sp-1 : quadrate dorsal spine-1
d. sp-2 : quadrate dorsal spine-2
en: egg nucleus
es : esophagus.
fe: fertilization envelope
gpl : genital plate
hl: hyaline layer
hy : hydrocoel
I: intestine
i. amb. z: inter-ambulacral zone
l.a.cil.b (d): left anterior ciliated band of the dorsal side
l.a.cil.b (v): left anterior ciliated band of the ventral side
l.ala: left antero-lateral arm
l.alr: left antero-lateral rod
lar. sp: larval spine in the echinus rudiment

l.br: left body rod
l.cv:: left coelomic vesicle
l. mic: larger micromere
l. pcv: left posterior coelomic vesicle
l. pda: left postero-dorsal arm
l. pdr: left postero-daosal rod
l. poa: left post oral arm
l. po.cil.b: left posterior ciliated band
l. por: left post oral rod
l. proa: left preoral arm
l. pror: left preoral rod
l. vtr: left ventral transverse rod
m: mouth
mac: macromere
me: mesenchyme cell
meso: mesomere
mic: micromere
m.p: madreporic pore
n: zygotic nucleus
pig : pigment cell
pmc: primary mesenchyme cell
po. cil. b: posterior ciliated band
pr. cil.b : preoral ciliated band
proa: preoral arm
pror: preoral rod
pt f: primary tube foot
r.a.cil.b (d): right anterior ciliated band of the dorsal side
r.a.cil.b (v): right anterior ciliated band of the ventral side
r. a.cv: right anterior coelomic vesicle
r. ala: right antero-lateral arm
r. alr: right antero-lateral rod
r.br: right body rod
r. cv: right coelomic vesicle
r. p.cv: right posterior coelomic vesicle
r. pda: right postero-dorsal arm
r. pdr : right postero-dorsal rod
r. poa : right post oral arm
r. por : right post oral rod
r. po.cil.b: right posterior ciliated band
r. proa: right preoral arm
r. pror: right preoral rod
r. vtr: right ventral transverse rod
ru. da: rudiment of dorsal arch
ru. l. pdr: rudiment of left postero-dorsal rod
ru. ptf: rudiment of primary tube foot
ru. r. pdr : rudiment of right postero-dorsal rod
s: spicule
smc: secondary mesenchyme cell
s. mic: smaller micromere
sp: spine
st : stomach
to : tooth or teeth
v. sp : ventral spine
v t r: ventral transverse rod

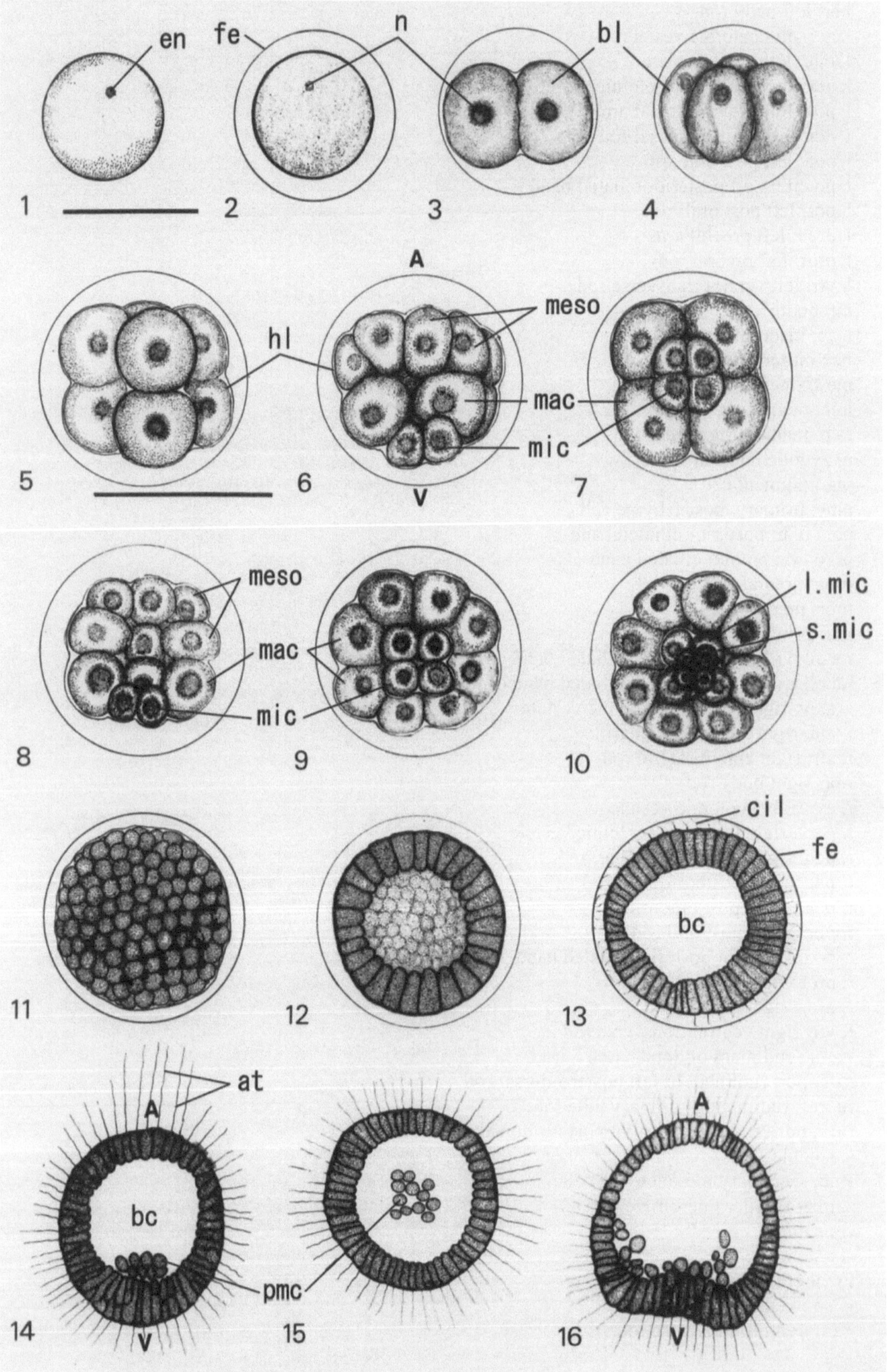
en
fe
n
bl
1
2
3
4
A
meso
hl
mac
mic
5
6
7
V
meso
l. mic
s. mic
mac
mic
8
9
10
cil
fe
bc
11
12
13
at
A
A
bc
pmc
14
V
15
16
V

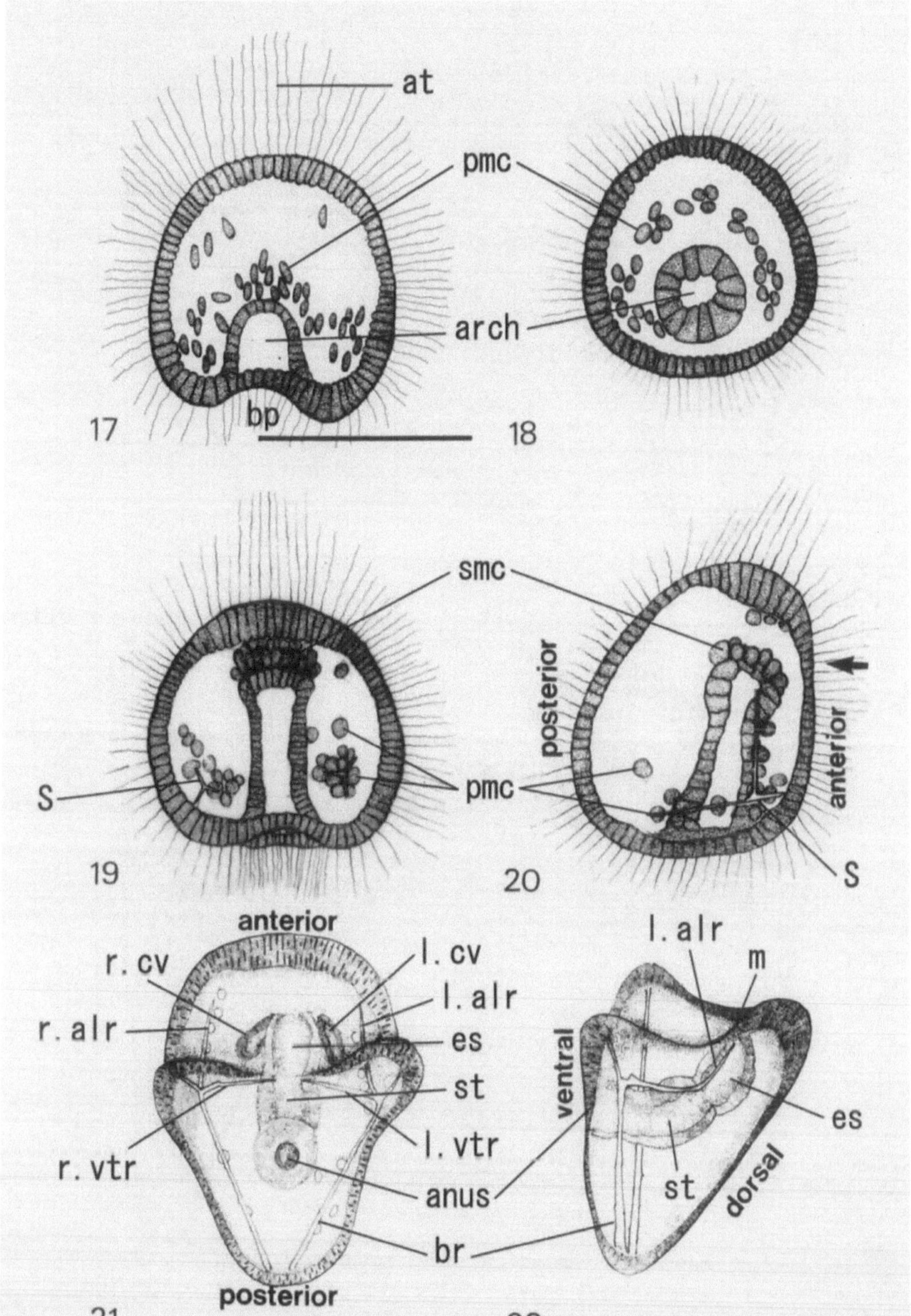
at
pmc
arch
bp
17
18
smc
posterior
anterior
S
pmc
19
20
S
anterior
r. cv
l. cv
r. alr
l. alr
es
st
r. vtr
l. vtr
anus
br
posterior
21
l. alr
m
ventral
es
st
dorsal
22

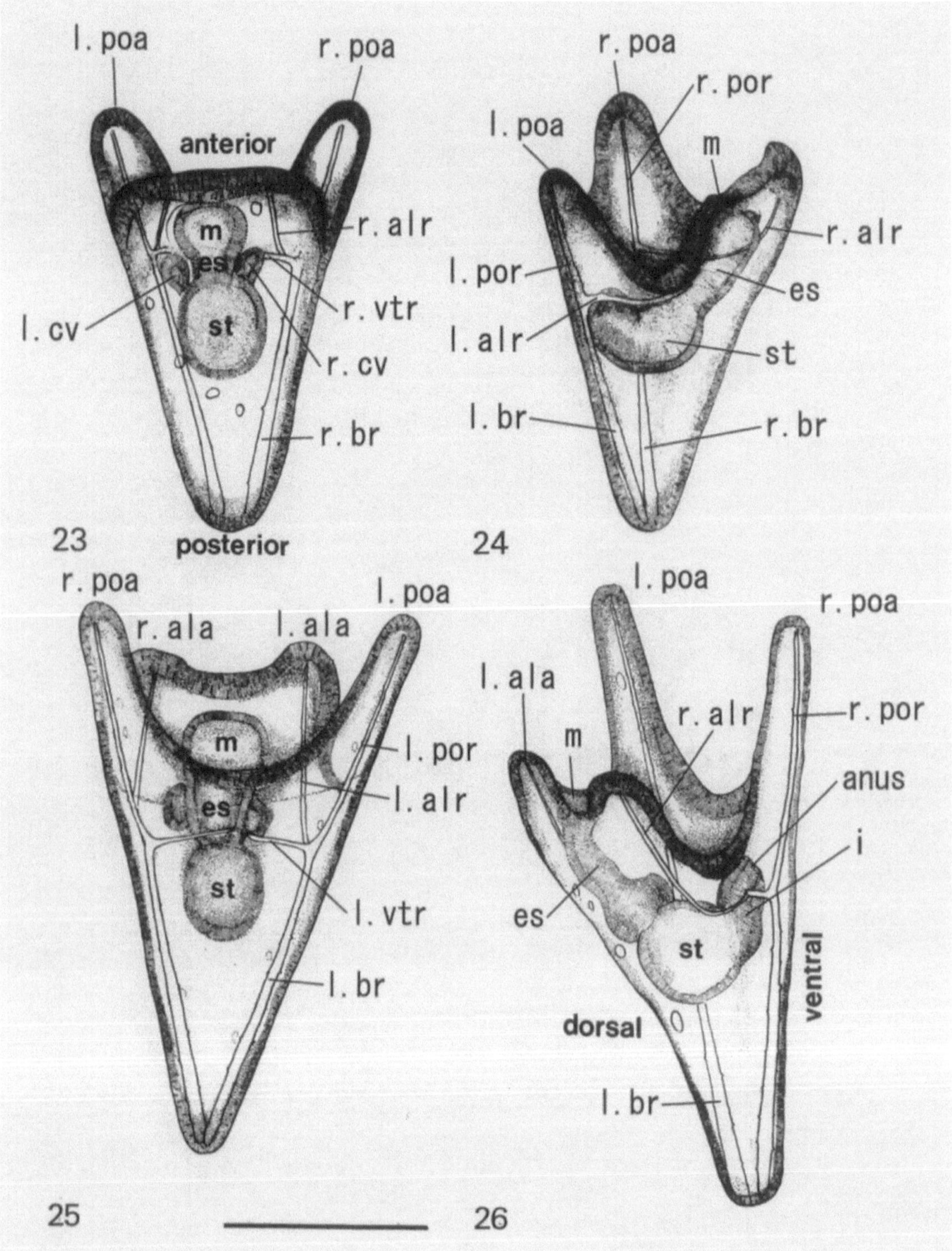
l. poa
r. poa
anterior
r. alr
m
es
l. cv
r. vtr
st
r. cv
r. br
23
posterior
r. poa
r. por
l. poa
m
r. alr
l. por
es
l. alr
st
l. br
r. br
24
r. poa
l. poa
r. ala
l. ala
m
l. por
l. alr
es
st
l. vtr
l. br
25
l. poa
r. poa
l. ala
r. alr
r. por
m
anus
i
es
st
ventral
dorsal
l. br
26

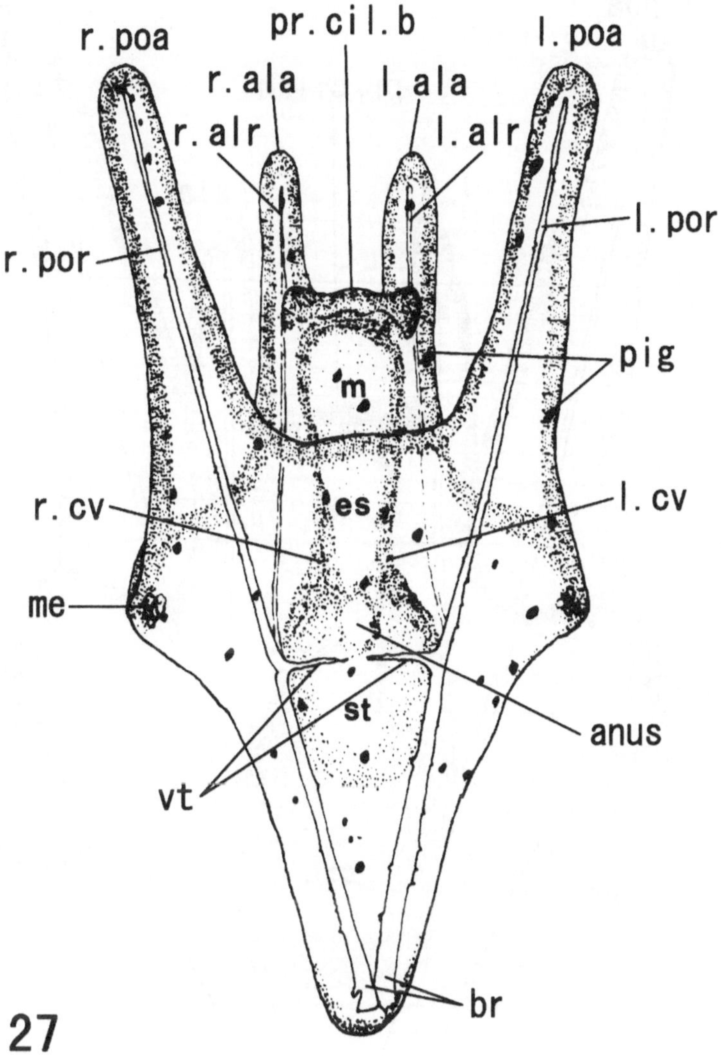

r. poa
pr. cil. b
l. poa
r. ala
l. ala
r. alr
l. alr
l. por
r. por
pig
m
es
r. cv
l. cv
me
st
anus
vt
br

27

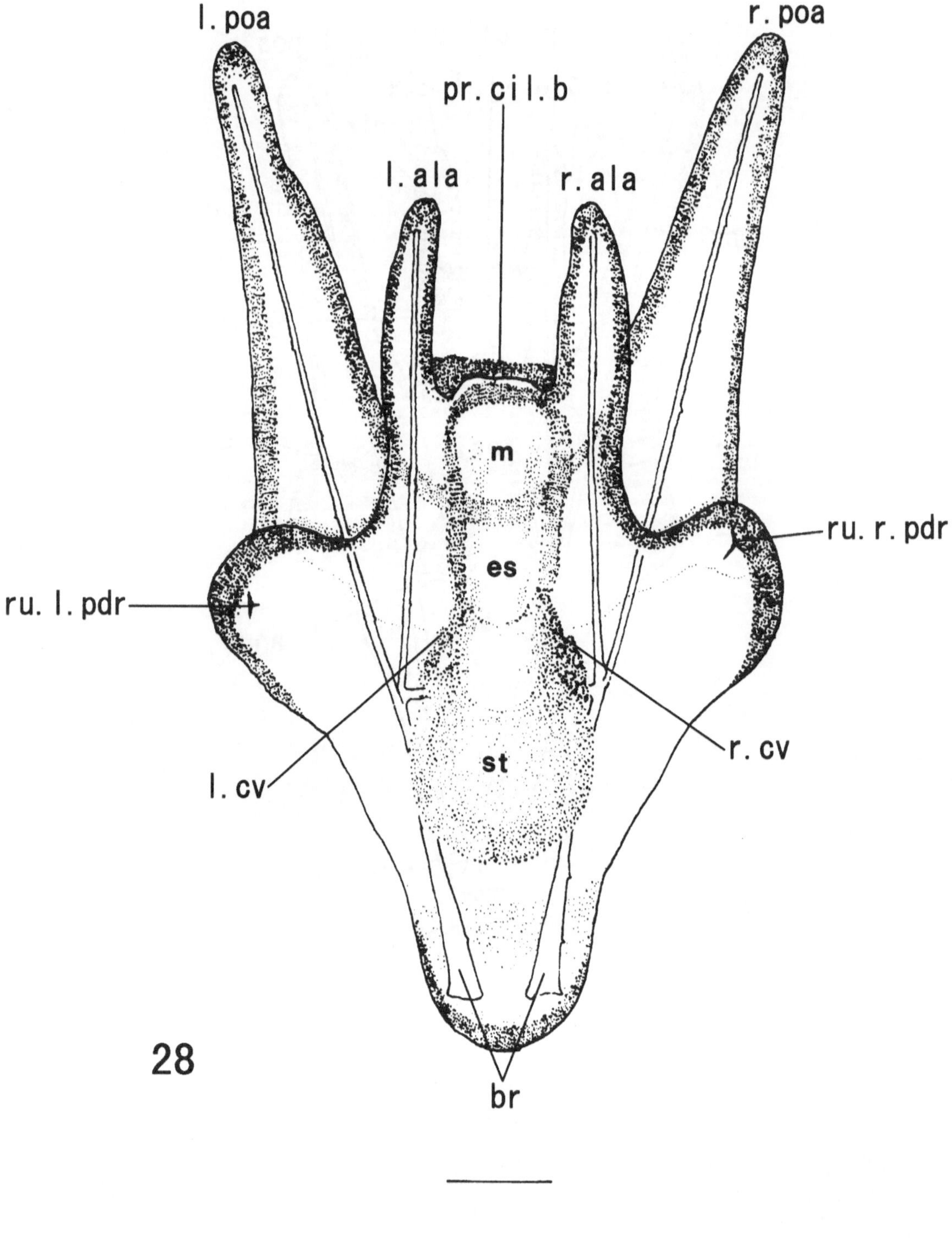

28

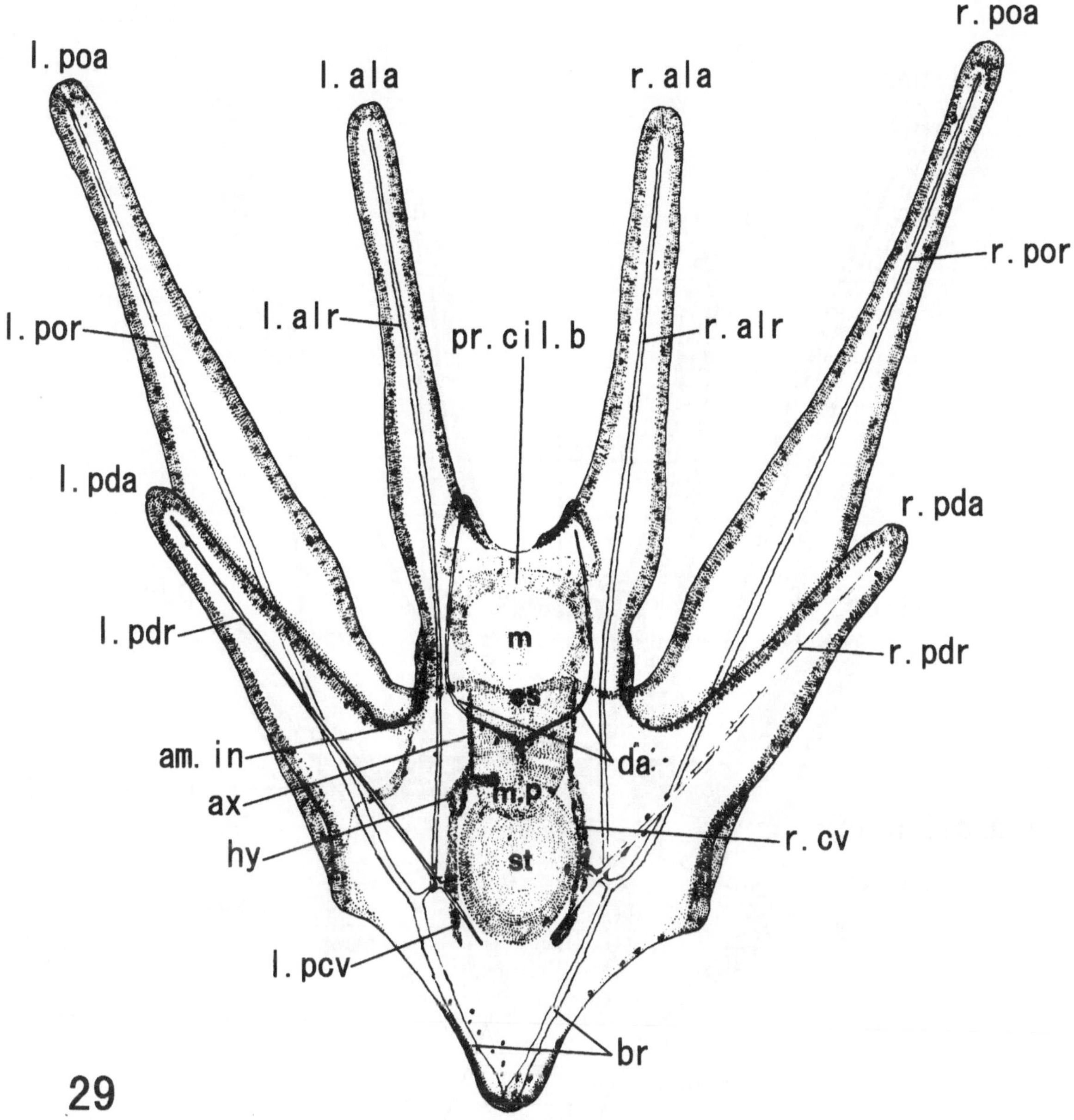
l.poa
l.ala
r.ala
r.poa
r.por
l.por
l.alr
r.alr
pr.cil.b
l.pda
r.pda
l.pdr
m
r.pdr
es
am. in
da
ax
m.p
hy
r.cv
st
l.pcv
br

29

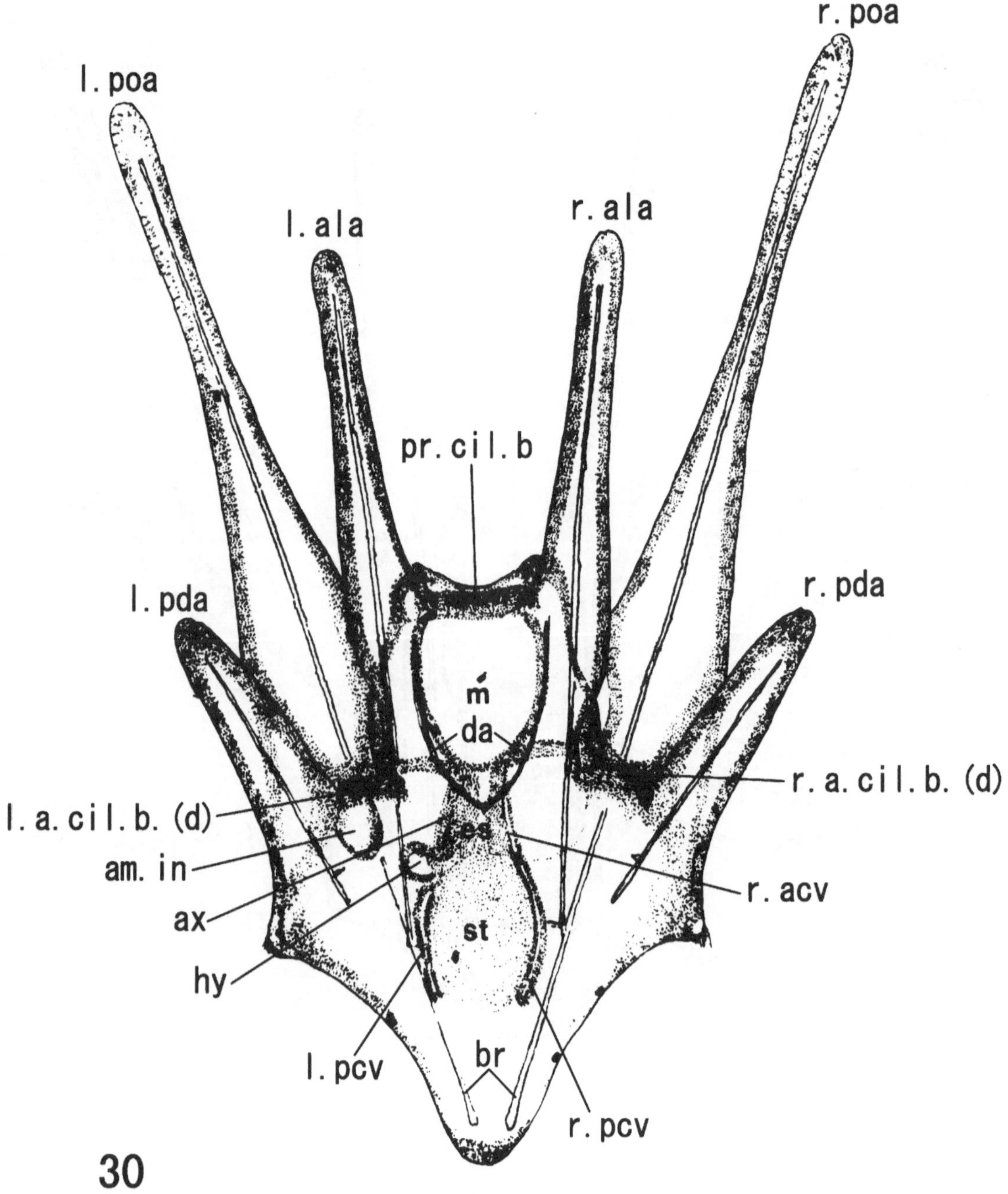
l. poa
r. poa
l. ala
r. ala
pr. cil. b
l. pda
r. pda
ḿ
da
r. a. cil. b. (d)
l. a. cil. b. (d)
es
am. in
r. acv
ax
st
hy
l. pcv
br
r. pcv

30

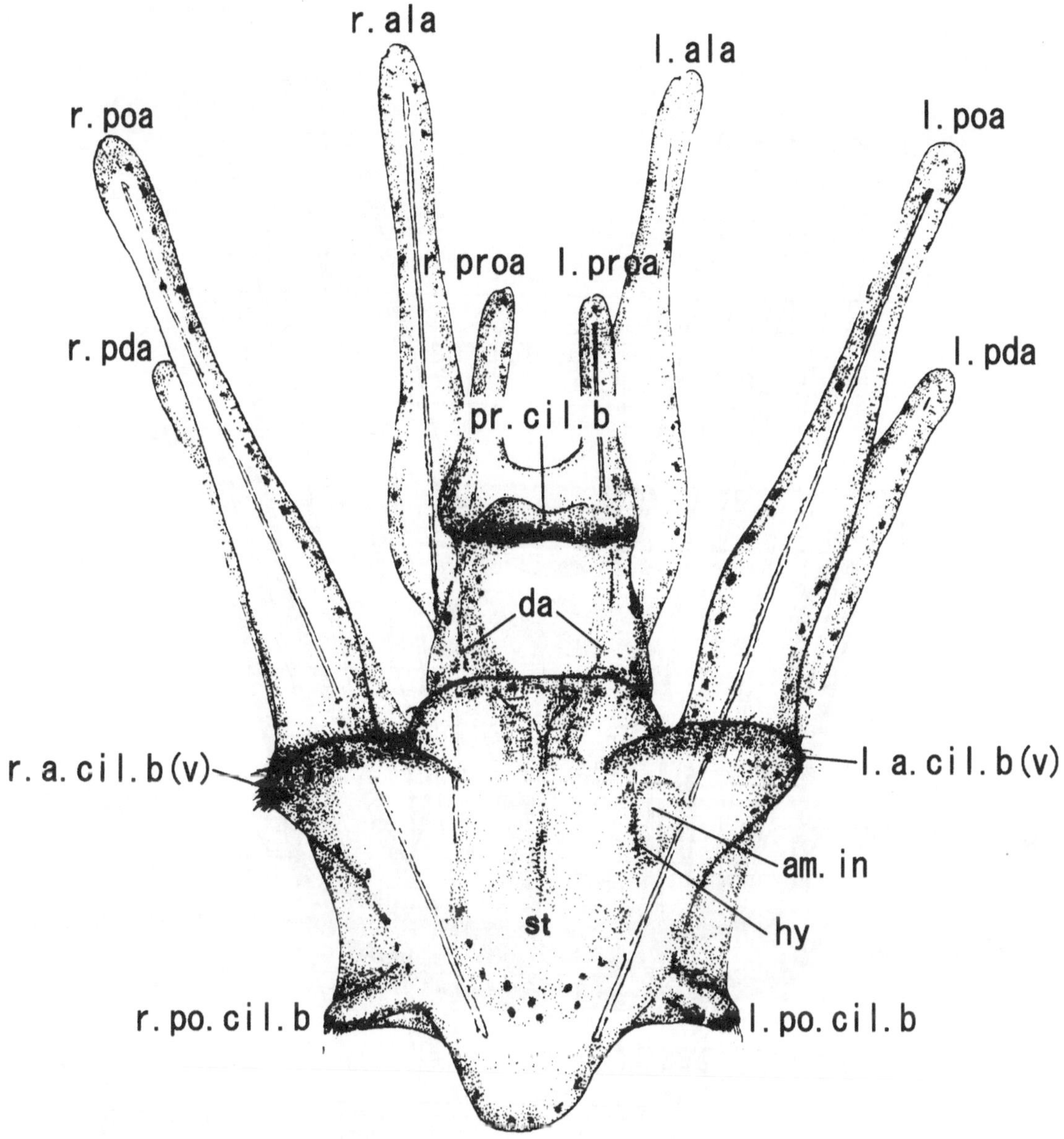
r. ala
l. ala
r. poa
l. poa
r. proa
l. proa
r. pda
l. pda
pr. cil. b
da
r. a. cil. b (v)
l. a. cil. b (v)
am. in
hy
st
r. po. cil. b
l. po. cil. b

31

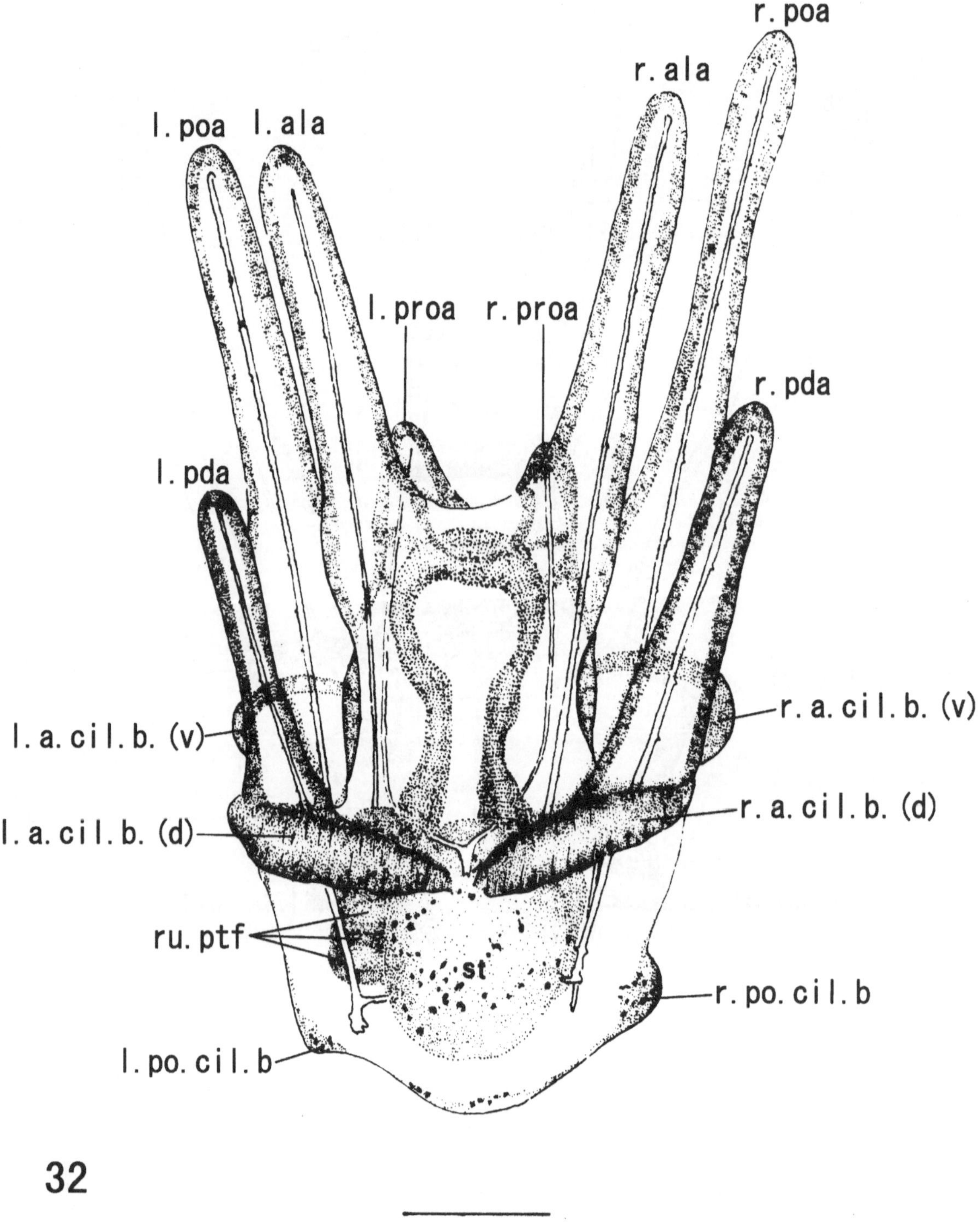
r. poa
r. ala
l. poa
l. ala
l. proa
r. proa
r. pda
l. pda
l. a. cil. b. (v)
r. a. cil. b. (v)
r. a. cil. b. (d)
l. a. cil. b. (d)
ru. ptf
st
r. po. cil. b
l. po. cil. b

32

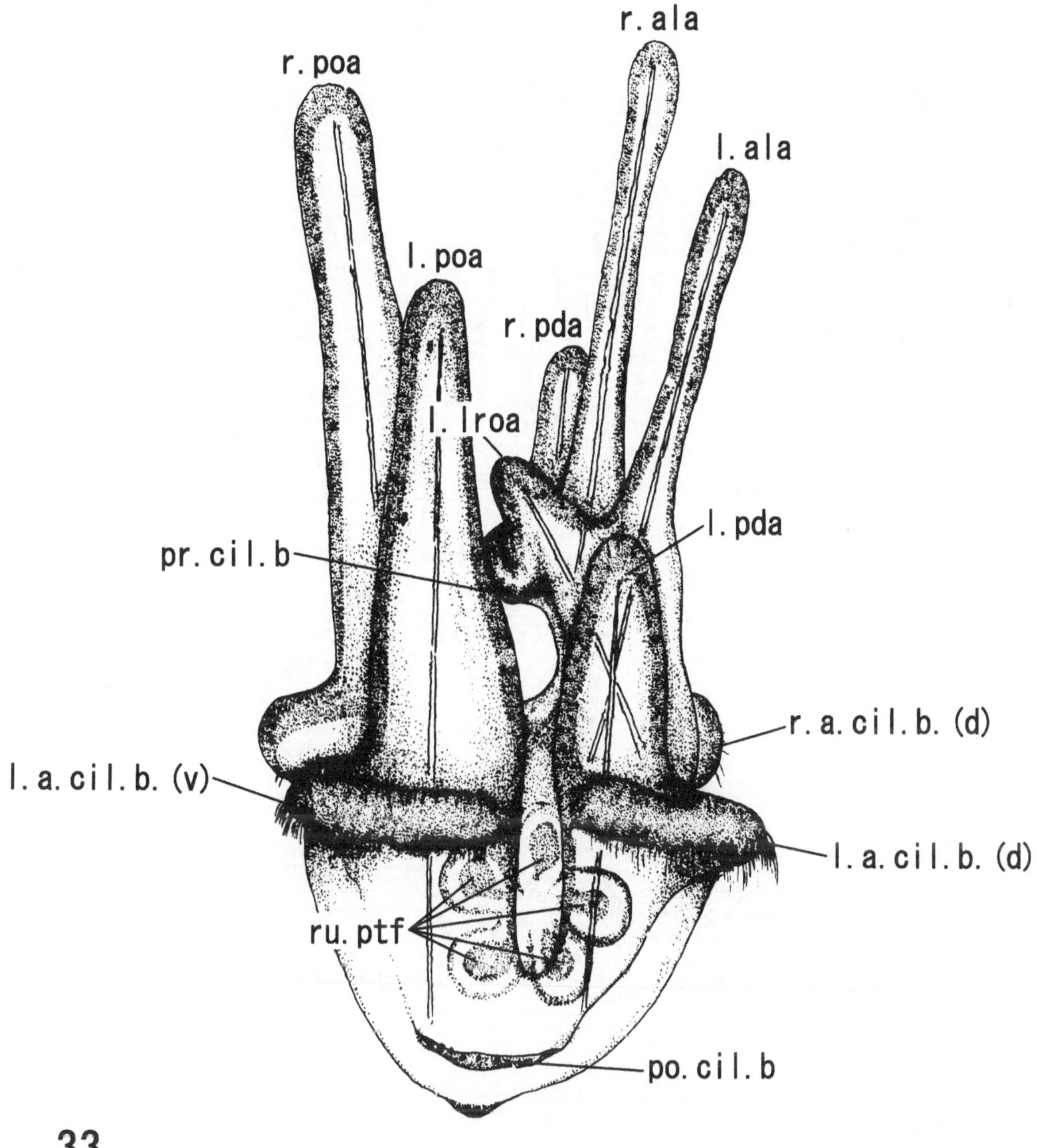

r. ala
r. poa
l. ala
l. poa
r. pda
l. lroa
l. pda
pr. cil. b
r. a. cil. b. (d)
l. a. cil. b. (v)
l. a. cil. b. (d)
ru. ptf
po. cil. b

33

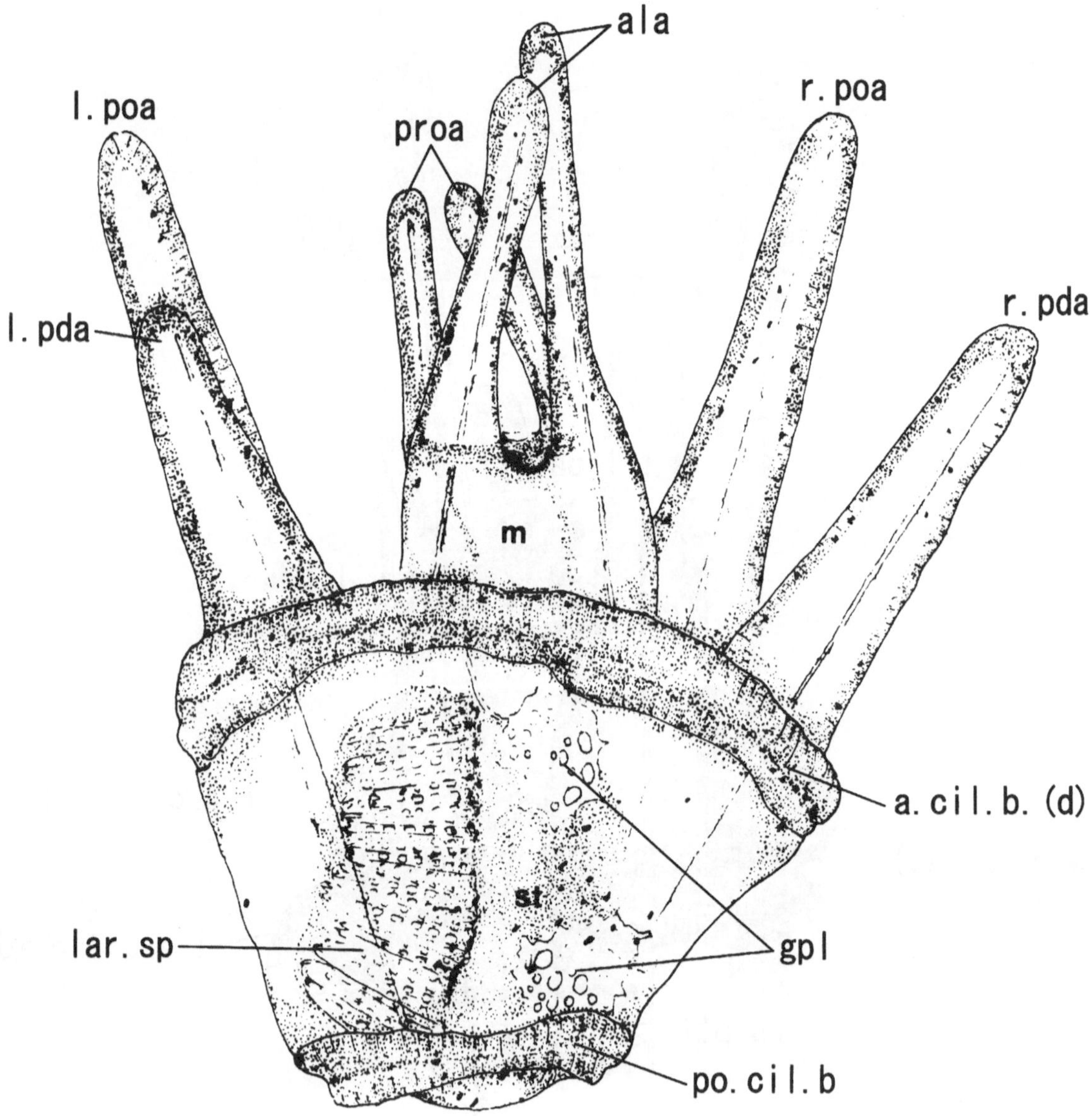

34

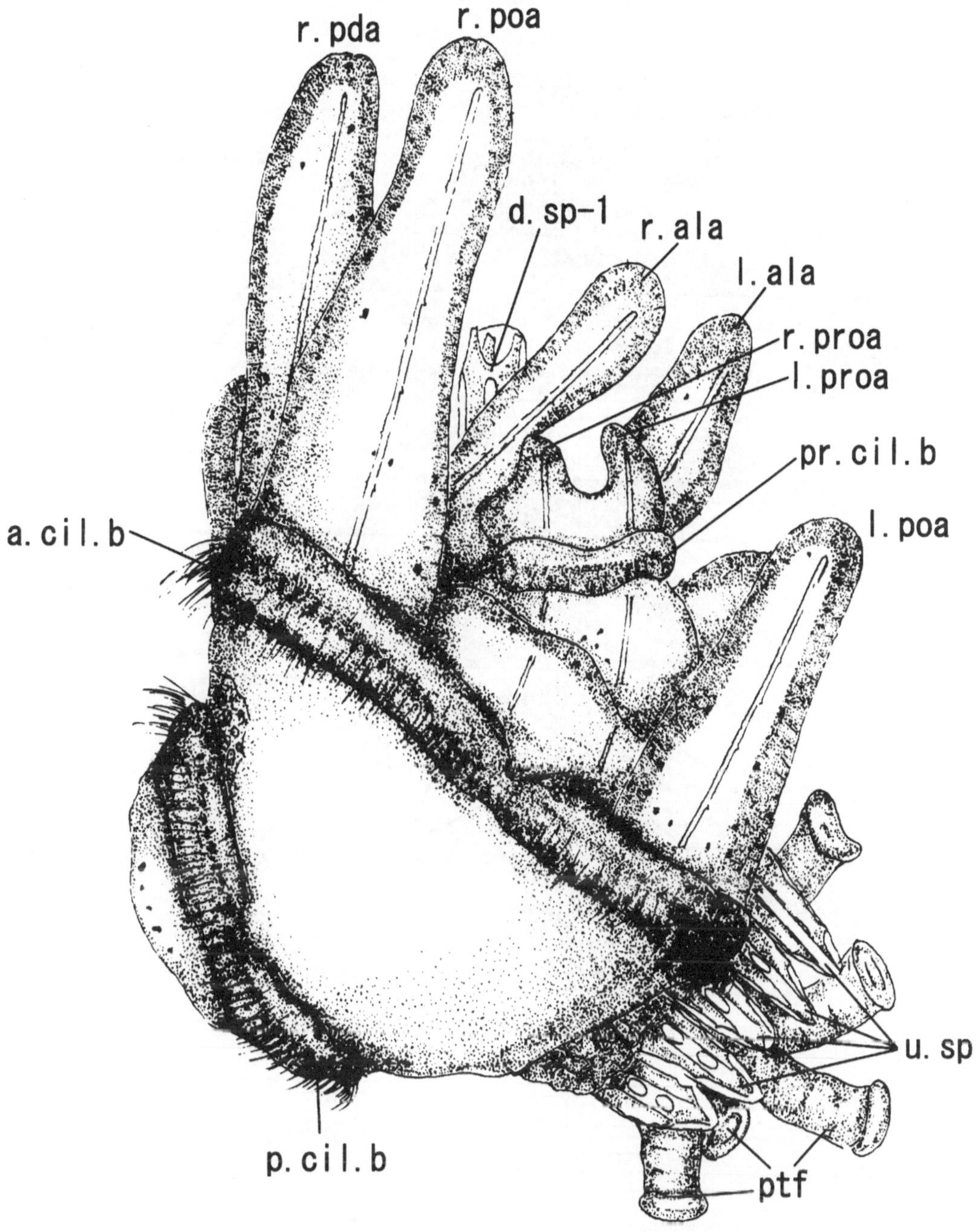
r. pda
r. poa
d. sp-1
r. ala
l. ala
r. proa
l. proa
pr. cil. b
a. cil. b
l. poa
u. sp
p. cil. b
ptf

35

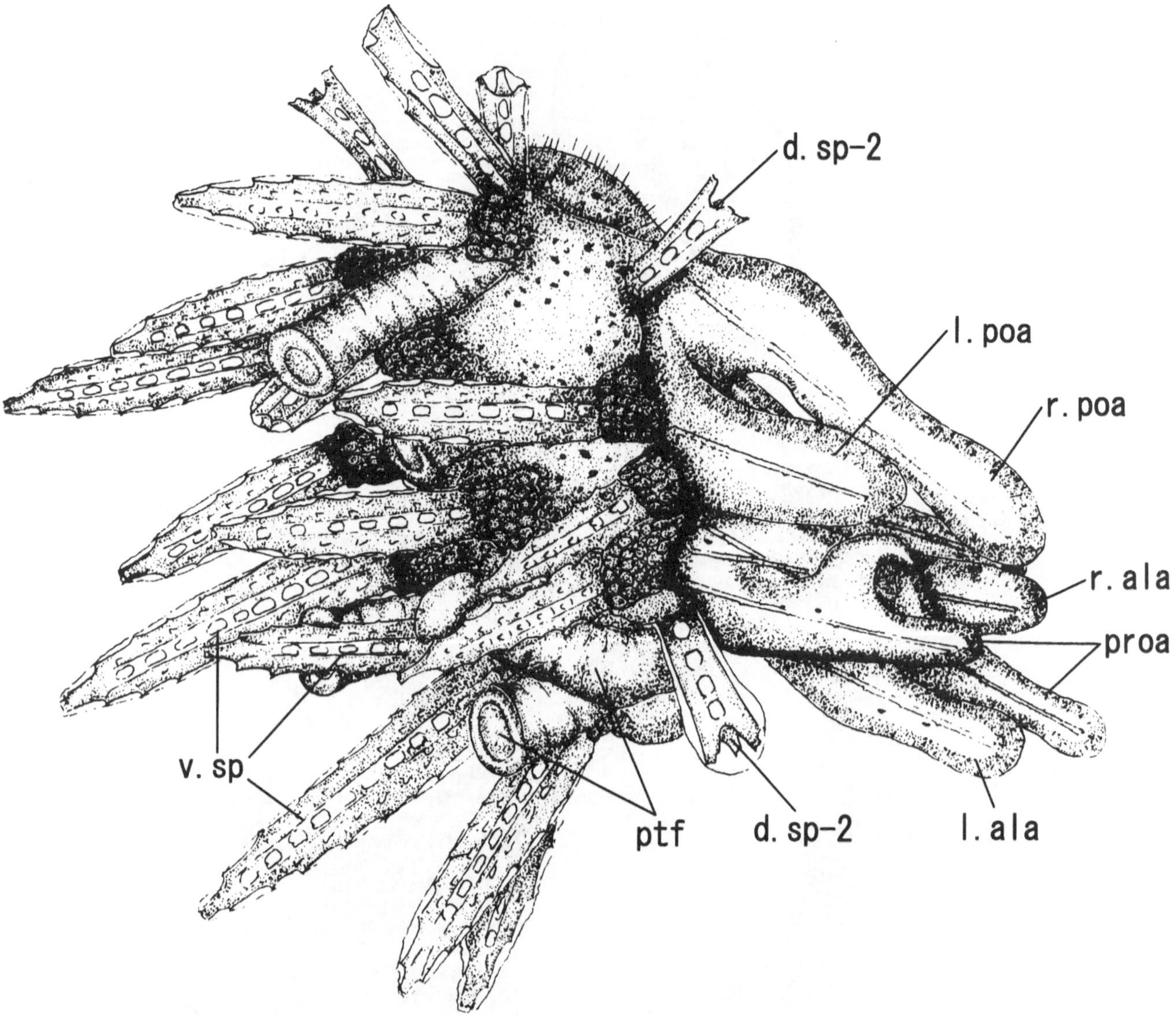

36

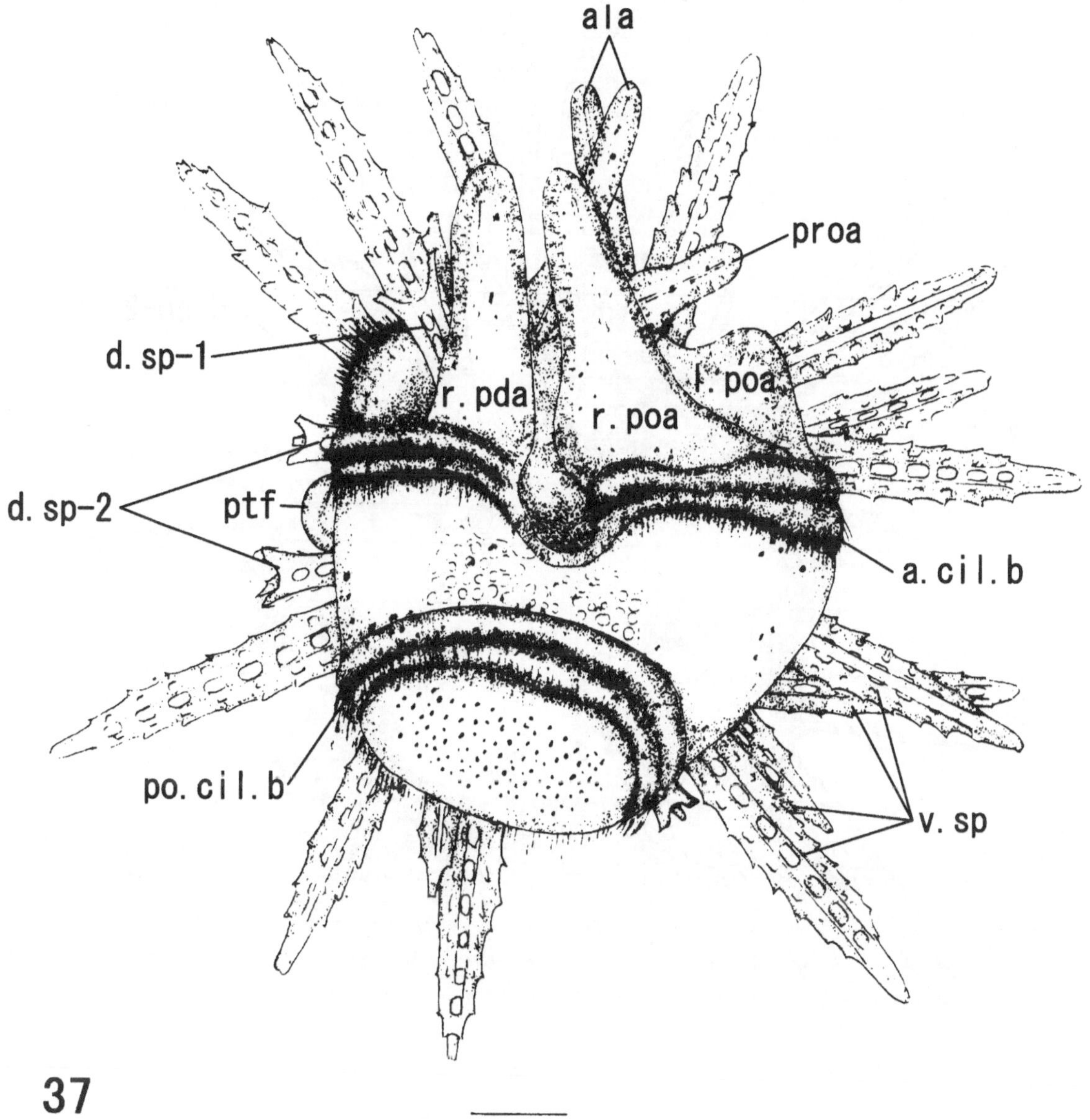

37

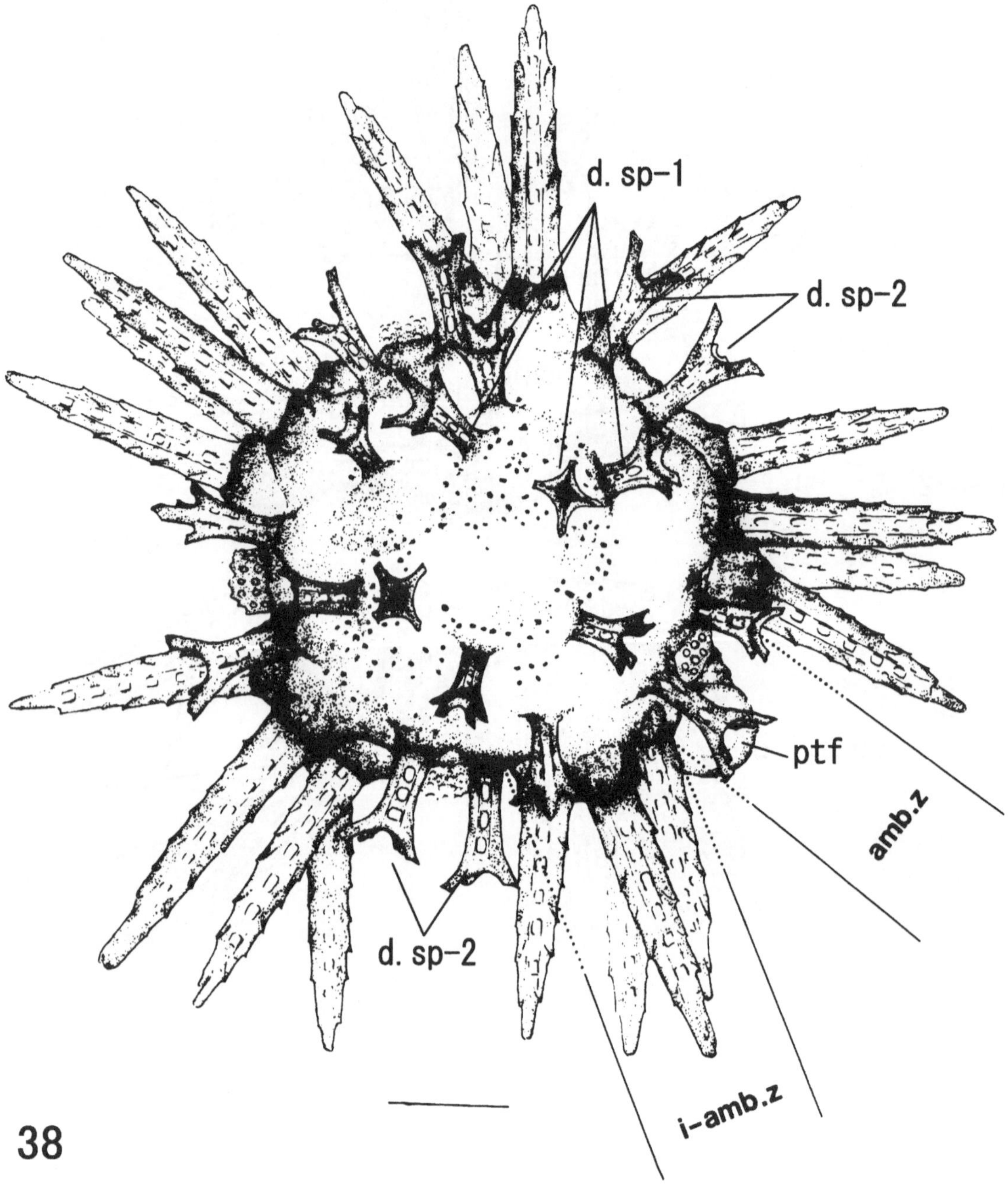

38

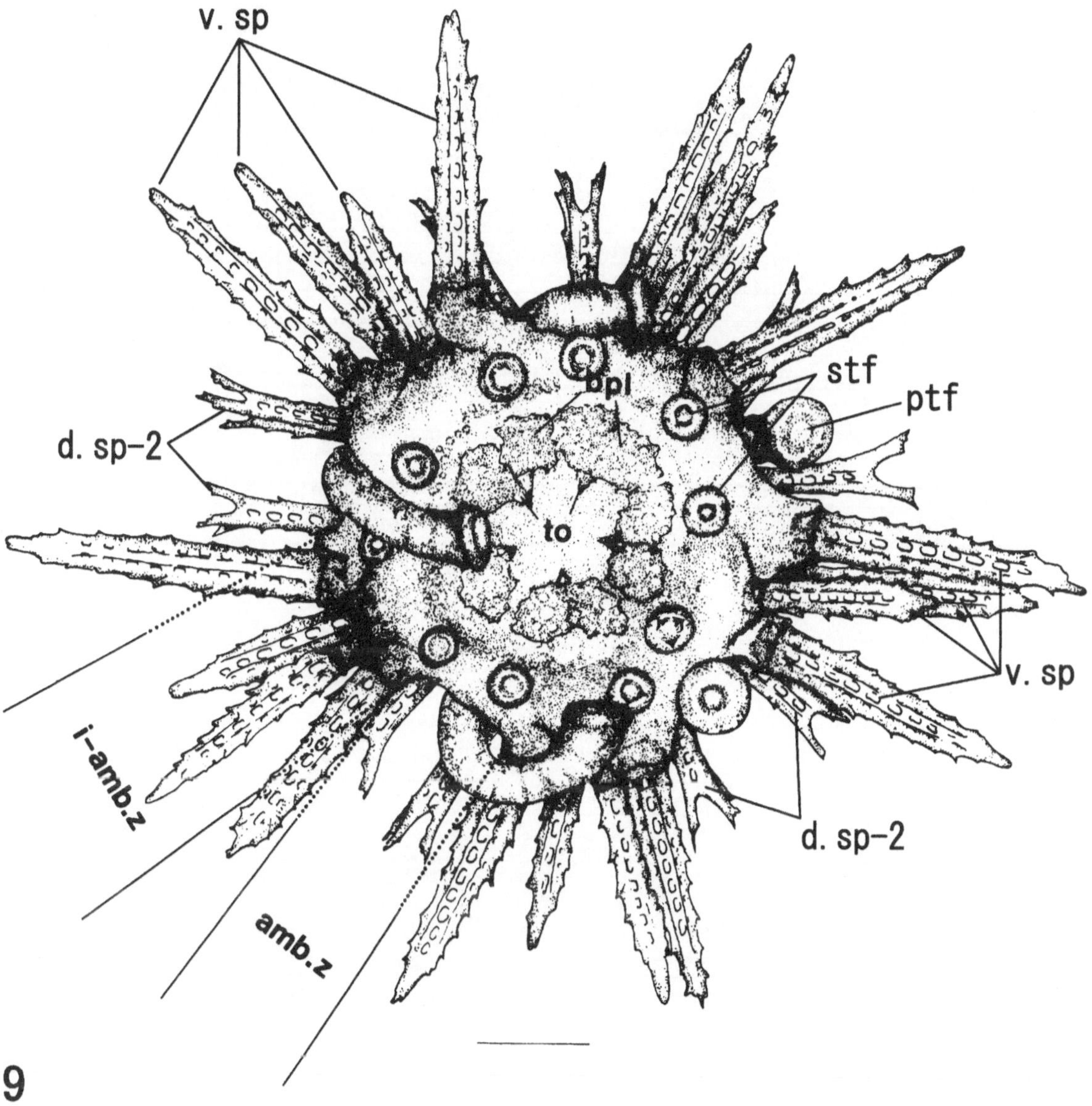

39

The Sea Urchin: From Basic Biology to Aquaculture, Yokota, Matranga & Smolenicka (eds)
© Swets & Zeitlinger, Lisse, ISBN 90 5809 379 4

Development of *Paracentrotus lividus* embryos and larvae

V. Matranga & R. Bonaventura

DEVELOPING EMBRYOS

a) oocyte, b) unfertilized egg, c) fertilized egg, d) 2-cell stage, e) 4-cell stage, f) 8-cell stage, g) 16-cell stage, h) morula, i) early blastula, j) hatching blastula, k) swimming blastula, l) mesenchyme blastula, m) early gastrula, n) middle gastrula, o) late gastrula, p) late gastrula (vegetal pole view), q) early prism, r) early prism (lateral view), s) early pluteus, t) early pluteus, (lateral view) u) pluteus, (anal rods focus) v) pluteus (lateral view), w) pluteus (body roads focus). Bar = 50 μm.

FEEDING LARVAE

x) 4-arms pluteus, y) early 6-arms larva, z) 6-arms larva, A) early 8-arms larva with small size rudiment (arrowhead), B) 8-arms larva with medium size rudiment (arrowhead), C) 8-arms larva with large size rudiment, D, E, F, G, H, I) 8-arms larva with large size rudiment, J, K) hatching juvenile, L) juvenile (x-B, J). Oral side view, (C, E) anal side view, (D, F, G, K) left side view, H) apex view, I) right side view. Podia are clearly seen in Figure F and G. Bar = 100 μm.

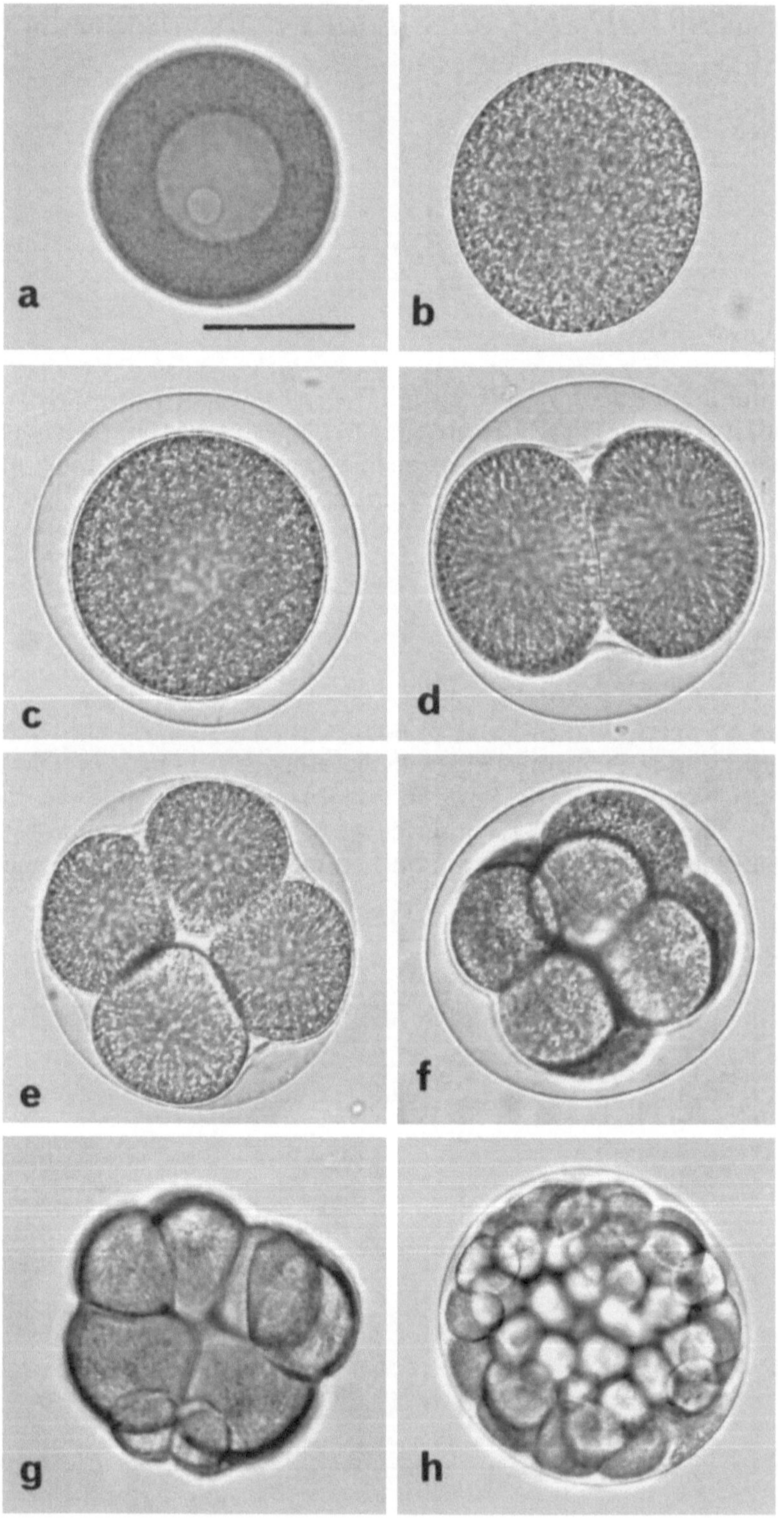
a
b
c
d
e
f
g
h

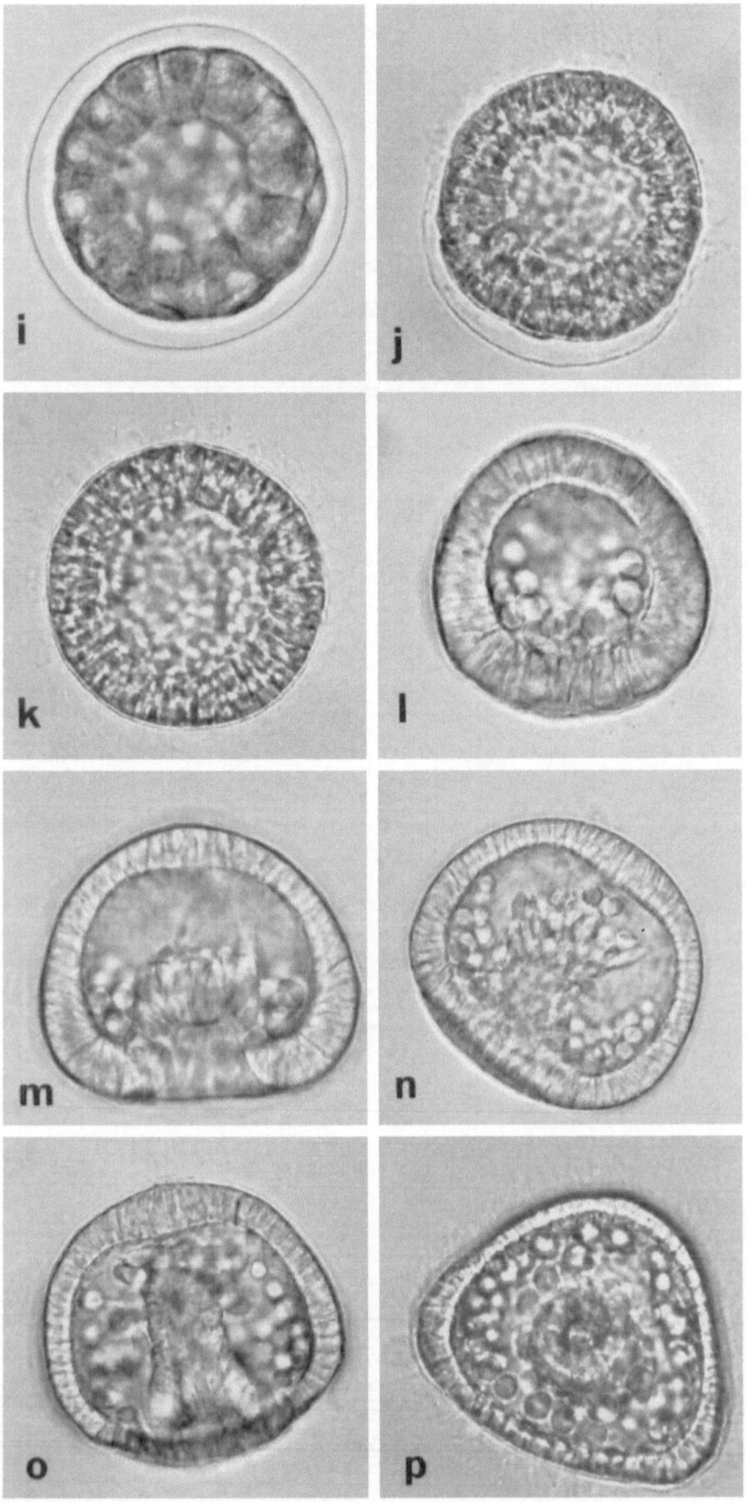

i
j
k
l
m
n
o
p

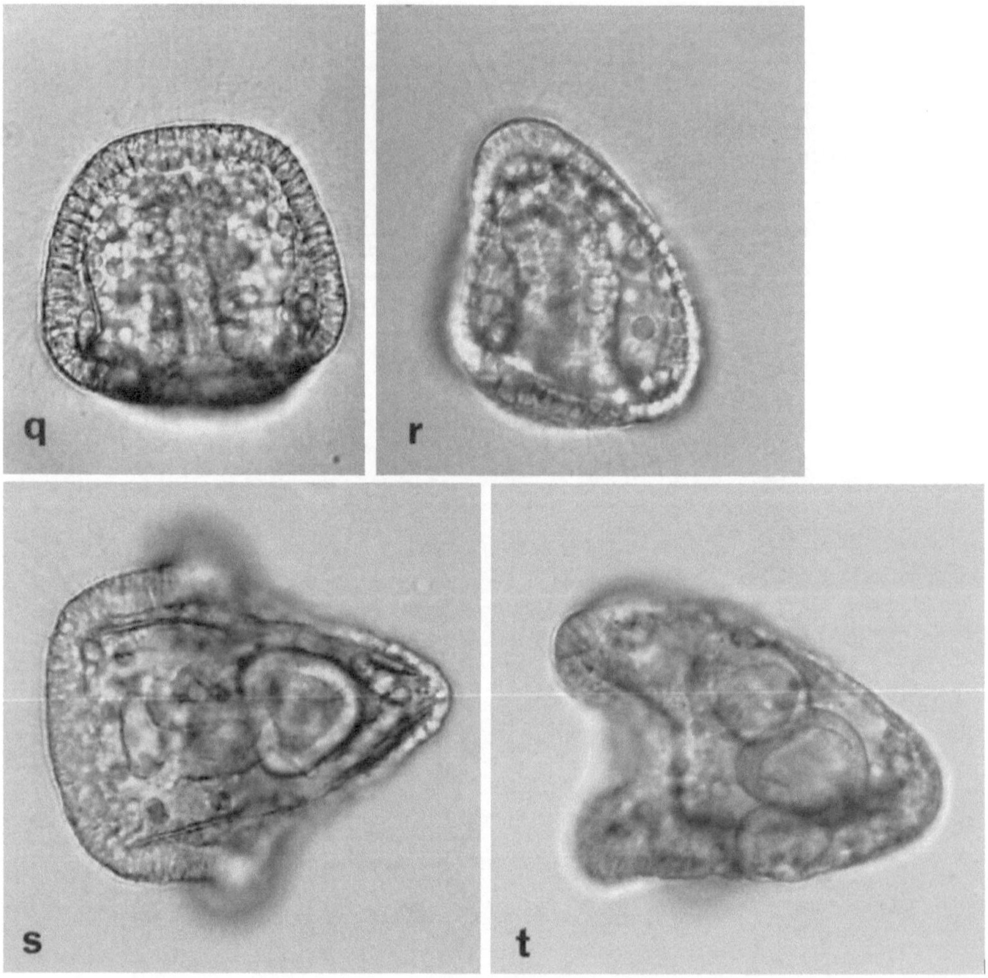
q
r
s
t

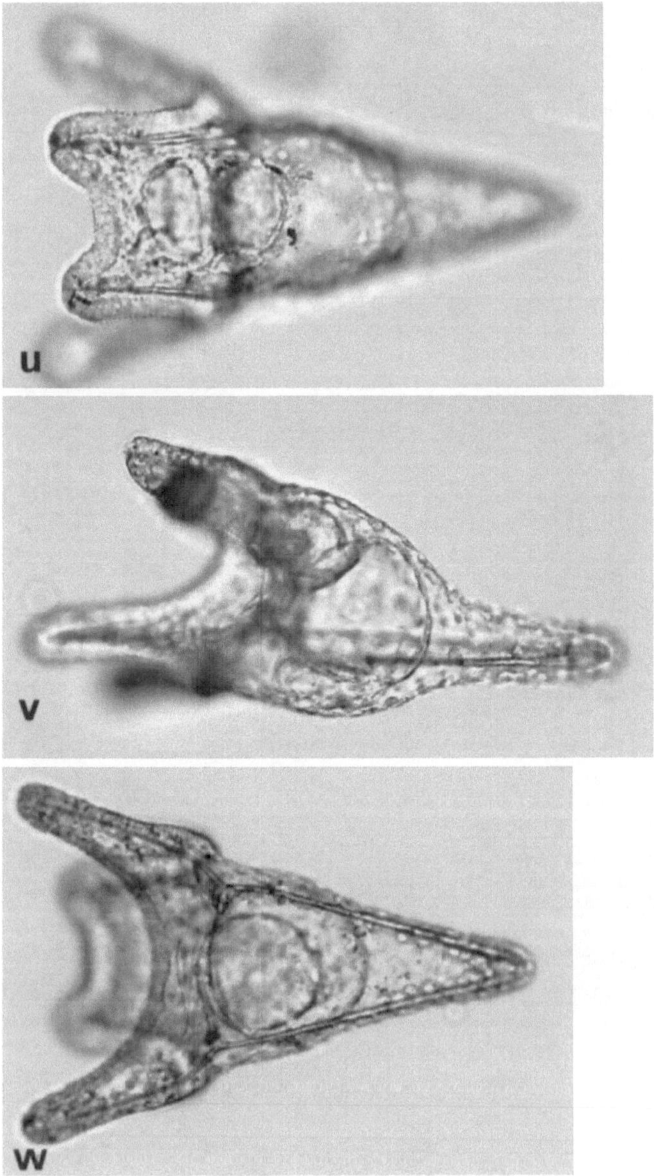
u
v
w

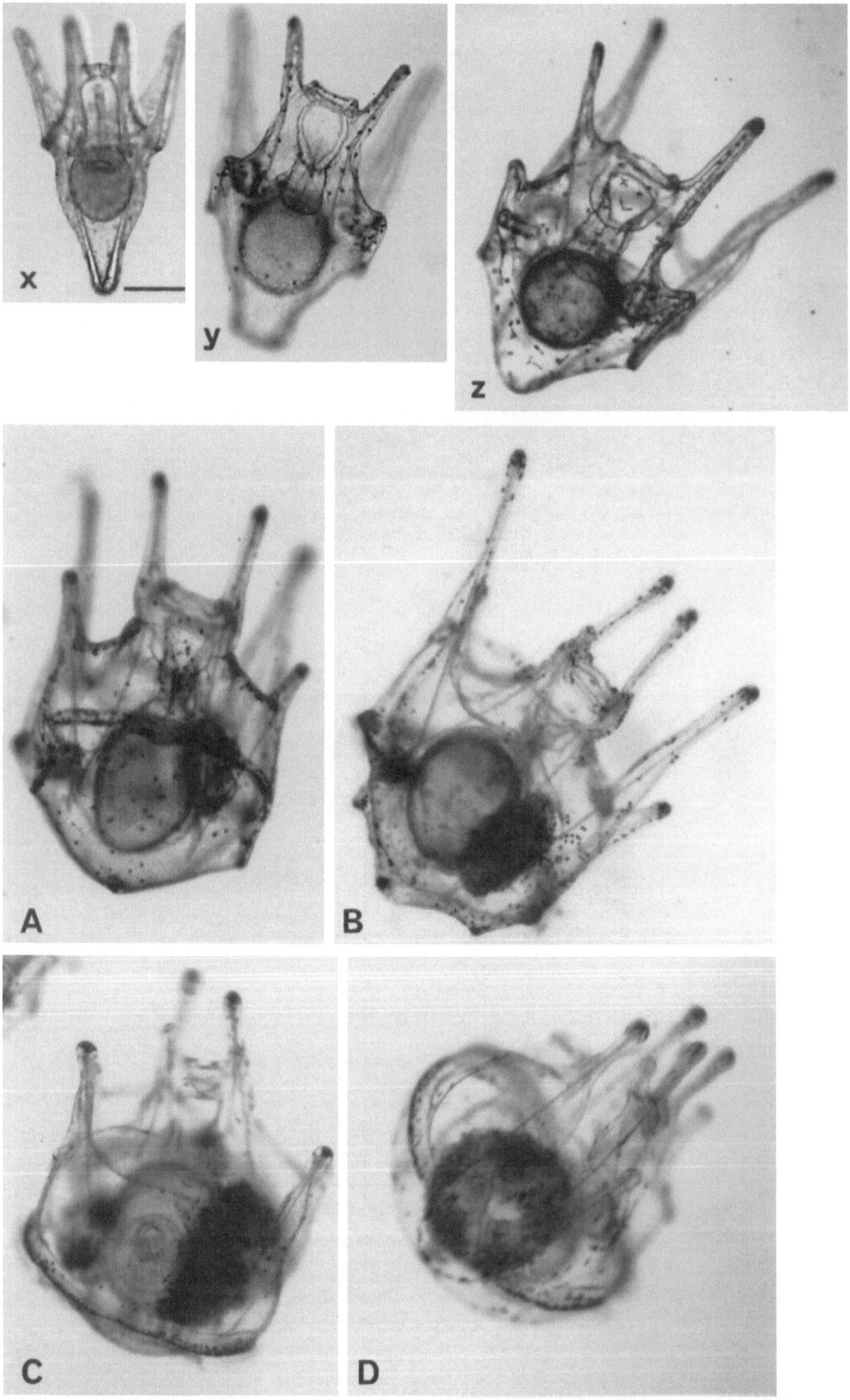
x
y
z
A
B
C
D

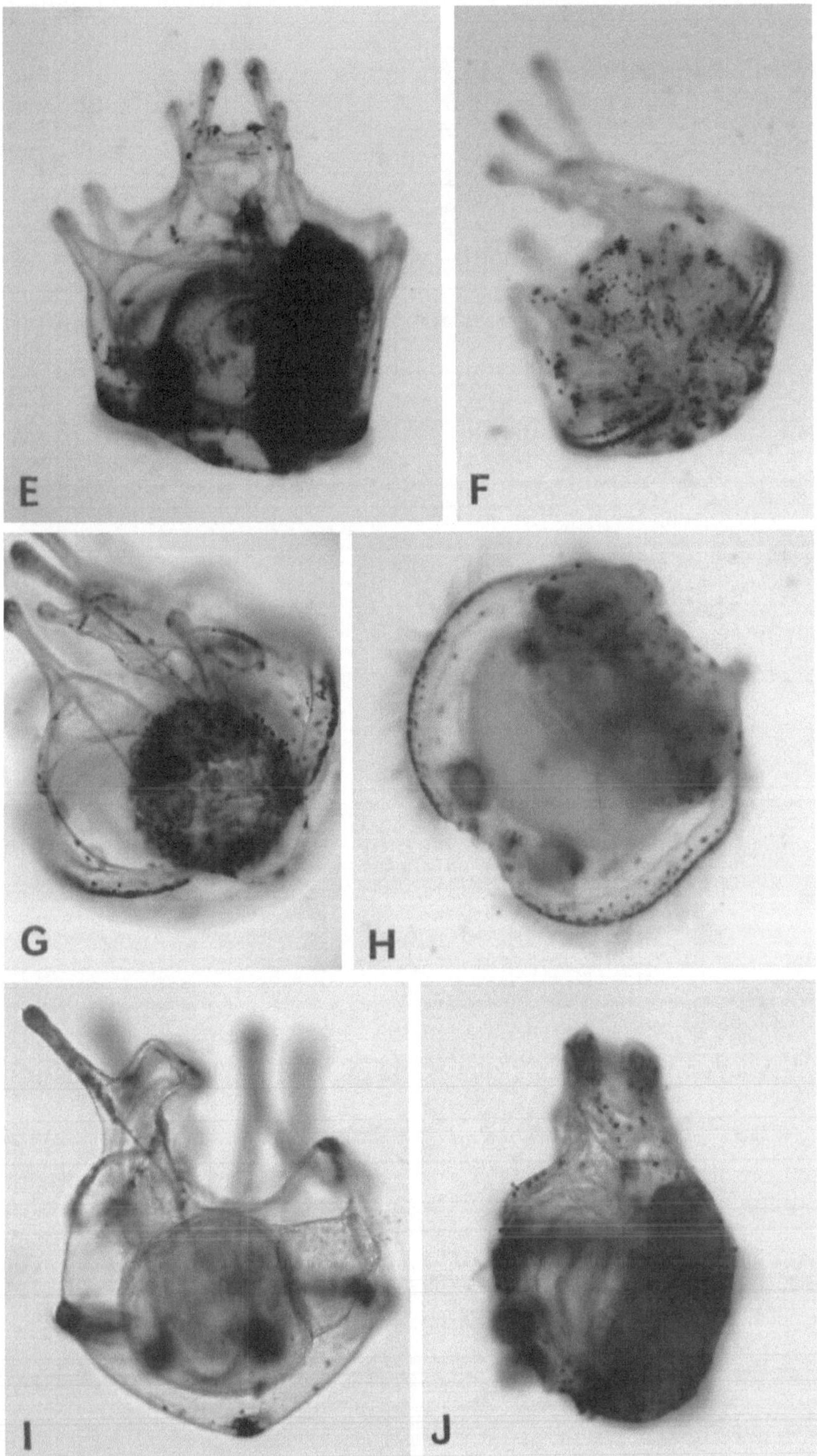
E
F
G
H
I
J

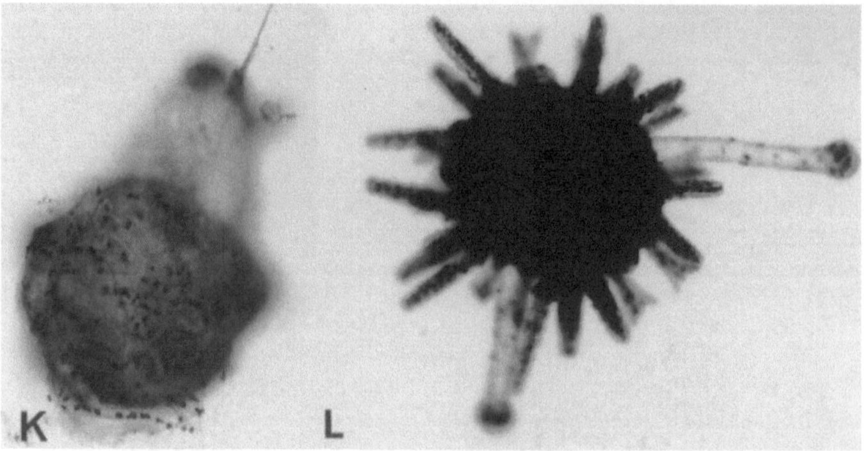

Author Index